Edited by
Lynn J. Frewer, Willem Norde, Arnout Fischer, and Frans Kampers

Nanotechnology in the Agri-Food Sector

Related Titles

Krug, H. (ed.)

Nanotechnology

Volume 2: Environmental Aspects

317 pages
2008
Hardcover
ISBN: 978-3-527-31735-6

Kaput, J.

Nutritional Genomics

Discovering the Path to Personalized Nutrition with Wiley Plus Stand-alone t/aNutrition: Everyday Choices Set

2006
Hardcover
ISBN: 978-0-470-13837-3

Castle, D., Cline, C., Daar, A. S., Tsamis, C., Singer, P. A.

Science, Society, and the Supermarket

The Opportunities and Challenges of Nutrigenomics

164 pages
2006
Hardcover
ISBN: 978-0-471-77000-8

Kaput, J., Rodriguez, R. L. (eds.)

Nutritional Genomics

Discovering the Path to Personalized Nutrition

496 pages
2005
Hardcover
ISBN: 978-0-471-68319-3

Brennan, J. G. (ed.)

Food Processing Handbook

607 pages with 189 figures and 41 tables
2006
Hardcover
ISBN: 978-3-527-30719-7

Schmidt, R. H., Rodrick, G. E.

Food Safety Handbook

approx. 864 pages
Hardcover
ISBN: 978-0-471-21064-1

Edited by Lynn J. Frewer, Willem Norde, Arnout Fischer, and Frans Kampers

Nanotechnology in the Agri-Food Sector

Implications for the Future

WILEY-VCH

WILEY-VCH Verlag GmbH & Co. KGaA

The Editors

Prof. Dr. Lynn J. Frewer
Wageningen University
Marketing & Consumer B. Group
Hollandseweg 1
6700 EW Wageningen
Niederlande

Prof. Dr. Willem Norde
Wageningen University
Dept. Physical Chemistry
Postbus 8038
6700 EK Wageningen
Niederlande

Arnout Fischer
Wageningen University
Marketing & Consumer B. Group
Hollandseweg 1
6700 EW Wageningen
Niederlande

Dr. Frans Kampers
University of Wageningen
Agrotechnology & Food Sc. Gr.
Dreijenplein 8
6703 HB Wageningen
Niederlande

Library of Congress Card No.: applied for

British Library Cataloguing-in-Publication Data
A catalogue record for this book is available from the British Library.

Bibliographic information published by the Deutsche Nationalbibliothek
The Deutsche Nationalbibliothek lists this publication in the Deutsche Nationalbibliografie; detailed bibliographic data are available on the Internet at <http://dnb.d-nb.de>.

Cover Design Adam Design, Weinheim
Typesetting Toppan Best-set Premedia Limited, Hong Kong
Printing and Binding betz-druck GmbH, Darmstadt

Printed in the Federal Republic of Germany
Printed on acid-free paper

ISBN: 978-3-527-33060-7

Contents

List of Contributors

Yuri Aleexev
Institute of Food Research
Norwich Research Park, Colney
Norwich NR4 7UA
UK

Aart van Amerongen
Wageningen University and
Research Centre
Food and Biobased Products
Biomolecular Sensing and Diagnostics
Bornse Weilanden 9
6708 WG Wageningen
The Netherlands

Remko M. Boom
Wageningen University and Research
Centre
Food Process Engineering Group
Agrotechnology and Food Sciences
Department
P.O. Box 8129
6700 EV Wageningen
The Netherlands

Hans Bouwmeester
Wageningen University and
Research Centre
RIKILT – Institute of Food Safety
Akkermaalsbos 2
6708 WB Wageningen
The Netherlands

Betty Bugusu
Institute of Food Technologists
525 West Van Buren
Suite 1000
Chicago, IL 60607
USA

Laurence Castle
The Food and Environment Research
Agency
Sand Hutton
York YO41 1LZ
UK

Qasim Chaudhry
The Food and Environment Research
Agency
Sand Hutton
York YO41 1LZ
UK

Enda Cummins
University College Dublin
Biosystems Engineering
UCD School of Agriculture
Food Science and Veterinary Medicine
Belfield
Dublin 4
Ireland

Arnout R.H. Fischer
Wageningen University and Research Centre
Department of Social Sciences
Marketing and Consumer Behaviour Group
Hollandseweg 1
6706 KN Wageningen
The Netherlands

John D. Floros
The Pennsylvania State University
Food Science Department
Food Science Building
University Park
PA 16802
USA

Lynn J. Frewer
Food and Society
Centre for Rural Economy
School of Agriculture, Food and Rural Development
Newcastle University
Agriculture Building
Newcastle upon Tyne NE1 7RU
UK

Wageningen University and Research Centre
Department of Social Sciences
Marketing and Consumer Behaviour Group
Hollandseweg 1
6706 KN Wageningen
The Netherlands

Frans W.H. Kampers
Wageningen University and Research Centre
Wageningen Bionanotechnology Centre (BioNT)
P.O. Box 8026
6700 EG Wageningen
The Netherlands

Hans Kastenholz
EMPA
Environmental Risk Assessment and Management Group
Lerchenfeldstrasse 5
9014 St. Gallen
Switzerland

Gijs A. Kleter
Wageningen University and Research Centre
RIKILT – Institute of Food Safety
Akkermaalsbos 2
6708 WB Wageningen
The Netherlands

Ursula Vanesa Lay Ma
The Pennsylvania State University
Food Science Department
Food Science Building
University Park, PA 16802
USA

Erik van der Linden
Wageningen University and Research Centre
Food Physics Group
Agrotechnology and Food Sciences Department
Bomenweg 2
6703 HD Wageningen
The Netherlands

Alan R. Mackie
Institute of Food Research
Norwich Research Park
Colney
Norwich NR4 7UA
UK

Bernadene A. Magnuson
Cantox Health Sciences International
2233 Argentia Road
Suite 308
Mississauga
Ontario, L5N 2X7
Canada

Hans J.P. Marvin
Wageningen University and Research Centre
RIKILT – Institute of Food Safety
Akkermaalsbos 2
6708 WB Wageningen
The Netherlands

E.N. Clare Mills
Institute of Food Research
Norwich Research Park
Colney
Norwich NR4 7UA
UK

Mark Morrison
Institute of Nanotechnology
Lord Hope Building
141 St James Road
Glasgow G4 0LT
UK

Willem Norde
Wageningen University and Research Centre
Laboratory of Physical Chemistry and Colloid Science
Chemistry Building, Dreijenplein 6
6703 HB Wageningen
The Netherlands

University of Groningen
University Medical Center
Hanzeplein 1
9713 GZ Groningen
The Netherlands
University of Groningen
Antonius Deusinglaan 1
9713 AV Groningen
The Netherlands

Bernd Nowack
EMPA
Environmental Risk Assessment and Management Group
Lerchenfeldstrasse 5
9014 St. Gallen
Switzerland

Niall O'Brien
University College Dublin
Biosystems Engineering
UCD School of Agriculture
Food Science and Veterinary Medicine
Belfield
Dublin 4
Ireland

Geertruida A. Posthuma-Trumpie
Wageningen University and Research Centre
Food and Biobased Products
Biomolecular Sensing and Diagnostics
Bornse Weilanden 9
6708 WG Wageningen
The Netherlands

Douglas K.R. Robinson
Institute of Nanotechnology
Lord Hope Building
141 St James Road
Glasgow G4 0LT
UK

Gene Rowe
Wageningen University and Research Centre
Department of Social Sciences
Marketing and Consumer Behaviour Group
Hollandseweg 1
6706 KN Wageningen
The Netherlands

Michael Siegrist
ETH Zürich
Institute for Environmental Decisions (IED)
Consumer Behavior
Universitätstrasse 22
8092 Zürich
Switzerland

Roger Strand
University of Bergen
Centre for the Study of the Sciences and the Humanities
P.O. Box 7805
N-5020 Bergen
Norway

Pieter Stroeve
University of California
Davis
Department of Chemical Engineering and Materials Science
1 Shields Avenue
Davis, CA 95616
USA

J.(Hans)C.M. van Trijp
Wageningen University and Research Centre
Department of Social Sciences
Marketing and Consumer Behaviour Group
Hollandseweg 1
6706 KN Wageningen
The Netherlands

Richard Watkins
The Food and Environment Research Agency
Sand Hutton
York YO41 1LZ
UK

Meike T.A. Wentholt
Wageningen University and Research Centre
Department of Social Sciences
Marketing and Consumer Behaviour Group
Hollandseweg 1
6706 KN Wageningen
The Netherlands

Introduction

Lynn Frewer, Arnout Fischer, Willem Norde, and Frans Kampers

As a basic science, nanotechnology has advanced considerably over the past decades. It has generally been agreed that nanotechnology deals with structures of size 100 nm (nanometers) or less in at least one dimension, and involves developing materials or devices within these size limits. Research in nanotechnology has resulted in applications across a wide range of areas, perhaps more so than for most areas of fundamental research in the natural sciences. The potential applications range from those within the medical and pharmaceutical sectors, the development of new materials, personal care products, to applications in agriculture and food (the focus of this particular volume).

In common with other emerging technologies, as well as existing technologies that are being re-evaluated, the opportunities for technological advancement are potentially profound. However, many scholars in the risk assessment community have raised concerns about the *toxicity* in regard to both human health and the environment. As a consequence, there is an ongoing discussion regarding whether specific measures regarding the regulation of nanoparticles are required. This is as true in the agri-food sector as in other nanotechnology application areas. Some academics have called for stricter application of the *precautionary principle,* with delayed marketing approval, enhanced labeling, and additional safety data development requirements in relation to certain forms of nanotechnology. From this, there have been discussions of the need to invoke the precautionary principle with regard to the application of nanotechnology. The precautionary principle states that, if an activity – for example, the application of an emerging technology – is potentially associated with health or environmental risks, the burden of proof that it is not harmful falls on those who advocate taking the action, if there is no evidence to suggest otherwise. On the one hand, the application of the precautionary principle allows policy-makers to take discretion in the absence of complete scientific proof of safety. On the other hand, this means that delays occur in marketing approval, and additional safety data may be needed in relation to specific applications of nanotechnology in the agri-food sector. In addition, appropriate labeling strategies may be needed to ensure informed consumer choice, given the emerging societal debate about nanotechnology and its applications to the agri-food sector.

Nanotechnology in the Agri-Food Sector: Implications for the Future, First Edition. Edited by Lynn J. Frewer, Willem Norde, Arnout Fischer, Frans Kampers.

The editors have attempted to address these (and related) issues in the current volume. In Part One, the fundamentals of nanotechnology applied to the agri-food sector are discussed, in particular, intermolecular interactions and self-assembly of macromolecules. In Part Two, the basics applications of nanotechnology in the agri-food sector are identified. Novel techniques such as encapsulation, diagnostics and sensing, and packaging are presented in detail, and their applications to food production are described. Part Three, on specific applications to food, deals with application in crop and livestock production, the application to improving the food supply (in terms of quality, safety, and security), discussion of functionality, and commercialization. Finally, Part Four, which deals with nanotechnology and society, focuses not only on the potential benefits of nanotechnology, but also on potentially emerging risks and what needs to be done to ensure safety. As well as a chapter focusing on the toxicology of nanomaterials in the agri-food sector, there is additional consideration of what this implies in terms of putative changes to risk regulation and governance. As part of this, it is essential to take account of the views and preferences of society, in terms of risk–benefit perceptions and preferences for co-development. These issues are addressed from both a theoretical and a practical perspective. The question that is asked is how should consumers and citizens be *effectively* involved in the societal debate about the development, application, and commercialization of food nanotechnology. As part of this, emerging ethical issues need to be addressed, and a chapter has been dedicated to discussion of these.

Part One
Fundamentals

1
Intermolecular Interactions

Willem Norde

1.1
Introduction

Nanotechnology may be broadly defined as the study, fabrication, and application of systems by manipulating structures or objects having nanoscale dimensions (say, between 1 nm and 100 nm). Of course, molecular scientists, in both chemistry and biology, have been dealing with nanoscopic (polymer) molecules and biological cell components for decades. So, what's new? New is that, with the advent in the 1980s of new instrumentation, in particular scanning probe microscopes – for example, atomic force microscopy (AFM) – *individual* nano-objects can be observed and manipulated (see Figure 1.1).

Using AFM, the positions of molecules and nanoparticles, relative to each other, may be rearranged in a controlled way. AFM furthermore allows the measurement of interaction forces between nanoparticles as well as between nanoparticles and macroscopic objects. Other recently developed devices, the so-called optical tweezers and magnetic tweezers, also enable the controlled motion of, and the determination of forces between, nanoparticles.

Manipulation on the nanoscale may be done in two "directions", referred to as *top-down* and *bottom-up*. In the top-down approach, structures are made increasingly smaller by progressively removing matter, usually by etching. Perhaps the most well-known example of a top-down structure is the electronic chips present in various devices. Another example is the micro- or nano-sieve, a solid wafer punctured with equally sized micro- or nanopores. Nano-sieves are in particular relevant for food processing and water treatment. Because various agricultural and dairy products are of heterodisperse particulate nature, that is, emulsions, foams, and dispersions of solid particles, they may be fractionated using a series of sieves of varying pore size. The separate components thus obtained may be recombined to give newly composed products of superior quality. Also, nano-sieves could be used in (cold) sterilization by filtering out microbial cells.

In the agri-food sector, however, bottom-up nanostructures are more often encountered. Bottom-up implies that atoms or molecules are distributed and rearranged to build new, functional nano-objects. Nature itself is full of bottom-up

Nanotechnology in the Agri-Food Sector: Implications for the Future, First Edition. Edited by Lynn J. Frewer, Willem Norde, Arnout Fischer, Frans Kampers.

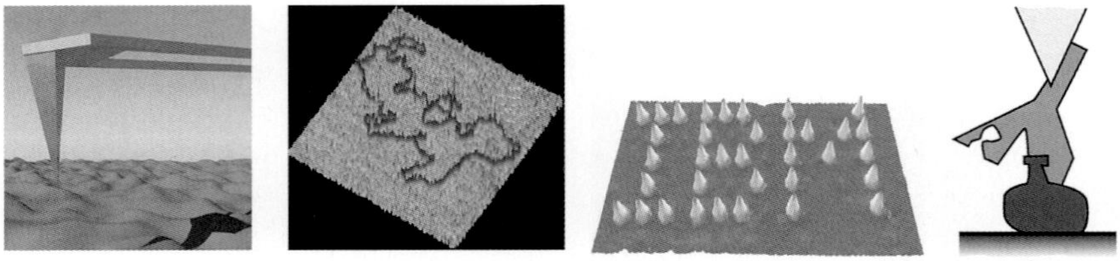

(a) (b) (c) (d)

Figure 1.1 Atomic force microscopy. (a) The topography of a surface is scanned with subnanometer resolution, so that nano-sized objects can be (b) observed and (c) manipulated. (d) Atomic force microscopy may also be applied to determine the interaction between two objects.

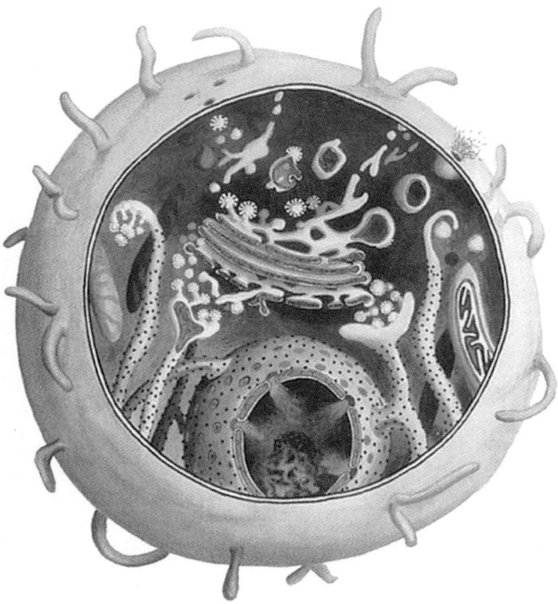

Figure 1.2 Cartoon of a biological cell showing a variety of nano-sized subcellular structures.

nanostructures, especially in living species. Think of viruses, where nucleic acids and proteins are arranged and interact such that viral activity results. Think of microbial, plant, and animal cells in which the various nano-sized organelles and membranes are complex bottom-up assemblies of precisely arranged building blocks (Figure 1.2).

Although nature is capable of making structures far more complicated and sophisticated than the ones that scientists can – for the time being – achieve in their laboratories, it may not be a surprise that nano-engineers are strongly inspired by nature. A few examples come to mind: in making addressable biocompatible nanoparticles to be used for the encapsulation and delivery of nutriceuticals and pharmaceuticals, nature provides clues as to how the surface of such particles should look; viruses may serve as a model in the design of particles carrying deoxyribonucleic acid (DNA) fragments to be used in gene therapy; non-fouling surfaces may be mimicked from the outer composition and structure of cell surfaces; and the texture of foodstuff may be optimized by imitating nanostructures as they occur in nature, for example, fibrillar protein aggregates in meat replacers and three-dimensional polymer networks in mousses.

To achieve the specific architectures related to the desired (biological) function of the nanostructure, the physicochemical interactions between the building blocks should be tuned with high accuracy. Needless to say, understanding the mechanisms underlying the various types of interaction is a prerequisite for successful tuning.

In this chapter an introduction to the main types of interactions that may play a role in bottom-up nanotechnology is given. These are physicochemical interactions more or less sensitive to changing environmental conditions and therefore result in the formation of annealed, responsive structures. The discussion here may not be the most rigorous one, as, in view of the scope of this book, the scientific language of chemistry and physics that involves formulas and equations will be avoided as much as possible.

In natural systems, including those of the agri-food sector, most nano-objects exist by virtue of their interaction with an aqueous environment. Not only their existence but also their shape and spatial structure are to a large extent determined by their interaction with water. It is, therefore, essential first to pay attention to some physicochemical properties of water.

1.2
Water

Water is one of the most abundantly occurring chemical compounds on Earth (although very unevenly distributed). Because of its ubiquity, we are inclined to think of water as a trivial, common, and normal liquid. However, from a physicochemical point of view, water is a highly extraordinary substance. By virtue of its unique properties, water is the medium in which life has evolved and is sustained. Which properties make water so special, and how can these properties be explained and understood at the molecular level?

Water, H_2O, has a molar mass of $18\,g\,mol^{-1}$. Under ambient conditions, water boils at 100 °C. Among other components of comparable molar mass, this is an exceptionally high temperature. For instance, the boiling points of methane ($14\,g\,mol^{-1}$) and ethane ($30\,g\,mol^{-1}$) are −162 °C and −88 °C, respectively. Also, the heat of vaporization, which is essentially the energy required to separate the molecules when they go from the condensed liquid phase to the gas phase, is extremely high for water, $2255\,J\,g^{-1}$, whereas for methane and ethane this is a little more than $500\,J\,g^{-1}$. Another interesting property is the heat capacity, the amount of heat needed to increase the temperature of a substance by one degree Celsius. For the sake of fairness, equal amounts of the substances should be compared and at the same temperature and pressure. While the heat capacity of water at 20 °C and 1 atm amounts to 4.18 J $K^{-1}g^{-1}$, the values for other liquids are much lower (cf. for chloroform it is 0.90 J $K^{-1}g^{-1}$ and for ethanol 2.49 J $K^{-1}g^{-1}$). These anomalously high values for the boiling point, heat of vaporization, and heat capacity (and, in this context, further extraordinary characteristics of water could be presented) originate from the phenomenon that water molecules attract each other. They attract each other so much that they strongly attach to one another. In scientific terms, water shows a strong internal coherence. Hence, to evaporate the liquid, favorable interactions between the water molecules have to be disrupted (which explains the large heat of vaporization), and this does not occur before the molecules have attained strong thermal motion (explaining the high boiling point). The large heat capacity reflects not only that heat is used for increasing thermal motion (corresponding to a one degree temperature rise) *per se*, but also that loosening of the internal coherence is necessary to increase thermal motion, which is the major energetic cost.

To understand the strong internal coherence, we zoom in on the molecular structure of water. Figure 1.3a shows a model of the molecular architecture of an H_2O molecule. The hydrogen (H) atoms are very small relative to the oxygen (O) atom. Hence, the H_2O molecule is nearly spherical, having a radius of about 0.14 nm. Atoms consist of a positively charged nucleus around which negatively charged electrons are orbiting. Hydrogen atoms have the tendency to donate their electrons for sharing with oxygen, which eagerly accepts that donation. Hydrogen is an electron donor and oxygen is an electron acceptor. Because of the positions of the H atoms relative to the O atom, the charge in the (overall electrically neutral) H_2O molecule is not evenly distributed. Positive charges (+*q*) are centered on each

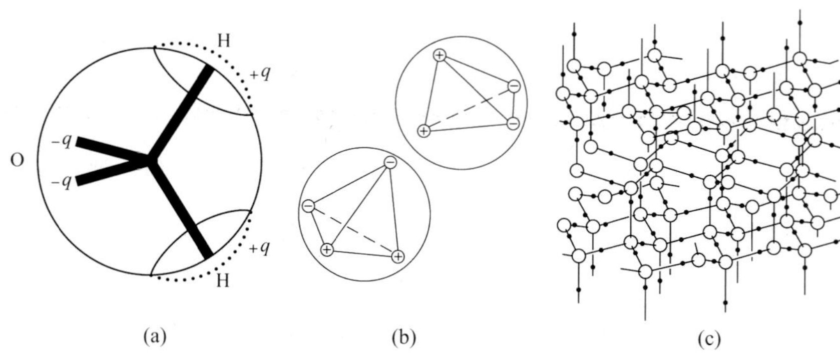

Figure 1.3 Water. (a) Model of a water, H_2O, molecule showing the positive charges +*q* on the hydrogen atoms and the negative charges −*q* on the oxygen atom. (b) Polar interaction occurs through so-called hydrogen bonds between water molecules. (c) The three-dimensional structure of a water lattice in which all potential hydrogen bonds are realized (i.e., ice).

H atom, and compensating negative charges ($-q$) are on the opposite sides of the O atom.

Because of those positive and negative sides, H_2O is said to be a polar molecule. It is the polar character of the molecules that causes the strong internal coherence in water: the positive and negative poles attract each other (Figure 1.3b), so that each H_2O molecule tends to be connected to four other H_2O molecules via so-called hydrogen bonds. In its solid state, ice, the water molecules are in more or less fixed positions, with all four hydrogen bonds realized. Owing to the positions of the poles on the H_2O molecules this results in a relatively open structure (Figure 1.3c) with an H_2O volume density of 55%. When put under pressure some hydrogen bonds may become disrupted and the regular ice structure will be distorted: ice melts under pressure, and in the liquid state H_2O has a somewhat less open structure or, in other words, a higher density than in the solid state. This is another peculiarity of water. In the liquid state the H_2O molecules are still strongly associated in clusters and participate in about three out of the potential four bonds per molecule. Contrary to what one would expect for an associated liquid, the viscosity, that is, the fluidity, of water is not strikingly different from that of other, non-polar, liquids. The mobility of the individual water molecules in the clusters, underlying the macroscopic fluidity of the liquid, is retained, because the molecules readily rotate and hop about every 10^{-11} s from one partner to another, while having most, but not all (as in ice), hydrogen bonding potentialities satisfied.

Another property of water that deserves attention is the dielectric constant. Without going into detail, the dielectric constant is a measure of the ability to screen the electrostatic interaction between two charges at a given separation distance. Water has a high dielectric constant: in water, electrostatic interaction is 20 times weaker than in chloroform and about five times weaker than in ethanol. It is for this reason that salts in water dissociate into their oppositely charged ions. For the same reason, (bio)polymers, such as proteins, DNA, and polysaccharides, as well as synthetic polymers that contain ionizable groups, acquire charge in an aqueous medium. And so do the surfaces of (solid) materials when they are exposed to water.

Now, having gained some insight into some relevant characteristics of water, we may be able to understand the crucial role of water in shaping bottom-up nanostructures.

1.3 Hydrophobic and Hydrophilic Interactions

It has been illustrated and discussed in Section 1.2 that water is a strongly associated liquid because of favorable polar intermolecular interactions. Addition of another substance (referred to as "solute") will disturb the coherence between the water molecules (Figure 1.4). If the solute molecules are also polar or have a net charge (ions), the polar water molecules interact favorably with the solute as well, just as they do with other water molecules. In that case the solute readily dissolves

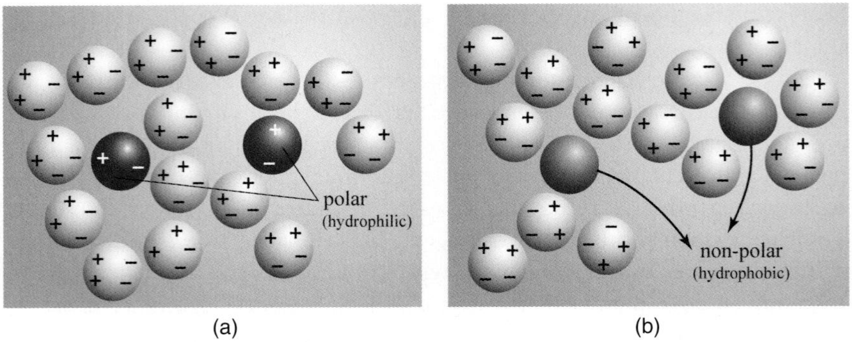

Figure 1.4 (a) Polar and (b) non-polar molecules immersed in water.

in water. The polar substance is called hydrophilic. Salts, sugar, and alcohol are examples of hydrophilic substances. However, if the solute is uncharged and non-polar (i.e., does not have an uneven charge distribution over its molecule), the water molecules prefer to stay attached to each other rather than to the non-polar solute molecules. This results in the non-polar molecules being expelled from the water and driven together to form a separate phase. Oils and fats therefore do not mix with water. For the same reason, the surfaces of plastics, Teflon, and so on, are poorly wetted by water. Such substances, disliked by water, are referred to as hydrophobic.

The description given above of water bordering other substances is highly simplified, especially in the case of non-polar, hydrophobic materials. There are still controversial issues to be solved. Nevertheless, theoretical and experimental studies indicate that, at hydrophobic surfaces, reorientation of water molecules imposes a higher degree of structural order in the adjacent water layer (the so-called hydration layer). Obviously, water molecules bordering non-polar surfaces tend to arrange themselves in a preferred orientation that allows them to form as many H bonds as possible with water molecules in the nearest-neighboring layer, and so on. As nature strives for disorder (in scientific terms, for maximum entropy), hydration of non-polar material is unfavorable and, consequently, the contact area between water and non-polar substances tends to be as small as possible. As a result, non-polar substances spontaneously associate in water because this leads to reduction of the contact area between water and the non-polar substance. Such association between non-polar, hydrophobic substances in an aqueous environment is known as *hydrophobic interaction*. Clearly, the hydrophobic interaction scales with the water-accessible surface area of the non-polar moieties involved. Hydrophobic interaction is one of the major types – if not *the* major type – of interaction occurring in biological systems. If interference by any other type of interaction does not occur, hydrophobic interaction leads to a featureless two-phase system: an "oil" (popular indication of non-polar) phase separates out from an aqueous phase. However, a different and more interesting pattern presents itself when solute molecules possess a non-polar and a polar part. Such molecules are referred to as *amphiphilic*. They occur abundantly in nature, for instance in lipids and proteins.

A typical lipid molecule has a non-polar tail and a polar head-group, as depicted schematically in Figure 1.5a. In water, such molecules show dual behavior. The hydrophobic tails, rejected by the water, assemble together, but the polar heads are preferably hydrated. The exposure of the heads to the aqueous medium prevents phase separation on a macroscopic scale. Instead, microscopic or nanoscopic non-polar phases, surrounded by the polar, hydrated head-groups, are spontaneously formed. Depending on the dimensions of the tail and the head, relative to each other, the amphiphilic molecules aggregate in spherical, worm-like or bilayer structures that close to form spherical objects called vesicles or liposomes. Such structures that are spontaneously formed are named *self-assembled structures* (see Figure 1.5b–d).

Self-assembly plays an important role in shaping the spatial structure of proteins as well. Proteins may be viewed as hundreds of units (the amino acids) linked together to form a long chain, illustrated in Figure 1.6a. There are 22 different amino acids available and, therefore, an almost infinite number of sequences of amino acids along the chain is possible. Some amino acids are polar, others are non-polar. This makes the protein amphiphilic. In an aqueous environment, the chain will try to fold such that the non-polar groups are shielded from contact with water and the polar ones are exposed to water. Of course, other types of interaction, such as electrostatic ones (to be discussed in Section 1.5) may interfere with these tendencies. Still, globular proteins possess a relatively non-polar interior and a polar exterior (see Figure 1.6b). It may be clear that the exact folding pattern depends on the composition and, even more so, on the sequence of the various amino acids along the chain. Many different proteins are active in biological systems, each one having its own specific structure and function.

It should be emphasized that the lipid assemblies shown in Figure 1.5 and the folded protein structure in Figure 1.6 represent highly ordered structures. Their spontaneous formation demonstrates the constructive power of chaos: such ordered structures exist because of increased disorder in the water that is released from hydrophobic hydration.

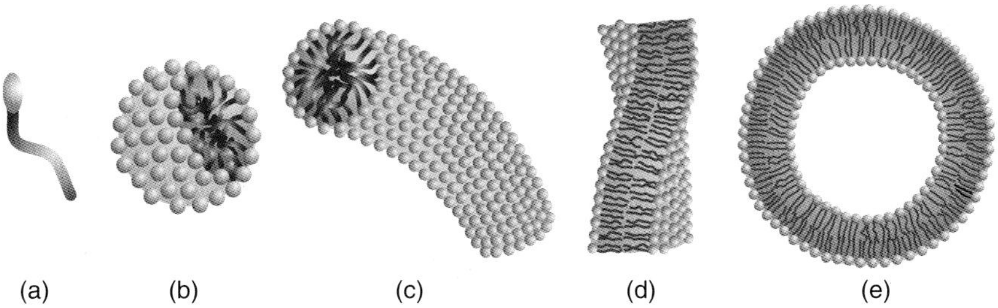

Figure 1.5 (a) The typical structure of an amphiphilic (= "lipid") molecule consisting of a polar and a non-polar part. (b)–(d) In an aqueous environment, amphiphilic molecules aggregate to form supramolecular structures of various geometries.

(a)

(b)

Figure 1.6 Folding of (a) a poly(amino acid) chain, containing polar (black) and non-polar (gray) amino acids, into (b) a compactly structured globular protein molecule, where the non-polar amino acids prefer to reside in the protein's interior, shielded from contact with water, and the polar amino acids prefer the periphery, exposed to the aqueous environment.

It is no surprise that (bio)nano-engineers are strongly inspired by the phenomenon of self-assembly. They design and synthesize tailor-made amphiphilic molecules to build their desired supramolecular assemblies and nano-objects.

1.4
Dispersion Interaction

Dispersion interaction is the most generic one among the different types of physical interactions. Wherever there is matter, irrespective of its properties, there are dispersion interactions.

The origin of dispersion interaction between two atoms may be understood as follows. In any atom, negatively charged electrons orbit around a positively charged nucleus. Although the time-average position of the electrons with respect to the nucleus may be concentric, at any point in time their positions deviate from that average, which gives rise to a small dipole moment. This instantaneous dipole generates an electric field, which, in turn, induces a dipole moment in a neighboring atom, and so on. Thus dispersion interactions are larger between atoms that have a larger polarizability. When the interaction occurs across a medium, the excess polarizabilities (i.e., the polarizability of the atoms in vacuum minus that of the atoms of the medium in vacuum) should be taken into account. Dispersion interaction across a medium is therefore strongly reduced, as compared to vacuum. Dispersion interaction between two atoms diminishes very strongly with separation distance (say, within 0.5 nm), but between particles, containing a large ensemble of atoms, they are more long-ranged and may be effective over (tens of) nanometers.

The strength and the range of dispersion interaction between particles depend not only on the polarizability of the constituent atoms but also on the density (number of atoms per unit volume) and the size of the particles. Dispersion interactions may play significant roles in aggregation of particles, in coating of surfaces with particles, and in folding polymeric molecules into condensed structures, such as, for example, globular proteins. Because dispersion interactions are non-directional, do not require special properties of the interacting species, and occur always and everywhere, they hardly provide a handle to direct bottom-up fabrication of nanostructures.

1.5 Electrostatic Interactions

Electrostatic interactions occur between electric charges. We may distinguish between interactions involving (i) atoms and small molecules, and (ii) polymers, particles, and surfaces.

1.5.1 Atoms and Small Molecules

Atoms and small molecules interact electrostatically when they have a net charge (i.e., when they are ions), and also when they are uncharged but polar. Obviously, interaction between positive and negative charges is attractive, whereas charges having the same charge sign repel each other. In Section 1.2 it was mentioned that interaction between charges across a medium is inversely proportional to the dielectric constant of that medium. It was further pointed out that the dielectric constant is directly related to the polarity of the molecules of the medium.

Charged groups are almost always surrounded by water, but sometimes they reside in a non-aqueous environment. This occurs, for instance, when ionic amino acids become trapped in the interior of a globular protein molecule. In such a low-dielectric environment, ions can only exist when they form a pair: two oppositely charged ions close together, as depicted in Figure 1.7.

The question arises whether ion pairing would contribute to the folding of the protein molecule. An ion pair in the compact structure is electrostatically favored over a wider separation in the unfolded hydrated chain (where the dielectric constant is almost that of water), but the pairing goes at the expense of favorable hydration of the ions in the unfolded structure. These two effects more or less balance each other. Hence, ion pairing as a factor promoting either a compact or unfolded structure in an aqueous medium is highly uncertain. Yet, if ionic groups are forced into a non-aqueous, low-dielectric environment – for instance, due to hydrophobic bonding of adjacent non-polar moieties – pairing between ions of opposite sign is strongly promoted.

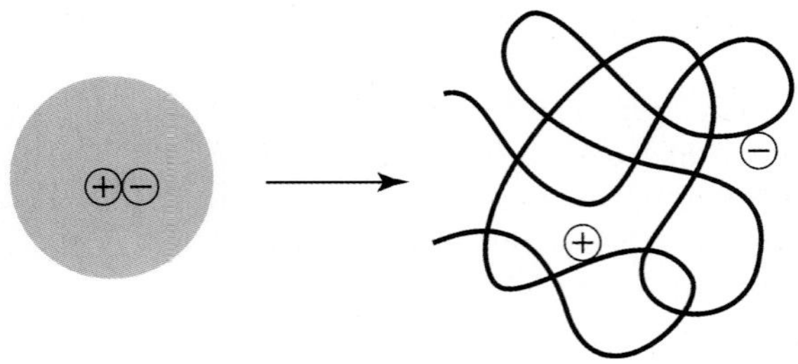

Figure 1.7 Ion-pair disruption in an unfolding protein molecule.

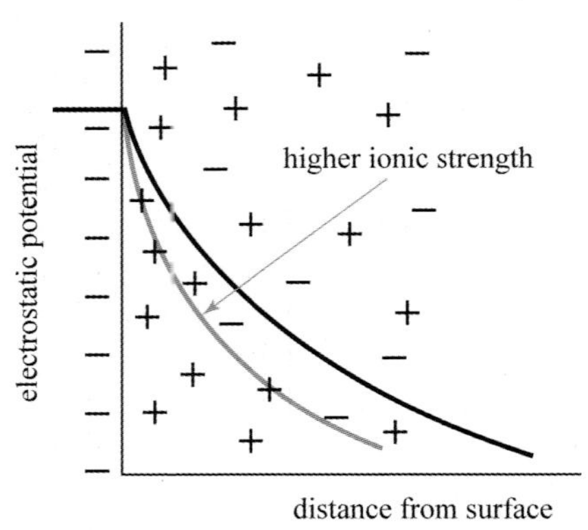

Figure 1.8 Electrostatic potential profile across an electrical double layer.

Similar reasoning applies to the weaker electrostatic interactions between an ion and a dipole, and to the even weaker dipole–dipole interaction.

Because of the relative strength of electrostatic interactions, inserting charged groups at selective locations in a polar and/or non-polar environment can help to direct construction of a desired nanostructure.

1.5.2
Polymers, Particles, and Surfaces

Polymers, particles, and (macroscopic) surfaces in an aqueous environment are often charged, in most cases due to groups that dissociate or associate with hydrogen ions. The charge is balanced by an uneven, diffuse distribution of counter-ions in the adjoining solution in a so-called electrical double layer (Figure 1.8). Electrostatic interactions involving polymers, particles, and surfaces are therefore referred to as *electrical double layer interactions*.

The diffuse distribution of ions in the electrical double layer gives rise to an electrostatic potential that drops off with increasing distance from the charged object. Without explaining the details here, it should be mentioned that the potential decays more steeply with increasing salt concentration in solution (= higher ionic strength) and the separation distance over which the two charged bodies interact decreases correspondingly. For example, in tap water that distance would be about 15 nm, in milk 1.3 nm, in blood 0.8 nm, and in sea water 0.3 nm. For nano-engineers electrical double layer interactions are convenient to work with because of the possibility to modulate their magnitude by just varying easily adjustable parameters such as pH and ionic strength.

In bottom-up nanotechnology, charged soluble polymers, named polyelectrolytes, are essential building blocks. Natural polymers, such as proteins, deoxyribonucleic acid (DNA), ribonucleic acid (RNA), and polysaccharides, are well-known examples, but synthetic polyelectrolytes are regularly used in nanotechnological applications as well. When the polyelectrolytes have the same charge sign, they repel and try to avoid each other; but when they are oppositely charged, interesting phenomena occur. Oppositely charged polyelectrolytes form complexes mainly because of the release of counter-ions from the electrical double layers, which implies increasing disorder and therefore a higher entropy in the system. Here, too, as described for amphiphilic molecules in Section 1.3, the formation of macroscopic aggregates is prevented if at least one of the two polyelectrolytes is linked to an uncharged hydrophilic polymer in a so-called block copolymer. During complexation of the two polyelectrolytes, the uncharged polymer block tends to remain dissolved in the surrounding water, thereby preventing the complexes from growing to macroscopic dimensions. This results in the formation of polyelectrolyte micelles, as shown in Figure 1.9, that are generally referred to as *polyion condensates* or *complex coacervate core micelles*. Recently these structures have attracted a lot of attention as potential nano-containers for the encapsulation and controlled release of pharmaceuticals and nutriceuticals, as well as for their application as nano-bioreactors.

Complexation between oppositely charged polyelectrolytes can also be used to regulate the consistency of certain foodstuffs. The texture of dairy products such as (drink) yogurts may be optimized by adding positively charged polysaccharides that associate with, for instance, negatively charged milk proteins.

Particles dispersed in water carry their charged groups exclusively at the aqueous periphery (except, perhaps, for a few ion pairs formed inside the non-aqueous

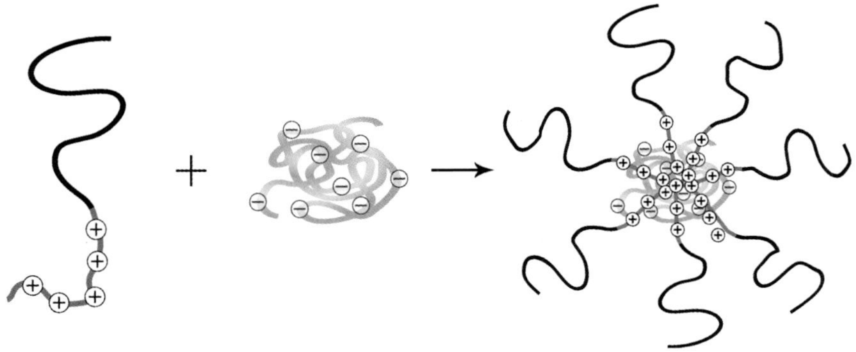

Figure 1.9 Formation of a polyelectrolyte micelle.

interior of the particle; see Section 1.3). When charged particles approach each other, the electrical double layers overlap, giving rise to repulsion between like-charged particles, and attraction if the particles are oppositely charged. A repulsive electrical double layer overlap helps to keep the particles apart and therefore contributes to the stabilization of the dispersion. Even when the particle surface is predominantly hydrophobic, at not too high ionic strength, electric double layer repulsion operating over a larger distance than attractive dispersion and hydrophobic interaction (see Sections 1.3 and 1.4) prevents the particles from aggregating. In media of higher salt concentration, typically beyond, say, 1% sodium chloride, electrical double layer repulsion is too short-ranged, and attractive dispersion interaction takes over and the like-charged particles aggregate. It goes without saying that oppositely charged particles readily aggregate under all conditions of ionic strength (see Figure 1.10).

The stability of nanoparticles against aggregation is a requirement in various food products, and in pharmaceutical and cosmetic formulations. Destabilization of particulate dispersions may be required when clearing, for example, wine or fruit juices and in (waste)water purification processes.

Electrostatic interactions between polymers and particles, between polymers and surfaces, and between particles and surfaces follow the same principles as described above for polymers and particles. Thus, attachment of polymers onto particles and planar surfaces, and of particles onto planar surfaces may be electrostatically favored or disfavored. As in nature, in nanotechnology, polymer–particle interactions often determine whether or not particles connect to each other. This will be further discussed in Section 1.6.

Surfaces can be covered by a layer of nanoparticles to provide the surface with special properties with respect to, for example, wetting or scratch resistance. Finally, polymers are applied at surfaces to render them resilient against (bio) fouling by suppressing the deposition of proteins, biological cells, and microbes. Non-fouling surfaces are a requirement for successful operation of equipment in food processing and water purification equipment, such as separation membranes, heat exchangers, and so on.

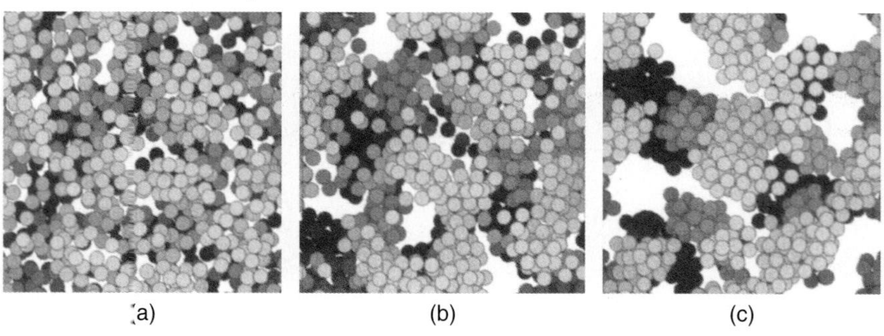

Figure 1.10 Different packing patterns of aggregated particles.

1.6 Steric Interactions Involving Soluble Polymers

In both naturally occurring and man-made nanoscopic constructs, polymers are present in solution and/or attached (= adsorbed) onto (particle) surfaces. The polymers could be uncharged or charged (polyelectrolytes), strongly hydrated or more compact. Polymers, either attached to a surface or not, influence the behavior of nanosystems through steric interactions. These interactions are intricate and delicate, and may change from attractive to repulsive by subtle variations of environmental conditions.

1.6.1 Depletion Aggregation of Particles by Non-adsorbing Polymers

The center of a non-adsorbing polymer molecule cannot approach a particle surface closer than its own radius. Hence, around each particle there is a zone where polymer molecules cannot be accommodated. When particles, by diffusion or otherwise, come closer than twice the radius of the polymer, the polymer will be moved aside and the gap between the particles will be depleted of polymer and just be filled with solvent. This situation is schematically illustrated in Figure 1.11. Because solvent and soluble polymers tend to mix homogeneously, solvent molecules flow from the gap into the solution, which drives the particles together. This type of aggregation induced by non-adsorbing polymers is referred to as depletion aggregation. Depletion aggregation is enhanced by increasing the concentration and the size of the polymer molecules.

1.6.2 Bridging Aggregation of Particles by Adsorbing Polymers

Most polymers tend to adsorb at surfaces. This may be due to binding of hydrophobic segments in the polymer chain (to escape from the hostile water), to hydrogen bonding with surface groups, or, in the case of polyelectrolytes and

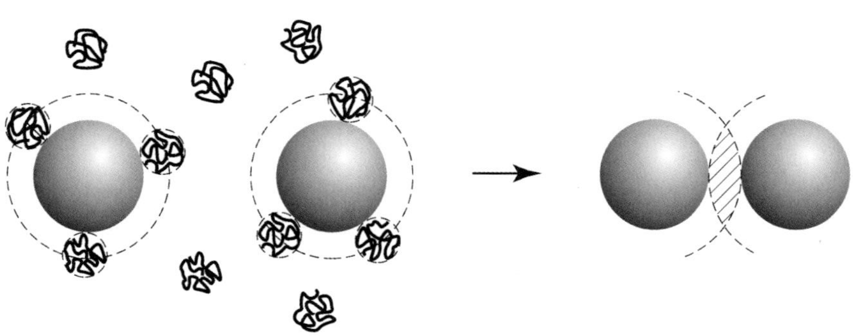

Figure 1.11 Aggregation of particles driven by overlapping polymer depletion zones.

charged surfaces, to favorable electrostatic interactions. When a polymer adsorbs onto a surface, it usually does not lie flat on that surface but adopts a "loopy" structure with loose tails at the terminal ends, as shown in Figure 1.12. If the particle surface is in excess of the polymer, the surface will only become partially covered by the polymer. This allows one and the same polymer molecule to bridge between different particles (see Figure 1.13). Similar bridging aggregation can occur if bare particles are added to particles that are covered with polymer molecules, fully or not.

Bridging aggregation requires that the polymer adsorbed on one particle extends into the surrounding solution far enough to reach another particle. When both particles are charged and repel each other electrostatically (Section 1.5), polymer bridging can only take place if the extension of the polymer exceeds the range over which electrostatic repulsion operates. For this reason, bridging aggregation is more likely to happen with longer polymers and at higher salt concentrations in solution, where the range of electrostatic interaction is reduced. The aggregates formed are loosely structured, with the particles not in direct contact with each other. Closer approach of the particles is detrimental, as it would impose unfavorable deformation of the polymer bridges between the particles.

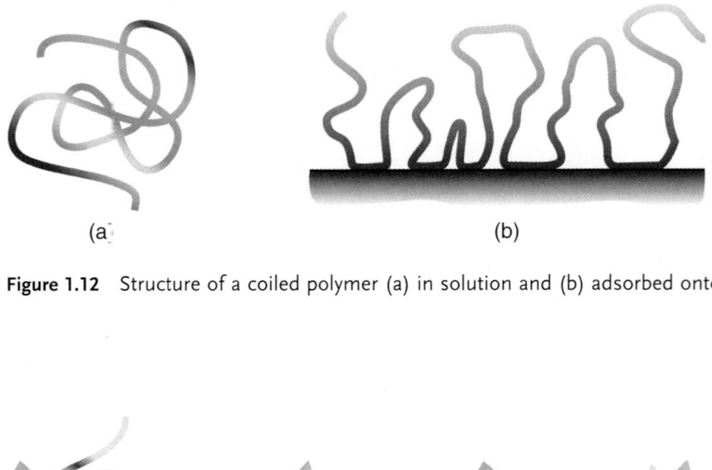

Figure 1.12 Structure of a coiled polymer (a) in solution and (b) adsorbed onto a surface.

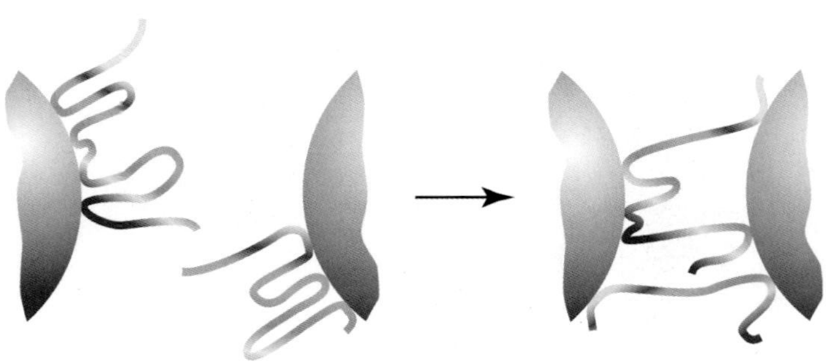

Figure 1.13 Particle aggregation by polymer bridging.

1.6.3
Stabilization of Dispersed Particles by Adsorbing Polymers

When there is an excess of polymers, each particle is saturated with a layer of adsorbed polymer. Then, upon approach of the particles, the outermost fringes of the loops and tails anchored to the respective particles begin to interpenetrate. This leads to unfavorable confinement of the polymer and, hence, to repulsion between the particles that is effective over roughly twice the thickness of the adsorbed polymer layer (Figure 1.14).

Such polymer coatings may be applied to stabilize dispersions under conditions where electrostatic repulsion is insufficiently strong to keep the particles apart, that is, in the case of oppositely charged and uncharged particles, but also of like-charged particles in media of high ionic strength (Section 1.5). To protect the particles from aggregation, the separation distance at which steric repulsion becomes effective (to be regulated by the thickness of the polymer coatings) should be large enough to suppress attractive dispersion interaction (Section 1.4).

Thus, the impact of polymers on interparticle interaction strongly responds to environmental conditions and can therefore readily be manipulated. For instance, by varying the quality of the solvent (e.g., by changing pH, ionic strength, temperature, additives, etc.), the polymer behavior may be adapted from adsorbing to non-adsorbing and vice versa. In this way, bridging may be eliminated, steric stabilization converted into depletion aggregation, and so on. When solvent quality is reduced to below the solubility of the polymer, steric repulsion between fully polymer-covered particles changes into attraction.

1.6.4
Polymer Brushes to Prevent Particle Aggregation and Particle Deposition at Surfaces

A very effective method of steric stabilization can be achieved by grafting polymers at one end onto a (particle) surface, leaving the rest of the molecule dangling in solution. This is best obtained by using diblock copolymers, of which one block has a strong affinity for the surface and the other for the solvent. The profile of

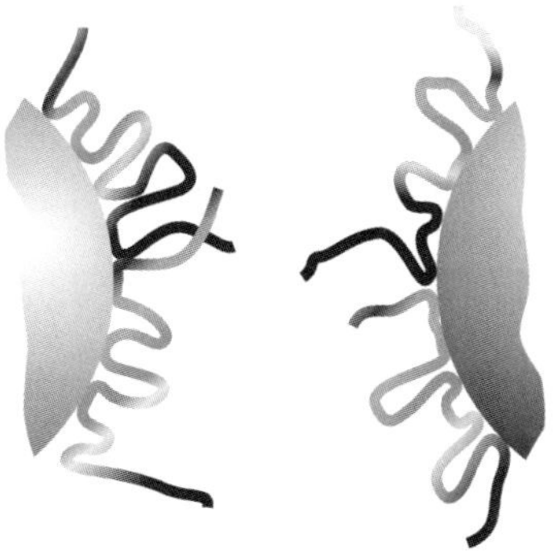

Figure 1.14 Steric stabilization of polymer-coated particles.

the polymer at the surface depends on the grafting density, as shown in Figure 1.15. Obviously, steric stabilization against particle aggregation or deposition improves with increasing grafting density, and is extremely effective when the polymer molecules are stretched out from the surface in a so-called polymer brush. The density of the brush prevents indwelling particles entering, and the strong hydration and high mobility of the polymer chains cause a strong resilience against compression. According to this principle, liposomes (see Section 1.3) used in drug delivery systems may be protected against removal by the immune system, allowing them a longer circulation time in the body. Similarly, the corona polymers of polyion micelles (see Figure 1.9) form a brush that stabilizes the micelles, possibly loaded with some functional ingredient, against external attack by, for example, enzymes or immunoproteins.

Polymer brushes may also be applied to planar surfaces to make them resistant to protein adsorption and microbial adhesion, which, in turn, suppress biofouling. An example is given in Figure 1.16. Besides use in various biomedical applications,

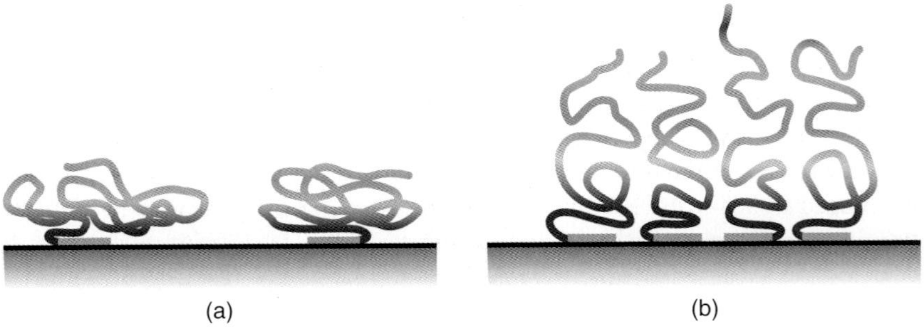

Figure 1.15 End-grafted polymers in (a) a mushroom and (b) a brush conformation.

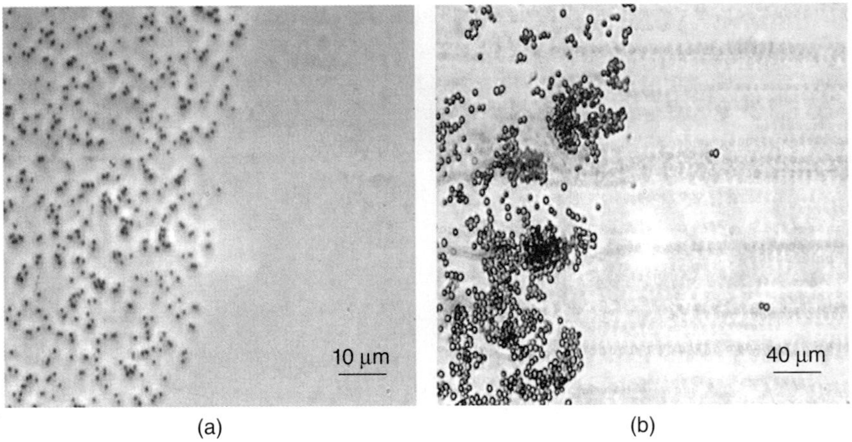

Figure 1.16 Effect of applying a polymer brush (right-hand side of each image) on the adhesion of micro-organisms: (a) *Staphylococcus epidermidis* and (b) *Candida albicans*.

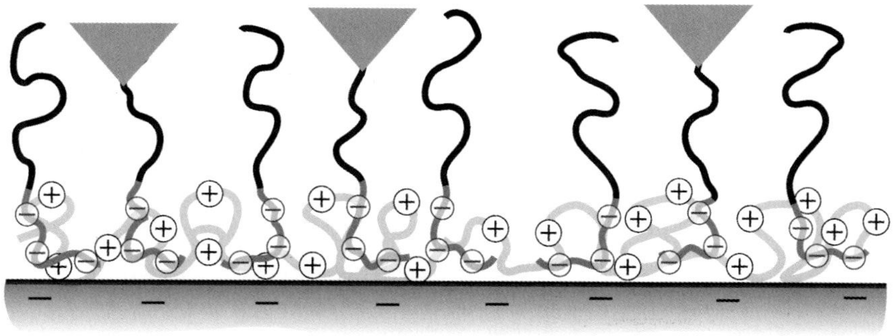

Figure 1.17 Cartoon of a functionalized bio-selective surface.

non-fouling surfaces are of the utmost importance in the food industry to avoid – or, at least, to retard – the formation of a biofilm that will reduce heat and mass transfer and increase frictional resistance and, moreover, may be a source of microbial contamination.

Nature uses its own polymers, usually polysaccharides, to keep cell surfaces clear of unwanted particles or molecules. Nature even goes a step further. At the far end of some polymer chains, receptor molecules may be attached that bind specific target molecules with high affinity. In this way, a bio-selective surface (Figure 1.17) is obtained that binds target molecules but prevents non-specific deposition of other species. Bio-nanoresearch is under way to mimic this principle for application in highly specific and sensitive biosensors and solid-state diagnostics. Such devices have great potential to be used for quality control in the agri-food sector.

1.7 Epilog

For (bio)nanostructures to be functional, they have to respond to an external trigger or signal. Internal bonds in such structures should therefore not be permanent but be sensitive to changing environmental conditions. The physical interactions discussed in this chapter provide such flexibility. For the sake of simplicity, the different types of interaction were presented separately. However, it should be realized that they rarely occur separately, but are usually interdependent. For instance, hydrophobic interaction may be accompanied by ion pairing in a non-polar environment. Conversely, like-charged ions may prevent hydrophobic interaction from occurring as in the case of stabilization of hydrophobic particles by electrical double layer interaction. Another example of interweaving interactions is the combination of electrostatic and steric effects induced by polyelectrolytes.

Hydrophobic, electrostatic, and steric interactions depend on different properties, that is, polarity, charge, solubility, and polymer adsorption behavior, whereas dispersion forces are less specific. Knowledge of the origin, characteristics, and mutual dependence of the various types of interaction provides the nano-engineer with clues to design the building blocks to be used in bottom-up nanostructuring. If tailor-made building blocks are brought together in a well-defined, usually aqueous, medium, they may self-assemble to yield the desired nano-object. Isn't it miraculous? It is almost magic! It is like a car spontaneously emerging from the proper blend of its parts, screws, and bolts.

Further Reading

For more specific and detailed information the reader is referred to:

Norde, W. (2011) *Colloids and Interfaces in Life Sciences and Bionanotechnology*, 2nd edn, CRC Press, Boca Raton, FL (forthcoming).

Israelachvili, J.N. (2004) *Intermolecular and Surface Forces*, 3rd edn, Academic Press, New York.

2 Supramolecular Structures

Pieter Stroeve

2.1 Introduction

Well before the current fascination with nanoscience and nanotechnology, scientists were studying phenomena on the nanoscale. For example, colloid and surface scientists have been interested in colloidal dispersions, micelles, vesicles, and surface modification by a layer of molecules for more than 150 years. Cell biologists have studied the organized structures existing in living cells since the nineteenth century. These structures are now known to have intricate geometry on the nanoscale, with very specialized molecular functions such as transport, synthesis, and energy generation. Plant cells have very complex internal structures similar to living cells. Most of our food is ultimately derived from plants, and studies of plant cells on the nanoscale are giving microbiologists, plant scientists, food scientists, and engineers new information about how to modulate plant growth, plant diversification, harvesting, food processing, and food preservation. For food science, nanotechnology has a different meaning from that encountered in other disciplines, such as the fabrication of integrated circuits for high-speed computers. For food, nanotechnology can be defined as the understanding of food on the nanoscale and translating this knowledge into new processes for food modification and enhancement of food value and preservation. This approach is one of the greatest challenges in food science and engineering.

The understanding of plant cells on the nanoscale is the fundamental basis for developing the nanoscience and nanotechnology to produce new and improved foods. By way of example, pulsed electric field (PEF) processing may be mentioned. PEF processing has been used to increase the rate of dehydration of water from fruits to produce dried fruit. It is well known that for certain conditions PEF can irreversibly open nanopores in the plant cell membranes (electroporation) and water can then escape more rapidly from the plant cell. The cell membrane is made up of a lipid bilayer, and lipids have self-assembly properties. The optimal conditions for PEF to increase dehydration for different fruits are not known. Likewise, the optimum nanopore size in the cell plant membrane and the number of openings per unit area of membrane are not known. Although there

Nanotechnology in the Agri-Food Sector: Implications for the Future, First Edition. Edited by Lynn J. Frewer, Willem Norde, Arnout Fischer, Frans Kampers.

is now some knowledge about the response of the lipid bilayer to PEF, in a cell membrane proteins are associated with the lipid bilayer and they change the properties of the cell membrane significantly. Thus the study on PEF and its effect on plant cell membranes on the nanoscale can give important information that would be useful for making decisions on PEF processing of plant materials used for food.

In addition to the drying of fruits and vegetables, PEF is also useful in killing bacteria in process water streams by making the membranes of bacteria leaky. Another application of PEF that has been explored is enzyme deactivation. Further, reversible electroporation may provide an opportunity to introduce desirable components (color, flavor, nutrients, antioxidants) or remove valuable components while damage to the plant is reversible.

Another process that is receiving considerable interest is high-pressure processing of foods. In this process, food is treated at elevated pressures of the order of 6000 atm. The purpose of the treatment is to inactivate bacteria and to change the food quality. The precise effects of high-pressure processing on plant cell structures and properties are not known but are currently under study by a number of groups.

The study of the properties and functions of nanostructures in plant cells and their changes due to processing is of utmost importance to develop the database for devising new food processes. The molecules in plant cells inherently can self-assemble into structures and this process of self-assembly will be treated in this chapter.

2.2 Self-Assembly

Self-assembly is the process in which a disordered system of molecules spontaneously forms an organized structure or pattern that is at equilibrium or in a quasi-equilibrium state. A typical example is when surfactant molecules dissolved in water self-assemble to form micelles. A typical surfactant is sodium dodecyl sulfate (SDS), shown in Figure 2.1, which is an anionic surfactant. A surfactant contains both an "oil-loving" hydrocarbon chain and a "water-loving" hydrophilic headgroup. This gives the surfactant molecule amphipathic ("being of two kinds") properties, in that the surfactant can be in either an aqueous or a hydrophobic

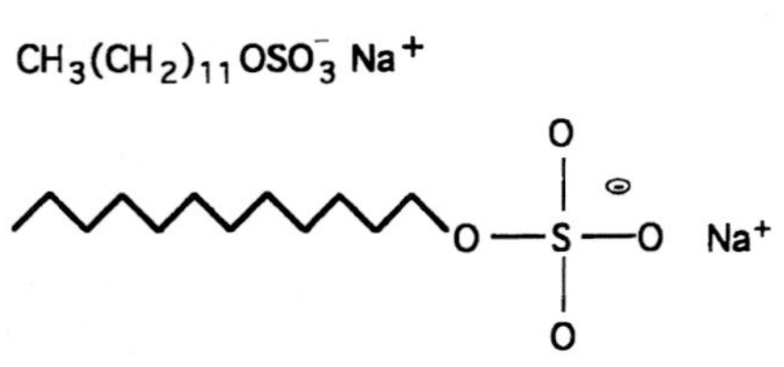

Figure 2.1 The chemical structure of sodium dodecyl sulfate.

environment. In an aqueous environment, SDS can self-assemble its hydrocarbon chains into a micellar aggregate.

In self-assembly, the organized structure or pattern formed has a reduced free energy compared to the initial state of the disorganized molecules. The specific molecular interactions existing between the molecules, before and after the spontaneous change, lead to the lowering of the overall free energy. Self-assembly processes in plant cells often take place at constant temperature and ambient pressure in an aqueous environment. For a spontaneous process to take place at constant temperature and pressure, the second law of thermodynamics states that the Gibbs free energy change between the initial state of a mixture of molecules, and the final state of some organized molecules, must be equal to or less than zero:

$$\Delta G \leq 0. \tag{2.1}$$

Since the Gibbs free energy is related to the change of enthalpy and entropy, Equation 2.1 can be changed to

$$\Delta G = \Delta H - T\Delta S \leq 0, \tag{2.2}$$

where H is the enthalpy, S is the entropy, and T is the absolute temperature. It can be inferred from Equation 2.2 that a decrease in enthalpy and an increase in entropy favor the occurrence of the process. However, if one of the terms, that is, ΔH or $T\Delta S$, opposes the process, the other one must have an (over)compensating favorable contribution to allow for a spontaneous process. The process of self-assembly of molecules to form organized structures does lead to an overall increase of entropy if both the organizing molecules and the solvent molecules are considered. This phenomenon can be further understood by the fact that, for the initial state, the water molecules in contact with an individual organic molecule are organized around the organic molecule like a cage. The water cage is similar to a clathrate and has a crystal-like structure. Upon self-assembly of the organic molecules (with themselves), the initial state and order of the water molecules around the individual organic molecules is changed, and the water molecules now in their final state are more random than before and are more like bulk water. Thus, even though the organic molecules have increased order and decreased entropy, the *net* entropy of the system has increased due to the increase in disorder and greater increase in entropy of the water molecules.

The spontaneous and reversible organization of molecular units into ordered structures occurs by non-covalent molecular interactions. The molecular interactions include van der Waals forces, hydrogen bonding, π–π bonds, and ionic interactions. These molecular interactions are often called weak interactions because their energies are considerably weaker (by a factor of 10 or more) than covalent or other bonds. Nevertheless, weak interactions play a very important role in nature. Weak interactions are responsible for the state of a pure component, such as the liquid or solid state versus the gaseous state. Obviously, weak interactions are of great importance in biological systems, such as the self-assembly of organized structures in plant cells. Examples of organized structures in biology are self-assembled monolayers, lipid bilayers, micelles, and vesicles. The folding of peptide chains into functional proteins and enzymes is another example of structures in biology and occurs in plant cells. The weak interactions are also responsible for the possibility that the organized structures can undergo changes due to a change in thermodynamic variables, and return to the original structure if the thermodynamic conditions return to the original values, that is, the structures can show a degree of reversibility. The weak forces allow the structure to change, effectively seeking a new minimum free energy, depending on the existing thermodynamic conditions. Thus the properties of organized structures can change depending on the thermodynamic conditions. In terms of applications, such as food processing using plant materials, it makes external control of induced changes by processing steps such as temperature, pressure, electric field, and so on more difficult and troublesome because one needs to fully understand not only the structural changes that occur on the nanoscale but also the change in the properties on the nanoscale level and the change in the overall properties on the macroscopic level. On the other hand, it opens exciting new opportunities to change the properties of foods and the possibility to create new value-added foods.

It is possible for chemical reactions to cause molecules to self-assemble. An example is the chemisorption of molecules on a surface to form an organized monolayer. The driving force for self-assembly is the change in enthalpy due to the chemical transformation. This type of self-assembly is not reversible.

Before considering plant cells and self-assembled structures, one should understand that the complexity of molecules that can self-assemble is of great importance. The chemical composition of molecules, their size and shape play important roles in what type of organized structures can be created. Because of this complexity, a wide variety of nano- and mesoscopic structures can be formed depending on the type of molecules involved. Further, it is not necessary that the molecules are all the same: mixtures of different molecules (composition, molecular weight, shape, size, and charge) can self-assemble into organized structures. Properties associated with the structures depend on the nature of the structure and the types of molecule that make up the structure. Nature has learned to exert fine control on the formation of cellular structures with specific functions and properties, by choosing the appropriate precursors to self-assemble into the desired structure. These structures can be called supramolecular assemblies.

From a historical point of view, researchers in the discipline of chemistry were the first to explore the field of supramolecular chemistry, the assembly of synthesized molecules that can arrange into precise, well-organized structures. The synthesis of special molecules that can give rise to supramolecular assemblies is known as supramolecular chemistry. Supramolecular assemblies include molecular self-assembly, folding, molecular recognition, and host–guest chemistry (enzyme–substrate). The principles of supramolecular chemistry and supramolecular assembly are similar to what is done by nature in biological systems. In both situations, molecules can interact because of weak molecular interactions. The folding of two single-stranded deoxyribonucleic acid (DNA) chains into a double helix is a supramolecular assembly and it is a consequence of weak interactions. The base-pairing in the DNA double helix formation is an example of molecular recognition that is solely due to weak molecular interactions. The study of non-covalent interactions is important in many biological systems. Self-assembly is crucial to the function of cells. An example is the self-assembly of lipids to form membranes, the formation of double helical DNA, and the assembly of proteins.

2.3 Plant Cells

Plant cells are eukaryotic cells that have distinctive, organized structures. A plant cell is shown in Figure 2.2. Like all cells, plant cells have a cell membrane, which has as a fundamental building block the lipid bilayer. It is known as the plasma membrane. However, plant cells contain a number of specialized structures. Plant cells have a cell wall, which provides the cell with structural support. A major function of the cell wall is to act as a pressure vessel to prevent over-expansion of the cell when water enters. Another role of the cell wall is to support the plant and to confer flexibility and tensile strength. Lignocellulose is the primary building block of plant cell walls. The cell wall is mainly composed of cellulose, hemicellulose, lignin, and smaller amounts of pectin, protein, and extractives (soluble non-structural materials such as non-structural sugars, nitrogenous material, chlorophyll, and waxes). The composition of these constituents may vary from one plant species to another. In addition, the ratio of the various constituents within a single plant varies with its age, stage of growth, and other conditions. Cellulose is the main structural constituent in plant cell walls and is found in an organized fibrous structure. This linear polymer consists of D-glucose subunits linked to each other by β-(1 $\rightarrow$ 4)-glycosidic bonds. Cellobiose is the repeating unit established through this linkage and it constitutes cellulose chains. The long-chain cellulose polymers are linked together by hydrogen and van der Waals interactions and

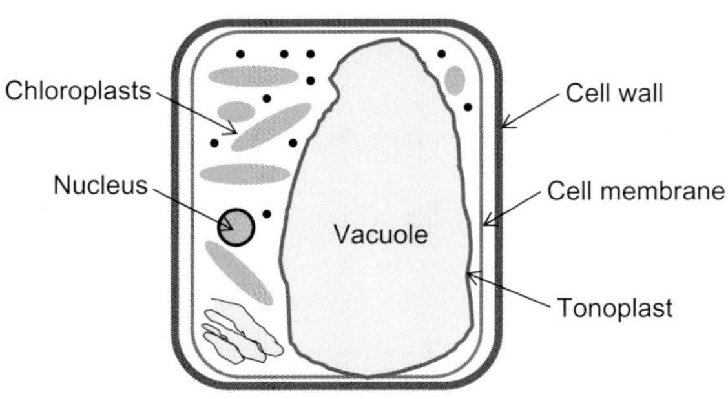

Figure 2.2 Schematic diagram of a typical plant cell. The vacuole can occupy from 70% to 90% of the interior of the cell.

cause the cellulose to be packed in microfibrils. Hemicelluloses and lignin cover the microfibrils. Cellulose in cell walls is present in both crystalline and amorphous forms. Crystalline cellulose forms the major proportion of cellulose, while there is a small percentage of non-organized cellulose chains, which form amorphous cellulose.

There are pores present in the cell wall, which are known as plasmodesmata. In the pores, the plasmalemma and endoplasmic reticulum of adjacent cells are continuous, which allows for cell-to-cell communication, which includes the transport of species.

Another specialized structure in the plant cell is a large central vacuole, which is completely enclosed by the tonoplast membrane. Like all biological membranes, the lipid bilayer is the building block that forms the tonoplast, although the tonoplast composition and properties are different from those of the cell membrane. The vacuole is filled with an aqueous mixture called the sap. The vacuole's membrane controls the movement of molecules between sap and the remaining fluid in the plant cell, the cytoplasm, which is known as the cytosol. The vacuole stores nutrients and digests waste materials and controls the cell's turgor.

Like many other cells, the plant cell contains mitochondria, Golgi vesicles, the Golgi body, small membrane vesicles, a nucleolus surrounded by a nuclear envelope containing nuclear pores, smooth endoplasmic reticulum, rough endoplasmic reticulum, ribosomes, and so on. The plant cell further contains starch grains, which serve as a storage of nutrients for the cell.

In the plant cell are plastids, which are organelles that serve as the site of synthesis and storage of important chemical compounds used by the cell. Plastids are responsible for photosynthesis, for the storage of products like starch, and for the synthesis of molecules, such as lipids, which are used as cellular building blocks and for the function of the plant cell. A structure of a typical lipid is shown in Figure 2.3, while the chemical structure of a phosphatidylcholine (lecithin) is shown in Figure 2.4. Lipids contain a hydrophilic head-group and two alkane chains. A double bond in a chain will cause a kink and has implications for the packing of the lipid in a lipid bilayer. Straight chains can pack tighter, leading to greater order of the bilayer, but less fluidity. Lipids with a kink in the chain cannot pack tightly in lipid bilayers, leading to more fluid-like behavior of the bilayer.

2.4 Organized Self-Assembled Structures

2.4.1 Langmuir Layers

The simplest organized structure is a monolayer of molecules organized at an interface. The study of monolayers at an interface can be done simply by using a Langmuir trough, also known as a Langmuir–Blodgett trough. Ultrapure water is placed in a Langmuir trough, and then a minute amount of a solution of a surface-

Figure 2.3 Structure of a typical lipid.

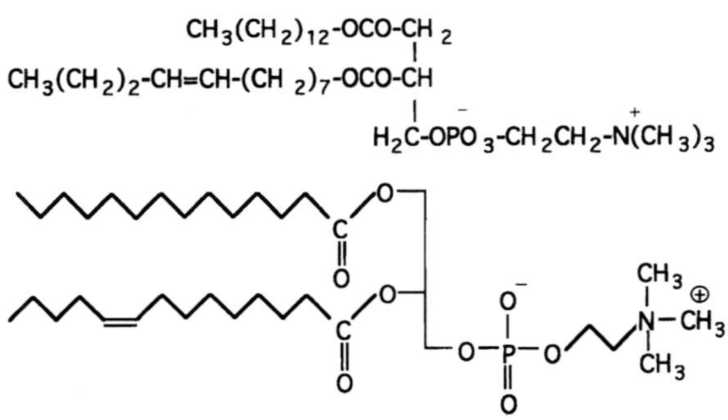

Figure 2.4 Chemical structure of phosphatidylcholine.

active species (e.g., surfactant, fatty acid or phospholipid) is administered to the air–water interface with a microsyringe. The solution is usually the surface-active species dissolved in a volatile, non-surface-active solvent such as hexane. By using the microsyringe, a known volume of the solution can be brought to the air–water interface by a series of small drops. The solution spreads rapidly (flashes) on the interface, due to the positive spreading coefficient of the solvent, which causes the surface-active species to be distributed on the air–water interface. At the same time, the solvent evaporates due to its high vapor pressure, so that the surface-active species is the only one left on the air–water interface.

The Langmuir trough contains a movable barrier that divides the air–water interface. By moving the barrier, the surface area of the air–water interface and consequently the surface concentration of the surface-active species can be changed. The monolayer of surface-active species is known as the Langmuir layer and it lowers the surface tension (surface energy) of the air–water interface. An increase in the surface concentration, due to the barrier movement, can further lower the surface tension. The surface tension can be continuously measured by a Wilhelmy plate as shown in Figure 2.5. On the molecular scale, when the barrier compresses the surface-active species at the air–water interface, the surface-active molecules will arrange in an ordered layer, with the head-groups associating with the water and the hydrophobic tails aligned and sticking out of the interface. The Langmuir layer can be transferred to a solid substrate by dipping the substrate through the air–water interface. Repeated transfer of the solid substrate through the air–water interface can transfer multiple organized layers of the surface-active species to the substrate. Transferred molecular layers on a solid substrate are known as Langmuir–Blodgett layers and these layers can be readily studied by surface analytical instrumentation such as infrared (IR) spectroscopy, surface plasmon resonance (SPR), and atomic force microscopy (AFM).

2.4.2
Lipid Bilayers

The lipid bilayer is a membrane made up from lipid molecules. The lipid bilayer is a fundamental component of all biological membranes and is essential to life. Its structure was discovered in 1925 by Gorter and Grendel, who compared the surface area of human red blood cells with that of a known amount of lipids in a Langmuir trough. Gorter and Grendel found that the area of lipids from a known

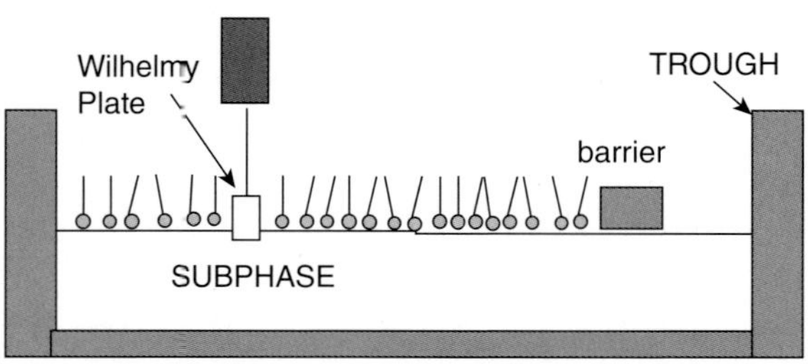

Figure 2.5 Cartoon of a Langmuir trough containing a water subphase and a surfactant spread at the air–water interface. The Wilhelmy plate is connected to an electronic balance. The movable barrier can be used to increase the surface concentration of the surfactant at the interface.

number of red blood cells, when spread out on the water of the trough, was twice the calculated surface area of the red blood cells. To explain their results, they concluded that the membrane is two lipid molecules thick and that the membrane is made of a bilayer.

The bilayer is composed of two layers of lipids arranged so that their hydrocarbon tails face one another to form an oily core held together by the hydrophobic effect, while their charged heads face the aqueous solutions on either side of the membrane. A phospholipid is an amphiphilic molecule that consists of a polar head-group and two non-polar fatty acid tails. The lipid bilayer forms a membrane matrix where other biomolecules such as proteins can be embedded.

The properties of the bilayer are determined by a number of factors, including the lipid composition, the lipid size and shape, and the temperature. The nature of the lipid head-groups and the length and degree of saturation of the hydrocarbon chains play an important role. The presence of a *cis* double bond in the carbon tail of a lipid produces a kink, which makes it more difficult to pack the tail with straight neighbors. Kinks effectively introduce disorder and lead to a more fluid behavior of the hydrocarbon region of the lipid bilayer. The more kinks there are, the greater the disorder and the more fluid the bilayer becomes.

The lipid bilayer acts as a barrier. The hydrophilic interfacial regions associate with water, while the inner hydrophobic core region contains essentially no water. Because of the oily nature of the bilayer, it is only permeable to small hydrophobic solutes. Hydrophilic molecules and ionic compounds have a very low permeability for transport through the lipid bilayer. Thus the lipid bilayer is permselective, allowing some molecules to pass through, but retaining others, thus regulating transport in to and out of the cell. The transport of species across the cell membrane can be by either passive diffusion, coupled diffusion, or active transport, which requires the expenditure of energy.

The cell membrane contains a wide variety of biological molecules, primarily proteins and lipids, which are involved in an array of cellular processes, such as transport, cell adhesion, and cell signaling. The plasma membrane also serves as the attachment point for both the intracellular cytoskeleton and the extracellular cell wall. The cell membrane surrounds the cytoplasm of the cell. In the plant cell, the cell wall forms the outermost boundary, but it plays mostly a mechanical support role rather than a role as a permselective boundary. The cell membrane anchors the cytoskeleton to provide shape to the cell, and in attaching to the extracellular matrix to help group cells together in the formation of tissues.

2.4.3
Solid-Supported Lipid Bilayers

Since a supported bilayer membrane was first used to investigate cellular immune responses, solid-supported lipid bilayers have been a widely studied topic of practical and scientific interest in recent years. Being well-defined models of biological membranes, phospholipid bilayers supported on solid substrates are important for their roles in fundamental biophysical research as well as in applications such as biosensors. Supported lipid bilayer membranes have been formed onto glass, quartz, and silicon surfaces, onto non-functionalized metal surfaces, or onto self-assembled alkanethiol monolayers. Methods for bilayer formation have included the Langmuir–Blodgett technique, vesicle fusion onto the substrate, spontaneous thinning of lipid–decane mixtures, and adsorption of charged lipids onto oppositely charged surfaces. A bilayer deposited directly on silica, glass or gold is a model membrane with a lack of functional integrity, as shown in Figure 2.6. Similar to lipid bilayers, supported lipid bilayers can have domains that depend on the composition of the lipids in the bilayer.

To yield space for accommodating large integral trans-membrane proteins in the supported lipid bilayer and to give lateral mobility to membrane components, a flexible polymer layer, preferably a hydrogel, can be inserted between the solid substrate and the bilayer, as shown in Figure 2.7. In Figure 2.7, the supported lipid bilayer is on top of a water-soluble polyion. The polyion itself is supported on a self-assembled monolayer (SAM) of an alkanethiol on a gold substrate. Hydrated polymer layers, self-assembled monolayers, and supported polyelectrolyte films have served as soft cushions for lipid bilayers.

For insertion of large membrane proteins into the supported lipid bilayer, a thicker polymer cushion is needed to lift the lipid membrane away from the solid substrate, which can be achieved by employing the layer-by-layer polyion adsorption technique. Additional layers of positively charged poly(diallyldimethylammonium chloride) (PDDA) can be adsorbed by interleaving with a polyanion such as negatively charged polystyrene sulfonate (PSS). The technique is very suitable to prepare polymeric films with well-defined thickness and homogeneity better than 1 nm. The dominant interaction, electrostatic attraction of opposite charges, can be used to deposit a bilayer on top of the multilayer polymeric film. Figure 2.7 is an example of a supramolecular assembly that can be fabricated using different self-assembly techniques: chemisorption, physisorption, and vesicular deposition. Exposure of the system shown in Figures 2.6 and 2.7 to a solution of membrane proteins from (for example) a cellular source may cause the proteins to penetrate into the lipid bilayer. A variety of methods, such

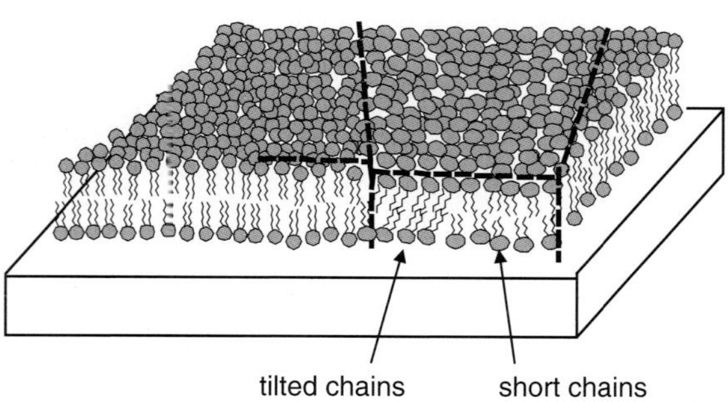

Figure 2.6 A bilayer supported on a solid support such as glass. The supported lipid bilayer is shown with two different domains. (Reproduced from Vidu *et al.* [1].)

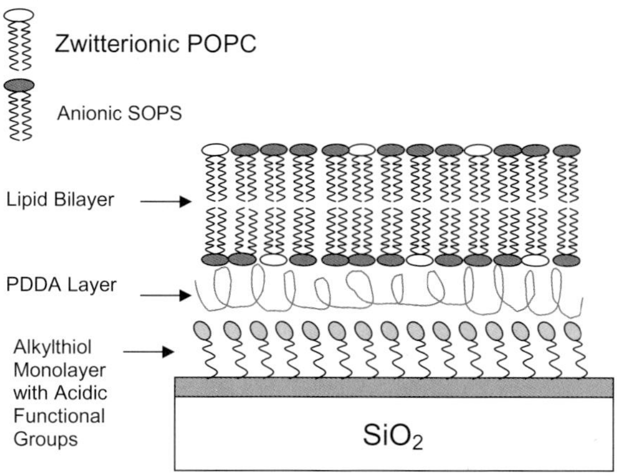

Figure 2.7 Schematic representation of the model membrane system. The alkylthiol 11-mercaptoundecanoic acid (MUA) layer is self-assembled on a gold surface. The negatively charged head-groups of MUA adsorb a cationic polymer (PDDA) layer. A lipid bilayer with negative charges is then deposited on the PDDA/MUA layer pair. (Reproduced from Zhang *et al.* [2].)

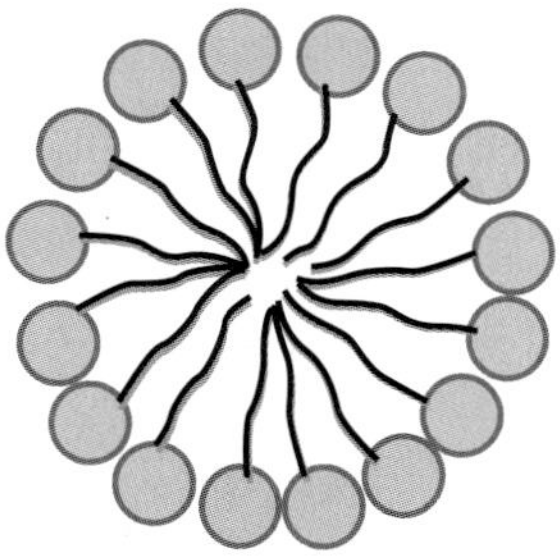

Figure 2.8 Schematic of a micelle in water.

as surface plasmon resonance, atomic force microscopy, cyclic voltammetry, and fluorescence, have been used to measure the uptake of proteins into supported lipid bilayers.

2.4.4 Micelles

A micelle is a colloidal self-assembled aggregate of surfactant molecules dispersed in a liquid and can form spontaneously from the monomer surfactant molecules if the surfactant concentration is sufficiently high. A micelle in an aqueous solution is a soft nanoparticle with the hydrophilic head-groups in contact with the surrounding water molecules and the hydrophobic tail regions sequestered inside the micelle center, as shown in Figure 2.8. The hydrophobic tails of the surfactant molecules have less contact with water when they are part of a micelle, and this

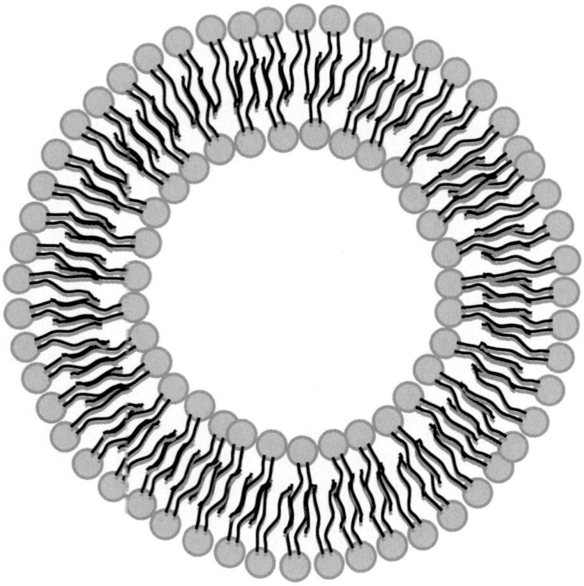

Figure 2.9 Diagram of a vesicle.

formation leads to a lowering of the free energy compared to the surfactant molecules being dispersed in the aqueous medium and interacting with water molecules.

In a hydrophobic medium, inverse micelles can form with the head-groups at the micelle center and the tails extending outward. Micelles are approximately spherical in shape depending on the solution conditions, such as surfactant concentration, solvent, temperature, pH, and ionic strength. Other micelle shapes include ellipsoids, cylinders, and bilayers. The shape and size of a micelle also depend on the composition and shape of the surfactant molecules besides the solution conditions. Micelles can form when the concentration of the surfactant is greater than the critical micelle concentration (CMC), and when the temperature of the system is greater than the critical micelle temperature, also known as the Kraft temperature.

Micelles composed of anionic or cationic surfactants have an electrostatic attraction to the counter-ions that surround the micelles in solution. The micelle charge affects the structure of the surrounding solvent at appreciable distances from the micelle. The distance of charge influence is known as the Debye distance, and it depends on the concentration of the ions in solution, the valences of the ions (but mainly the valence of the counter-ions), the dielectric constant, and the temperature. Ionic micelles can influence the properties of the colloidal mixture, including the electrical conductivity and the turbidity. The addition of salts to a colloidal solution of micelles decreases the strength of electrostatic interactions and can lead to the formation of larger ionic micelles.

2.4.5
Vesicles

A vesicle is an envelope of a lipid bilayer that forms a sac that encloses fluid and separates it from the continuous fluid. Vesicles can form naturally because of the self-assembly properties of lipid bilayers. A diagram of a vesicle is given in Figure 2.9.

Considerable research has been conducted on the use of lipid bilayers to understand the behavior of vesicles. The reason for this popularity is that the procedure of vesicle preparation is straightforward. Essentially, a phospholipid is first dissolved in a hydrophobic solvent with a high vapor pressure. The solution is then placed in a small flask or test tube and the container is rotated to allow the solution to wet the container walls. The solvent evaporates and the lipid deposits on the container wall. After all the solvent is evaporated, an aqueous medium is placed inside the container. At this point small vesicles can be produced by introducing an ultrasound tip and applying high-frequency mixing for a minute or so. Depending on the time and frequency of the ultrasound application, the vesicles will have a certain distribution of diameters. It is possible to produce vesicles with a diameter less than 1 μm. Alternatively, the container with the aqueous medium can be stored for several hours and over time the lipid on the walls becomes hydrated, separates from the container walls, and forms lipid bilayers that enclose to form vesicles. In the second process, very large vesicles (giant vesicles) are produced, with diameters in the order of from several to tens of micrometers. An image of a giant vesicle is shown in Figure 2.10.

Vesicles are used by the cell for organizing cellular substances. Vesicles can transport, store, and/or digest metabolites and waste products. They are involved in metabolism and enzyme storage, and can act as reaction chambers. Vesicles can fuse with the plasma membrane to release their contents outside of the cell. They can also fuse with the membranes of other organelles in the cell. Owing to transport mechanisms in the vesicle bilayer, the inside of the vesicle may be different from the cell interior.

There are a number of specialized vesicles in the plant cell. Lysosomes are vesicles that contain digestive enzymes used to break down substances in the cell. Food vacuoles are vesicles that contain mostly water and metabolic compounds. Food vacuoles fuse with lysosomes, which break down the components in the

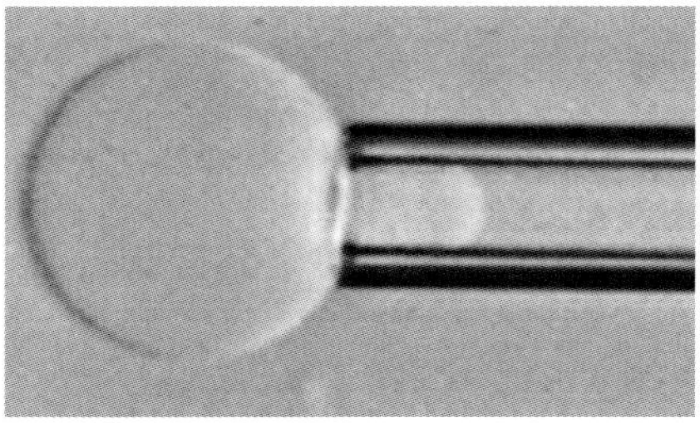

Figure 2.10 Giant vesicle with a diameter of 20 μm held (by suction) to a 7 μm micropipette. (Photo obtained courtesy of Dr. Henry Bouman.)

vacuole for further use in the cell. Lysosomes can also destroy defective or damaged organelles. The lysosomes fuse with the membrane of the defective organelle and then digest the organelle. Transport vesicles move molecules to different locations inside the cell. Secretory vesicles contain waste materials that need to be removed from the cell. The cellular control of the functions of the different types of vesicles is complex and not yet fully understood. The energy needed for these processes often comes from the metabolic reactions, which can also involve conversion of adenosine diphosphate (ADP) to adenosine triphosphate (ATP). ATP effectively stores chemical energy and can release it upon conversion to ADP. Study of cellular processes on the nanoscale seeks to understand the formation and functions of macromolecular assemblies and to couple this to what is known about the cell behavior on the molecular level.

2.5 Summary

Plant cell structures and functions are complex, and are determined on the nanoscale. Many of the structures can form by self-assembly. The understanding of how food processing changes these structures and their functions on the nanoscale is important to formulate new food products and to improve current processes. Nanoscience studies on how the processing of foods can cause favorable changes on the nanoscale can be explored to determine the optimum processing conditions to create value-added foods.

References

1 Vidu, R., Ratto, T.V., Longo, M.L., and Stroeve, P. (2003) Domains, cushioning, and patterning of bilayers by surface interactions with solid substrates and their sensing properties. *Membr. Sci. Technol.*, **7**, 887–915.

2 Zhang, L., Longo, M.L., and Stroeve, P. (2000) Mobile phospholipid bilayers supported on a polyion/alkylthiol layer pair. *Langmuir*, **16**, 5093–5099.

Part Two
Basic Applications

3
Nanotechnology in Food Production

Remko M. Boom

3.1
Introduction

The global supply of food is affected by a number of developments: growing world population, increased affluence of large groups, use of bioresources for fuel and chemicals production, intensification of agriculture, and monopolization of global food supply chains.

Nanotechnology may offer a way to produce very high-quality foods in a much more sustainable way, while offering better bioavailability of (micro)nutrients. An overview of current applications of nanotechnology in and around food products is given in this chapter. In addition, some examples are given on how the technology could contribute to the indicated problems: the strong improvement of bioavailability of lycopenes from nanocrystals or nanosized emulsions; the application of lipid-based delivery systems that may deliver components through the intestinal wall; the production of nanostructured plant-protein-based products; and the development of much better isolation and structuring methods.

The chapter concludes with a brief discussion of resistance to the use of the technology for food production. Even though rational arguments seem to favor the application of nanotechnology, human feelings are of prime importance in such an important subject as the supply of our food. They should not be ignored and may be sufficient reason for not applying the technology. However, this would leave the ethical issue of perpetuating our methods of food preparation at the expense of people who do not have sufficient supplies of food.

Nanotechnology in the Agri-Food Sector: Implications for the Future, First Edition. Edited by Lynn J. Frewer, Willem Norde, Arnout Fischer, Frans Kampers.

3.2
Food Production

3.2.1
Food and New Ways of Food Production

Food is among the most complex materials that we know. Biological tissues are generally structured down to the (macro)molecular level. When we convert living tissue into food, preservation ensures that our food will be safe, even if we do not consume it directly after harvesting, but can store it and prepare it later. The processing (preservation, storage, and preparation) induces changes that we often appreciate, and which may also improve the bioavailability of the nutrients inside the product.

Food is close to us. The expression "we are what we eat" is true in a literal sense, and in a metaphoric sense: our choice in the type and preparations of food, and the way we consume food, is closely connected to our personal and social identity. This implies that we do not like to consider large changes in the way we process foods. Our intuition tells us that we should use the same method as our grandparents used: what was good enough for them, must be good for us, as is claimed by Michael Pollan [1]. In this case, our intuition may not be reliable. Compared to 50–100 years ago, our food is now much safer, and no consumer would accept the risks of food consumption that were normal in the past. However, the statement by Pollan shows the power of our feelings for our food.

3.2.2
Why Do We Need New Processing and Preparation Methods?

If the introduction of new ways to prepare foods is so sensitive, why should we consider them? Why should we want to consider the use of new technology as nanotechnology for the processing of foods?

The main reason is the fact that food is becoming scarcer, expressed in higher market value. There are a variety of reasons for this. First, the world's population is still growing, and will reach approximately nine billion in 2050.

Second, at the same time, the population of large parts of the globe (mainly Asian countries, but also others) are quickly gaining affluence, which means that they are starting to consume more foods that have a high requirement for agricultural resources. For example, the production of meat is extremely costly in terms of usage of agricultural land or crops, use of water, and production of greenhouse gases by cattle. Growing feed for cattle consumes roughly 50% of all water in the USA, and 80% of the agricultural land. Cattle raised in the USA for food consume 90% of the soy, 80% of the corn, and 70% of the cereals.

With the increasing number of people in the world, and with increasing affluence in many regions, we will not be able to support meat production for all. If all the grain currently fed to livestock in the USA were consumed directly by people, the number of people who could be fed would be nearly 800 million [2].

Thus, it would make sense to try to produce high-quality, tasty protein foods directly from plants. Unfortunately, the structure of meat is so intricate that conventional structuring methods cannot nearly match nature; the quality of meat-replacing protein foods is considerably lower than the original.

A third factor is the increasing scarcity of fossil fuels. This makes the use of agricultural resources to produce biofuels (ethanol, biodiesel, and others) and biochemicals to replace chemicals from petrochemistry more attractive. However, this implies that the land used to produce these materials is not available for the production of food.[1)]

A fourth factor is that the increasingly intensive use of land leads to slow degradation of the agricultural land, via for example erosion or salination. This will slowly make the pressure on the remaining land even larger. And a fifth factor is the emergence of food as a political factor: the free, global market enables countries to swap their traditionally produced crops for high-value crops that yield more value.

3.2.3
More Efficient Fractionation of Crops

All these factors, and others, imply that it is important to consider processing technologies that can convert agricultural crops into as many useful (and edible) products as possible. The current technology is mostly aimed at the isolation and purification of a single product, or in some cases two products. Processing of sugar beet is optimized solely for the production of sugar, and the production process is only aimed at ensuring the quality of that product. The rest of the beet is thermally degraded, and is used as animal feed. In order to produce more than only sugar from such a crop, one needs to consider new ways of processing: for example, non-thermal methods, or methods that would enable very precise removal of components, while minimizing the change in the feedstock. Nanotechnologies may enable this, using for example molecular recognition techniques to isolate specific components.

3.2.4
More Efficient Product Structuring

A second issue requiring better and more sustainable processing technologies is the preparation or structuring of foods. Most people like meat as an important part of their diet–not only because of the nutritional value, but also because of the excellent taste of it. The fact that meat is a product that is fibrillar on a

1) This is at least true for the first-generation technology, which directly uses edible fractions such as starch and oil. The second-generation technology uses inedible fractions, such as cellulose and possibly lignin, which are now left on the land to fertilize and protect the soil. Using these for biofuels and biochemicals will reduce the nutrients left on the soil and may thus reduce productivity and promote land erosion in the long term.

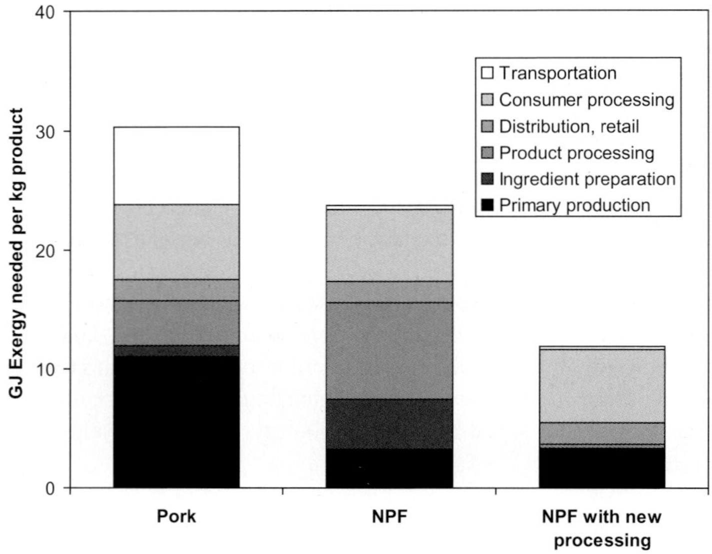

Figure 3.1 Exergy demand of producing 1 kg of pork meat and 1 kg of pea-protein-based meat replacer ("novel protein food", NPF) using conventional processing technology, and possible exergy demand when new and much better processing could be used (exergy demand for new processing estimated at 10% of the original processing). Partly based on [3].

nanometer scale is mainly responsible for this: the flavor components are only gradually released upon chewing, giving a good taste experience during the complete mastication.

One could envisage the preparation of a similarly nanostructured product, with the help of nanotechnology, but now based on plant proteins. The preparation of such a product would require considerably less agricultural resources, since the plant proteins are used directly for food preparation, instead of first being converted into animal protein (saving a factor 4–10). In addition, this direct plant-based product preparation would reduce animal suffering due to poor conditions during their life, and before slaughtering. This development would be welcomed for an additional reason. Especially in the Western world, there is resistance to the meat industry on animal welfare grounds. Nanotechnologies could offer techniques both to fractionate the biofeedstock efficiently into high-value fractions (such as plant protein) and also to nanostructure products that may replace meat in some of our meals.

The case below may serve to illustrate this. Apaiah *et al.* [3] calculated the total "exergy", that is, the potential to perform work (resources) needed to produce one kilogram of pork meat, and the same for a product based on pea protein, which was first purified from peas, and then converted into a meat-replacing product using extrusion (Figure 3.1).

The primary production of the pea-protein-based product is much more sustainable (i.e., requires less exergy), but the current technologies for purification and structuring of the protein product are not efficient. In addition, the product made with extrusion is clearly inferior in quality compared to the original and will therefore not be chosen by consumers. Overall, the new product is only marginally better in terms of sustainability and inferior in quality. It is therefore clear that new and better production technologies are required. If one could reduce the exergy needed both to fractionate the peas into valuable fractions and to structure the protein fraction into a good product, the figure shows that big steps could be taken, especially when the new technologies could lead to a product that is comparable in quality to meat, and would therefore be chosen by consumers, not only on idealistic grounds, but also because of the taste.

3.2.5
Optimizing Nutritional Value

A further driver would be to optimize products in terms of nutritional value. This is currently an important trend in the developed markets, usually referred to as functional foods. Nanotechnologies could enhance the nutritional value by increasing the rate of uptake of specific nutrients in the gastrointestinal tract. This can be done by using specific form of encapsulation, or by shaping the nutrient into nano-sized droplets or crystals.

3.2.6
Nanotechnology for Food Production?

Considering the above factors, it may be clear that, especially now, nanotechnology may play an important role in establishing a more sustainable supply of high-quality food products for the global population. However, one should not forget that food and food preparation represent more than just the rational supply of adequate food.

Perhaps surprisingly to many people, new methods for fractionation and structuring seem a logical match with biological (or sustainable) farming. A more sustainable primary production (e.g., in terms of soil use, and fertilizer and pesticide usage) combined with a more sustainable processing would yield much better sustainable food production. However, this would require a merger of the biological farming world, which currently is on a somewhat technophobic track, and the nanotechnology world, which is very much the reverse.

A last look at Figure 3.1 on the exergy demand for the preparation of protein products shows that, as soon as the fractionation and structuring steps have become much more sustainable, the main factor remaining is the preparation of the food at home. Using newly available technologies to make this process more sustainable would once more contribute to a more sustainable world. Whether this would be accepted by consumers is, however, unclear.

3.3
Nanotechnology and Food

3.3.1
What Is Nanotechnology?

Before continuing toward the developments in nanotechnology relevant for food production, it is important first to define what we mean by the term "nanotechnology", because there is a bewildering range of definitions available. While many say it is the technology that concerns itself with arranging molecules or clusters of molecules smaller than 100 nm, some argue that this could also be a description of colloid science for sizes between 1 and 100 nm or organic/inorganic chemistry for sizes between 0.1 and 1 nm.

Nanotechnology in general is characterized by a high degree of multidisciplinarity: chemical concepts such as self-assembly and molecular recognition are used, in conjunction with physical methods, such as use of atomic force microscopy, but also principles from biology, such as bilayer formation. One of the challenges for nanotechnology is to translate the level of control that, for example, organic chemistry has over the composition and conformation of molecules, to larger size scales. Next to the molecular scale is supramolecular chemistry, which concerns the assembly of a moderate number of molecules in well-defined clusters. Perhaps the ultimate challenge is to translate the concept into all levels of size, from the supramolcular to the colloidal, the mesoscopic, and all the way up to the level that humans can directly perceive, by touch, taste, or smell.

This is especially important for foods: they are characterized to a high degree by their internal hierarchical structure, that is, highly defined and well structured from the (macro)molecular (nanometers) level all the way up to the macrolevel (centimeters). While we will discuss this in more detail later on, we would like here to define nanotechnology for this chapter as the directed assembly of molecules or clusters of molecules into well-defined structures from the level of the clusters up to the larger scales relevant to direct human perception. This means that we will not strictly adhere to the size limit of 100 nm, but rather explore the hierarchical construction of product structure on several scales of size: while the fundamental building blocks are in the range of 1–100 nm, the structures built with them should be much larger in at least one dimension. The control over the structure should be there over all dimensions, from 100 nm to centimeter scales.

3.3.2
Nanotechnology in Food Production

Nanotechnology often makes use of the natural tendency of molecules to self-assemble into specifically shaped aggregates. Typical examples are the self-assembly of amphiphilic molecules (surfactants, some proteins) into micelles at lower concentrations, and into lamellar mesophases at higher concentrations. By somewhat changing the properties of the molecules, one can change the shape and morphology of the aggregates that are formed.

Lipids, such as phospholipids, self-assemble into bilayers, consisting of two layers of the molecules, with the hydrophobic parts of the molecules in between the layers, and the hydrophilic groups to the outside world. Phospholipids have the tendency to form vesicles (in fact, they form the membranes of living cells as well in this way) that are more or less spherical, and that have water inside and outside. Addition of ceramides, which are a different type of lipid, induces the bilayer to become less curved, and this induces the bilayer to form not spherical vesicles, but tubular ones. There are numerous shapes and morphologies of vesicles, but the shape is mostly governed by the molecular properties, plus the precise conditions in the suspending fluid, which determine the intermolecular forces acting between the molecules.

Vesicles and other structured aggregates can be used to encapsulate bioactive food ingredients, such as flavors, enzymes, prebiotics or even probiotics. They typically provide a barrier against the hostile environment in the stomach, and, due to their membrane-like structure, can deliver their contents to the cells of the intestinal wall. An example here is the use of cochleates, phospholipid–cation crystalline structures that form spiral lipid sheets with little or no internal aqueous space. They can encapsulate relatively hydrophobic components by taking them up in their bilayers. Cochleates that were loaded with a vaccine have been shown to give immune response after they were orally administered, which illustrates that they could deliver the active components into the cells; the vaccine in itself would have passed the intestinal wall [4].

There are many more examples of encapsulates for delivering ingredients to specific locations in the gastrointestinal walls. Since the techniques in the field of nanotechnology offer the possibility to precisely assemble a structure from individual molecules, it is clear that they can be applied for preparing these encapsulates. These encapsulates themselves are usually in the size range of 1–10 μm: smaller encapsulates would have an enormous interfacial area, complicating effective encapsulation, while for example probiotics consist of bacterial cells, which would make smaller encapsulates impossible. Encapsulates larger than 10 μm would make them perceptible to the human organoleptic system.

Vesicles can serve as encapsulates for specific ingredients to be dispersed into a food, but they are not a food matrix itself. A food matrix is characterized by the presence of structure on different scales. An example is the structure of meat, which is made of individual protein filaments only a nanometer thick, bundled into fibrils, which in turn are bundled into fibers, which in turn are bundled into fasciculi. The structure of meat cannot at this moment be imitated with plant proteins by existing technologies. Even though the flavors and the color of the meat can be easily matched, it is the complex structure that makes meat still unique.

Nanotechnology can contribute to this by providing ways to precisely position molecules into fibrils, bundling them together into fibers, and bundling them together into fasciculi-like structures. From this, it is also clear that the new technology should not stop at only arranging individual molecules into molecular strings; instead, it should enable us to arrange the clusters on several scales of magnitude at the same time. This challenge is present wherever we consider the preparation of food matrices, instead of food ingredients (e.g., encapsulates).

The use of nanotechnologies for food ingredients will be discussed, with the example of fibrils from protein that can serve as rheology enhancer even at very low concentrations, but which can also serve as building block of encapsulates. We will then continue with the discussion of the preparation of food matrices. This will be done with an example in which anisotropic structuring over a range of scales of magnitude was achieved.

There are many more examples possible in both categories; however, it is not the purpose to give a complete overview in this chapter; it rather attempts to sketch the challenges that future technology could and should aim at.

3.4 Applications of Nanotechnology in Foods

3.4.1 Sensing

Nanotechnology is associated with a range of applications in foods. One of the earliest was the development of sensors for detecting a specific molecule that is associated with the condition of a food product. One can think here of the detection of food spoilage by sensing metabolic products of spoilage bacteria, or direct detection of spoilage bacteria. While this is very challenging indeed, given the low concentrations of bacteria and their metabolic products, an even more challenging target is the detection of pathogenic bacteria, as their concentrations are even much lower. Not only would a sensor need to be sensitive to a single molecule, but even that might not be sufficient. One might need to concentrate a large amount of the product (liters), and then detect a single bacterium in the concentrate. The levels of sensitivity indicate clearly why one looks at nanotechnology to deliver these sensitivities. The sensors should probably be made in a very inexpensive way, to incorporate them into packaging of food products (see Chapter 5). This poses a further challenge to the technology, as expensive materials cannot be used, and the mode of production should be suitable for mass production, yet remain absolutely reliable, as the health of consumers may depend on it.

3.4.2 Packaging

A second application is the development of active packaging. A package with a build-in sensor showing the state of freshness of the product can be regarded as an active package, but there are other types. There has been a development in the incorporation of nano-sized particles into the packaging material itself. Incorporation of crystalline nanoparticles, such as nanoclays, can make the packaging material more impermeable to oxygen or modified-atmosphere gases such as nitrogen or carbon dioxide while improving its strength.

Another application is the incorporation of nano-sized silver particles, which give the material antibacterial properties. While the application of silver for this purpose is not new, the use of silver nanoparticles is; and it is claimed that the silver is more antimicrobially active in this form [5].

Another example is a packaging material composed of potato starch and calcium carbonate. This foam has good thermal insulation properties, is lightweight and biodegradable, and has been developed to replace the polystyrene "clam-shell" used for fast food [6].

3.4.3 Encapsulation

A third application is the nano-engineering of food ingredients and encapsulation. This is a wide field, which was initiated for medical purposes (delivery of active ingredients into the targeted area in the human body without spreading into other areas), but which may soon be a major application in the area of foods, especially for fortified or functional foods. Probiotic bacteria are, for example, at least partially inactivated by the adverse conditions in the stomach and other parts of the gastrointestinal tract. These bacteria may be protected until they have reached the large intestine, which is where they are supposed to be active. A similar argument holds for prebiotic ingredients (ingredients that cannot be directly digested by humans, but are nutrients to the probiotic micro-organisms in our gut): some of them will be digested partially even before reaching the large intestine.

An interesting development as a crossover between food and medicine is the development of oral vaccines. A major impediment in vaccination is the necessity for injection. In the developed world some groups do not want vaccination for religious reasons; in the developing world the costs involved and the assurance of hygiene with the injection is an important issue. Oral vaccination could alleviate some of these problems, but vaccines usually do not survive the conditions in the stomach, and when they do survive, they will not pass the intestinal wall, as this barrier evolved exactly to protect against the passage of pathogens. Encapsulation of the vaccine, such that it would be resistant to the stomach's conditions, and also would be delivered into the cells of the body, would make oral vaccination possible. Encapsulates that are (nano-)engineered to this end would therefore be a good development; there are developments in this area that are encouraging [4].

Apart from encapsulation, another development is the nano-engineering of food ingredients such that they become more bioavailable. This is achieved, for example, by preparing nano-sized crystals (in fact, the crystal size is not in the nanometer but in the submillimeter range) or emulsions that contain a supersaturated solution of a nutrient. The effectiveness of these routes in actual products and in the gastrointestinal tract is still under discussion, but it indicates the potential for nano-engineering ingredients to influence their destination in the human body.

Apart from nano-engineering food components, a new area of application could be the nano-engineering of food matrices. Food products in general are characterized by a very high degree of structure on a range of size scales (see also Section 3.3.1). Until now, nano-engineering has only been associated with engineering on the nanoscale. The successive arrangement of these structures on larger scales has not yet received much attention. Further, many of the nano-engineering procedures have been developed for use in a diluted environment, not in a concentrated, semi-solid matrix. However, the development of nano-engineering instruments to do that would have great value in the realm of food production.

The subjects of sensing and packaging are dealt with in different chapters and will therefore not be discussed further here. Therefore we will focus on some examples in the area of nano-engineering food ingredients. In addition, we will discuss an example of precisely structuring food matrices by combining directed assembly with well-defined process conditions, which results in the formation of a hierarchically well-defined structure on many size levels.

3.4.4
Nano-Engineering Food Ingredients to Improve Bioavailability

3.4.4.1 Nanocrystalline Food Ingredients

Many micronutrients and pharmaceutical components are poorly soluble in water, for example, lipids such as omega-3 fatty acids, flavors, antimicrobial components, antioxidants such as tocopherols, carotenoids such as β-carotene and lycopene, and also components like phytosterols. However, most foods have an aqueous continuous phase, as have the intestinal contents. The kinetics of uptake can therefore be slow, and in many cases the fraction taken up by the body is quite small. One way to improve this is by preparing so-called nanosuspensions or nanocrystals [7]. An example is the production of β-carotene nanocrystals [8]. Typical crystal aggregates with a size of 120 nm can be obtained, stabilized by gelatin for example, that contain crystallites around 30 nm (Figure 3.2). These particles can be created by forcefully mixing a solution of the carotene (in, e.g., an alcohol) into water usually containing a polymeric stabilizer. Tan and Nakajima [9] and Chu *et al.* [10] developed a method based on emulsification, where a β-carotene solution in hexane was emulsified in water containing sodium caseinate as stabilizer. By subsequent evaporation of the hexane, nanoparticles of 17 nm were created.

Co-precipitation with a biopolymer such as poly(lactic acid) can result in very stable nanoparticles, obviously at the cost of having a lower concentration of carotene, due to the presence of the biopolymer [11]. A strongly enhanced solubilization of the active components was noted, for example, by Trotta *et al.* [12] for nanoparticles of poorly soluble active components.

There is some evidence that particles in the range of 10 nm show a different structure than larger particles: their properties become different from the bulk properties. This, and the fact that their very large surface area allows their contents to be much more bioavailable, shows the potential of designing and producing

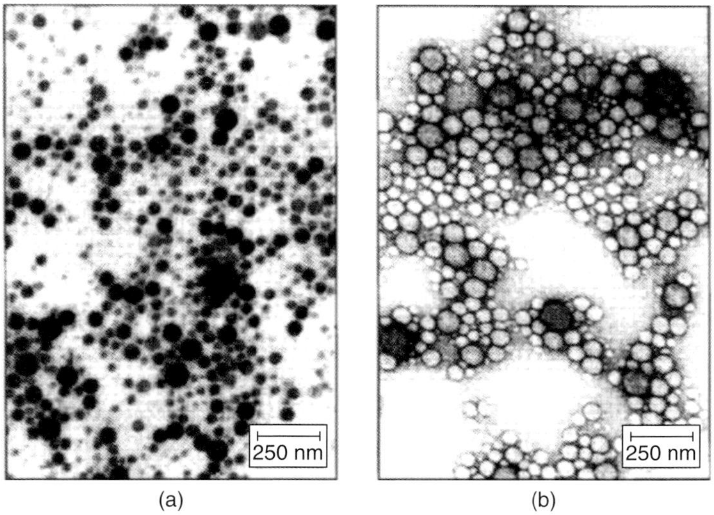

Figure 3.2 Transmission electron microscope (TEM) pictures of β-carotene nanoparticles: (a) stained with OsO_4, showing the β-carotene particles; (b) stained with uranyl acetate, showing the stabilizing gelatin coating. From [8].

particles of very small dimensions. Nanotechnology can help in the preparation of these particles, and in stabilizing them (e.g., by using microchannel emulsification methods, as has been developed by [13]), such that they have good shelf-life, and that they can be incorporated into complex food products.

3.4.4.2 Nano-Emulsions

Ribeiro *et al.* [14] have developed an interesting route for making carotene more bioavailable. Most lycopenes are completely insoluble in water and only slightly in oil. Therefore, only a small fraction of the lycopenes in our food is digested; most of it is excreted unused. The lycopene is typically at 180 °C in oil at a concentration of 15–30 wt%, which they then quickly emulsified into water using high-pressure emulsification. The resulting emulsion droplets are around 100–150 nm, which is so small that they do not contain sufficient material to form a critical nucleus; thus, the lycopene stays in solution and will be more available for digestion. It is obvious that the smaller the emulsion droplets are, the higher the lycopene concentration can be. Thus, engineering nano-emulsions would give added value. A similar system has been patented (Figure 3.3).

3.4.4.3 Nano-Engineered Protein Fibrils as Ingredient Building Blocks

Protein-Based Nanofibrils

Many proteins have the tendency to form aggregates when subjected to conditions under which they are less soluble. This may stem from a change in either solvent quality or the protein molecule itself (in fact, the two are not independent of each other: a change in solvent quality induces a change in protein conformation).

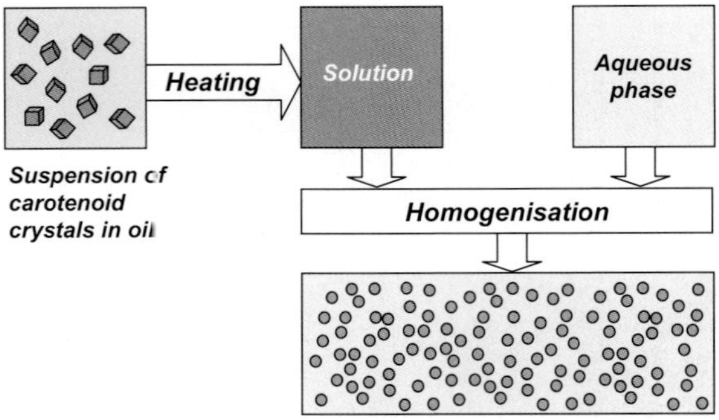

Figure 3.3 Procedure for making supersaturated emulsions of β-carotene. According to Schweikert and Kolter [15].

Owing to the conformational change, hydrophobic parts of the proteins become more available for interaction with other proteins. These interactions then lead to the formation of aggregates.

Aggregation due to protein unfolding (denaturation) is well known, and is one of the fundamental mechanisms underlying the preparation of food: boiling or frying an egg results in solidification of the egg white, due to the aggregation of the protein into a fine, random network. Meat, when cooked, becomes firmer, due to the (partial) denaturation of the proteins.

By selecting the conditions, one can tune the properties of the resulting solution. For example, when making a sauce or ice cream, one often heats the (egg protein) solution to a specific temperature (around 70 °C): at this temperature, the egg proteins do not denature completely, the interactions between the molecules remain mild, and the consistency of the solution becomes more viscous, without leading to large-scale aggregation and hence flocculation (curdling).

By partially denaturing a protein solution at conditions of low ionic strength and pH, one can make the aggregation process highly specific. At low pH the protein molecules have a high charge, while the low ionic strength ensures that the molecules will repel each other, although the partial denaturation enables hydrophobic interaction. Once at very specific spots the molecules will have the chance to interact and form a bond. This gives rise to string formation in the form of fibrils. The precise method of formation of such fibrils is not yet completely understood, but it is generally agreed that β-sheet formation plays an important role. Recent findings indicate that partial hydrolysis is necessary for some proteins to form fibrils [16]. After some time, the bonds between the individual molecules, that were initially reversible, become irreversible.

A system known for its tendency to form fibrils is β-lactoglobulin, a protein from whey. One typically heats a diluted solution of it (1–5 wt%) for 6–24 h at 80 °C and pH 2 and low ionic strength. This results in fibrils 1–8 μm long with a diameter of 4 nm, representing a thickness of one or two individual protein molecules (or significant fractions of them). The resulting fibrillar solution shows strongly increased viscosity (up to 10 000 times and more; see Figure 3.4) and shear thin-

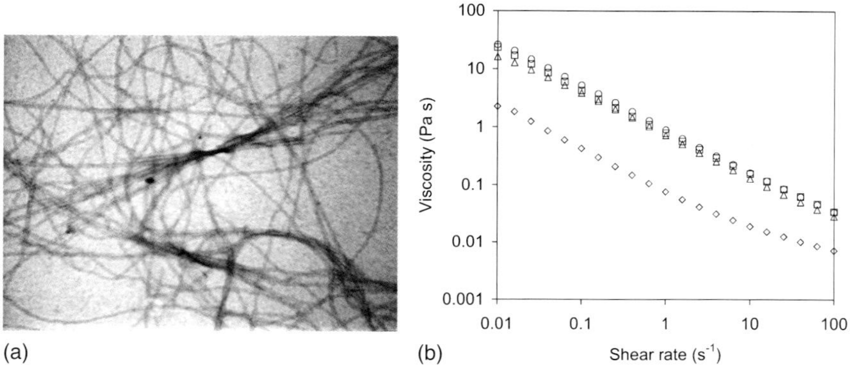

(a) (b)

Figure 3.4 (a) TEM picture of fibrils from β-lactoglobulin, made by 2 h heating at 90 °C while shearing at 200 s^{-1}. (b) Flow curves for solutions of 5.2 wt% whey protein prepared at different shear rates: ◇ 0 s^{-1}, □ 168 s^{-1}, ○ 337 s^{-1}, △ 673 s^{-1}. The viscosity of the solvent without protein is 0.001 Pa s. From Akkermans *et al.* [16].

ning behavior, assumed to be caused by percolation of the fibrils, forming a network, which is successively destroyed when applying shear.

A recent finding was that the self-assembly kinetics of this process can be influenced by applying mild shear during the assembly process [16]. Figure 3.4 shows that application of shear increases the viscosity by an order of magnitude. This is caused by a much higher yield. A higher shear rate does result in more fibrils, but this is not evident in the viscosities: at higher concentrations, the fibrils cannot assume a random orientation any more due to steric hindrance and they form nematic, liquid-crystalline domains, in which they align and hence have less influence on the viscosity. This was supported by the observation of birefringence in the solutions, indicating fibril alignment in the solution.

It is not clear how flow positively influences the self-assembly process. It may speed up the diffusion of proteins (or their fragments) toward the active tips of the fibrils. The Peclet number calculated with a shear rate of 168 s^{-1} is, however, only 3×10^{-5} (using a diffusion coefficient for β-lactoglobulin of $9.7 \times 10^{-11}\,m^2\,s^{-1}$ and a molecular size of 4 nm), which indicates that direct influence of the flow on the mass transfer should not be expected. However, when the fibrils start to form more densely packed, liquid-crystalline domains, diffusion from the surrounding solutions into these domains may become limiting. It is clear indication, however, that even a macroscopic parameter such as shear flow may influence a molecular process such as self-assembly of a protein. The fact that the fibrils start to line up into liquid-crystalline domains also indicates that larger-scale structure may be formed. Aligning these domains by applying shear flow (evident in the shear thinning behavior) may further yield options for forming the fibrils into a matrix, when one could fix the fibrils while aligned.

Using Nanofibrils for Microstructure Assembly Apart from their use as an ingredient to modify the rheological properties of a product, fibrils may have further use.

At low pH, the fibrils themselves are highly positively charged. This enables assembly of the fibrils into larger structures using electrostatic interactions. Recently, Sagis *et al.* [17] used this for structural reinforcement of encapsulates (Figure 3.5). They used emulsion droplets stabilized by β-lactoglobulin at low pH, and exposed them alternately to high-methoxyl pectin and fibrillized whey protein isolate (consisting mainly of β-lactoglobulin). The pectin was negatively charged and therefore formed a nanometer-thin layer on top of the positively charged droplet surface. The positively charged whey protein isolate fibrils then formed a layer on this negative surface, and so forth. Shells with only a few layers do not have much mechanical strength; however, the application of six layers gave the encapsulates considerable strength.

While the fibrils were created by using the combination of specific hydrophobic interaction with general electrostatic repulsion (to reduce random aggregation), the larger aggregates could be assembled by using electrostatic attraction.

There are many examples of using self-assembled or directionally assembled aggregates for encapsulation.

3.4.5
Preparation of Food Matrices

In Section 3.2.4 we discussed the relevance of being able to create food matrices with a well-defined hierarchical structure. Especially matrices with a fibrillar structure would be of relevance (e.g., to act as meat-replacing protein foods).

In principle, one could use the fibrillization procedures as discussed above for such a purpose. Recently, Akkermans *et al.* [16] showed that fibrils could be formed not only from proteins of animal origin, but also from proteins of plant origin: both soy glycinin and soy protein (a mixture of mostly glycinin and conglycinin)

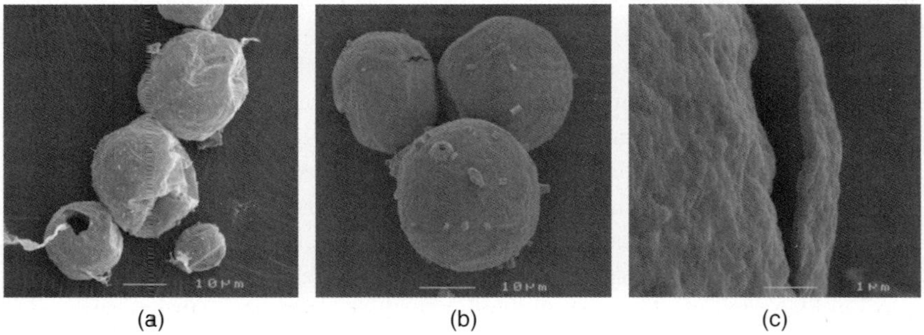

Figure 3.5 Shells made with alternating layers of high-methoxyl pectin and whey protein fibrils. Owing to the electrostatic interaction, the fibers do not stand out but are attached to the surface over their whole length. The fibres strengthen the encapsulates considerably. (a) These shells have only two layers (i.e., one pectin, one fibril layer). (b) Here the shells have six (three pectin, three fibril) layers. (c) A close-up of the wall of one of the six-layered shells, indicating a typical shell thickness of around 50 nm. From [17].

were shown to form fibrils. However, the fibrillization procedure is only effective with dilute systems: one cannot use concentrations higher than a few percent of protein: the system will form a dense gel already before fibrillization, while shear flow during fibrillization results in much shorter fibrils.

Formation of the fibrils at low concentrations, and subsequent concentration into a highly concentrated product, would not be sustainable: very large volumes would be necessary, and an excessive amount of (acidified) water would be needed. The same is true for many other methods to form fibrils; for example, wet spinning and electrospraying only yield low volumes of fibrils; thus one would need very large equipment to produce industrial volumes of protein product.

It is therefore important to start with systems having the same range of concentrations as would be required for the ultimate product. For a high-protein food such as meat, this would be in the range of 20–30 wt% of protein (raw meat).

Conventional technology in this range is the use of extrusion technology. The product ingredients are brought together, heated and compressed, and forced through a small die. Many protein-based systems will form a fibrillar structure under the influence of the extensional flow related with the focusing flow just before the die. The product is however not finely fibrillar, while the process consumes ample amounts of energy due to the intense process conditions. Other methods like wet or dry spinning produce limited numbers of fibers at the same time, and thus have limited scope for upscaling. An interesting process is the one used for the production of Valess, a product based on casein. By mixing the casein with a carbohydrate, and subsequently solidifying the carbohydrate phase, one can produce a product matrix that is fibrillar, down to a level of tens of micrometers. Even though this is a successful product, it cannot yet compete directly with meat, as its structure is still several orders of magnitude coarser.

A new process was proposed by Manski *et al.* [18, 19]. A 30 wt% calcium caseinate dispersion in water was subjected to plain shear flow in a special device. A cross-linking enzyme was added, solidifying the dispersion while it was being sheared. Under the right conditions, the caseinate was found to align into long fibrils with diameter of around 100–150 nm. The shear stresses applied were relatively low, so the process can be considered mild. The solid fibrillar product closely resembled meat in terms of structure (Figure 3.6).

The fibrillization was ascribed to what one might call directed self-assembly. Calcium caseinate is present in the dispersion as micelles of size around 125 nm, which is big enough to be aligned by the shear force. It is well known that particles in (non-Newtonian) suspensions have the tendency to align under plain shear flow (e.g., [20–22]). The calcium caseinate micelles have the tendency to cluster or "stick together", due to the divalency of the calcium ions, which can serve as a bridge between them. This is probably important in the alignment process. By concurrent cross-linking of the aligned micelles with the help of an enzyme, one can fix the strings of micelles, effectively creating a (soft) solid fibrillar matrix, because of the high concentration in the system.

The proposed mechanism was supported by the rheology of the dispersions. Calcium caseinate dispersions showed evidence of structure formation under

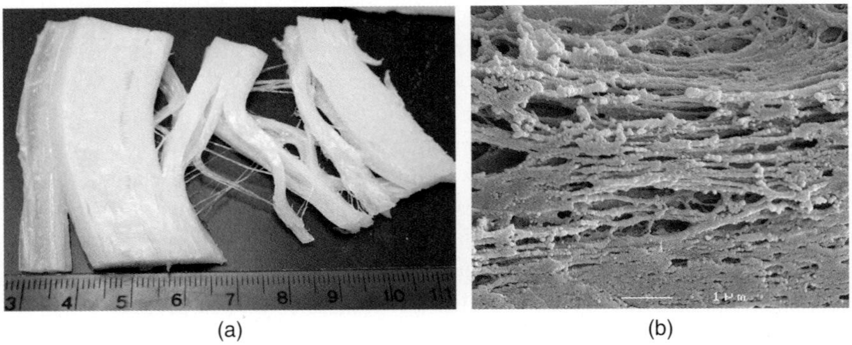

Figure 3.6 Hierarchically fibrillar protein structures made by shearing a 30 wt% dispersion of calcium caseinate that is slowly solidified by crosslinking with transglutaminase [18, 19]. The structure was shown to consist of fibrils of ~100 nm diameter, packed into bundles of ~100 µm thick (b), which themselves are arranged into larger-scale bundles evident in (a).

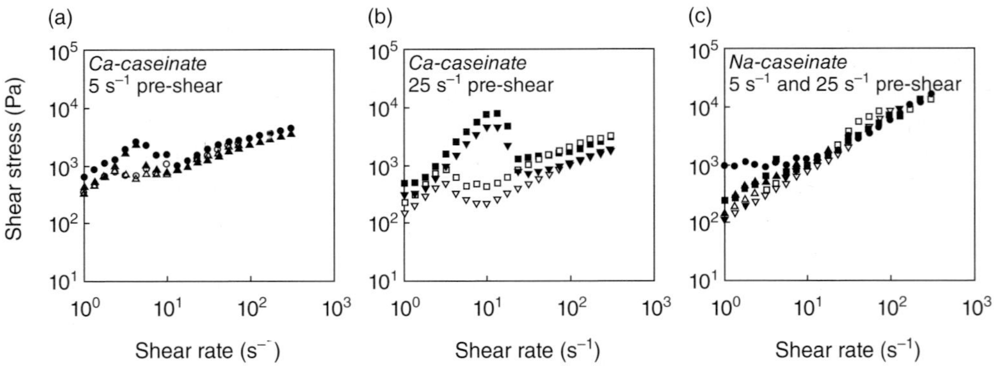

Figure 3.7 Rheology of calcium and sodium caseinate dispersions: closed symbols, up-sweep; open symbols, down-sweep; circles and squares, using pre-shearing directly before measuring; triangles, using pre-shearing, a rest period of 7 minutes, and then measuring. From Manski *et al.* [23].

shear flow, but sodium caseinate dispersions did not (Figure 3.7). Sodium caseinate micelles are smaller, therefore less easily aligned, and show no tendency to cluster or aggregate.

This example shows that a combination of self-assembly (well-defined protein clusters such as micelles) and positional assembly (alignment under shear flow) can yield better-structured food matrices. It seems probable that the range of examples can be extended when well-defined process conditions are used.

3.5
Concerns about Using Nanotechnology in Food Production

3.5.1
Risks of Nanotechnology

Nanotechnology in foods is usually associated in the media with the use of nanoparticles in foods, especially nanoparticles from non-biological origin, such as buckyballs (C_{60}) or carbon nanotubes. It is not yet clear how human physiology responds to these types of component, and one should therefore be very careful with application of this category of components.

Nanostructured components of biological nature such as fibrillized proteins (through β-sheet formation) may be safer, as these structures occur in nature, and thus our body has probably evolved in the presence of these components. The fact that some diseases seem to be correlated with the occurrence of amyloids may however warn us to be cautious. One should point out that there is no evidence that indicates that the consumption of, for example, fibrillized soy protein would in any way stimulate the formation of amyloids in the human body.

Another development is the use of nanoparticles of natural components that are smaller than those that occur in nature. It is unlikely that the use of nano-sized crystals or lycopenes would in any way be dangerous: these crystals have the tendency to disappear by dissolution quicker than the natural crystals do. Similarly, nano-sized emulsions will be broken down or absorbed very quickly, as they are less stable than larger emulsions. Emulsions with droplet sizes much smaller than 1 μm have been used extensively without any indication of any risk related to the droplet size.

Lipid-based encapsulates such as cochleates seem to be able to deliver deoxyribonucleic acid (DNA) or ribonucleic acid (RNA) through the intestinal wall, and therefore may be a bigger risk factor: any contamination by, for example, a virus or other harmful components might also be transported through the intestinal wall. These components may therefore have some risk involved.

The fibrillized food matrices as described will almost certainly be safe, as the structure is natural in size scales, and all the components are completely food-grade. The process merely influences the spatial arrangement of the elements that were present together.

So, the emergence of nanostructured food components shows no reason for caution any more than the development of any new component would merit, but of course every caution needed should be taken.

3.5.2
Rational Argumentation Versus Human Feelings

Many of the more popular discussions on the use of nanotechnology evoke statements such as the one of Pollan [1] (see above): we should use the same method for food preparation as our grandparents used – what was good enough for them, must be good for us. This type of statement is not part of a rational discussion, but has an intuitive and emotional background.

Even though, on a rational basis, there seem to be ample reasons to consider the use of nanotechnology for food production (surrounded by reliable precautions), the emotional side to it is just as relevant. Emotional repulsion from application of new technologies for our food is sufficient reason not to apply them, when there is no issue in obtaining sufficient food of high quality. The global situation however is different. Not considering the use of better technology may ultimately imply the deprivation of adequate food supply for many humans on our planet.

The application of the new technologies will therefore depend on the severity of the problems surrounding food and bioproduction. It seems important that engineers and scientists at least work on sustainable methods to produce high-quality food, at least to enable the societal choice between the different alternatives that will be apparent in the future.

References

1 Pollan, M. (2008) Quote from an interview with Michael Pollan. VPRO Television Noorderlicht, February.

2 Pimentel, D. (1997) Remark made during the July 24–26 Meeting of the Canadian Society of Animal Science in Montreal.

3 Apaiah, R.K., Linnemann, A.R., and van der Kooi, H.J. (2006) Exergy analysis: a tool to study the sustainability of food supply chains. *Food Res. Int.*, **39**, 1–11.

4 Gould-Fogerite, S., Mannino, R.J., and Margolis, D. (2003) Cochleate delivery vehicles: applications to gene therapy. *Drug Deliv. Technol.*, **3**, 40–47.

5 Rai, M., Yadav, A., and Gade, A. (2009) Silver nanoparticles as a new generation of antimicrobials. *Biotechnol. Adv.*, **27**, 76–83.

6 Stucky, G.D. (1997) High surface area materials. *Proceedings of the WTEC Workshop on R&D Status and Trends in Nanoparticles, Nanostructured Materials, and Nanodevices in the United States.* http://www.wtec.org/loyola/nano/us_r_n_d/07_03.htm

7 McClements, D.J., Decker, E.A., and Weiss, J. (2007) Emulsion-based delivery systems for lipophilic bioactive components. *J. Food Sci.*, **72** (**8**), 109–124.

8 Auweter, H., Haberkorn, H., Heckmann, W., Horn, D., Lüddecke, E., Rieger, J., and Weiss, H. (1999) Die supramolekulare Struktur ausgefällter, nanometergroßer β-Carotinpartikel. *Angew. Chem. Int. Ed.*, **38**, 2188–2191.

9 Tan, C.P., and Nakajima, M. (2005) β-Carotene nanodispersions: preparation, characterization and stability evaluation. *Food Chem.*, **92**, 661–671.

10 Chu, B.S., Ichikawa, S., Kanafusa, S., and Nakajima, M. (2007) Preparation of protein-stabilized β-carotene nanodispersions by emulsification–evaporation method. *J. Am. Oil Chem. Soc.*, **84**, 1053–1062.

11 Ribeiro, H.S., Chua, B.-S., Ichikawa, S., and Nakajimaa, M. (2004) *Preparation of Nanodispersions Containing β-Carotene by Solvent Displacement Method*, National Food Research Institute, Tsukuba, Japan.

12 Trotta, M., Gallarate, M., Pattarino, F., and Morel, S. (2001) Emulsions containing partially water-miscible solvents for the preparation of drug nanosuspensions. *J. Controlled Release*, **76**, 119–128.

13 Chuah, A.M., Kuroiwa, T., Kobayashi, I., Zhang, X., and Nakajima, M. (2009) Preparation of uniformly sized alginate microspheres using the novel combined methods of microchannel emulsification

and external gelation. *Colloids Surf. A Physicochem. Eng. Asp.*, **351**, 9–17.

14 Ribeiro, H.S., Ax, K., and Schubert, H. (2003) Stability of lycopene emulsions in food systems. *J. Food Sci.*, **68**, 2730–2734.

15 Schweikert, L., and Kolter, K. (1997) EP 0800824 A1.

16 Akkermans, C., van der Goot, A.J., Venema, P., van der Linden, E., and Boom, R.M. (2008) Formation of fibrillar whey protein aggregates: influence of heat and shear treatment, and resulting rheology. *Food Hydrocolloids*, **22**, 1315–1325.

17 Sagis, L.M.C., de Ruiter, R., Rossier Miranda, F.J., de Ruiter, J., Schroën, K., van Aelst, A.C., Kieft, H., Boom, R.M., and van der Linden, E. (2008) Polymer microcapsules with a fiber-reinforced nanocomposite shell. *Langmuir*, **24**, 1608–1612.

18 Manski, J.M., van der Goot, A.J., and Boom, R.M. (2007) Influence of shear during enzymatic gelation of caseinate–water and caseinate–water–fat systems. *J. Food Eng.*, **79**, 706.

19 Manski, J.M., van der Goot, A.J., and Boom, R.M. (2007) Formation of fibrous materials from dense calcium caseinate dispersions. *Biomacromolecules*, **8**, 1271–1279.

20 Lucey, J.A., Srinivasan, M., Singh, H., and Munro, P.A. (2000) Characterization of commercial and experimental sodium caseinates by multiangle laser light scattering and size-exclusion chromatography. *J. Agric. Food Chem.*, **48**, 1610–1616.

21 de Kruif, C.G. (1998) Supra-aggregates of casein micelles as a prelude to coagulation. *J. Dairy Sci.*, **81**, 3019–3028.

22 Dickinson, E., Semenova, M.G., Belyakova, L.E., Antipova, A.S., Il'in, M.M., Tsapkina, E.N., and Ritzoulis, C. (2001) Analysis of light scattering data on the calcium ion sensitivity of caseinate solution thermodynamics: relationship to emulsion flocculation. *J. Colloid Interface Sci.*, **239**, 87–97.

23 Manski, J.M., van der Zalm, E.E.J., van der Goot, A.J., and Boom, R.M. (2008) Influence of process parameters on formation of fibrous materials from dense calcium caseinate dispersions and fat. *Food Hydrocolloids*, **22**, 587–600.

4
Packaging

Frans W.H. Kampers

4.1
Introduction

People are very selective about the food that they eat. Evolution has taught us that contamination and spoilage are serious threats to consumer health, as well as food quality. Food must therefore be fresh and clean. Packaging is potentially an important way to deliver these requirements. Packaging has been used for centuries to contain foods, and to keep foods free of undesired contaminants. Food needs to be digestible by a biological organism, and so is by its very nature a perishable product. This means that food quality deteriorates over time to a level that consumers reject it. Again, packaging can help prevent or slow down deterioration of foods.

With the advent of nanotechnologies, innovative applications in the area of packaging are being developed, providing new opportunities to improve on the sometimes already very sophisticated packaging concepts that have been developed to fulfill the demands of the modern consumer. Improved packaging can deliver improved convenience and, at the same time, improve sustainability and reduce waste [1].

4.2
Reasons to Package Food Products

Apart from the obvious need to contain certain food products, such as beverages or powdery materials, to avoid contamination by dirt, and to keep foods free of rodents and other pests, the most important reason for packaging foods is to maintain the quality of the product for as long as possible. Quality deterioration of food products can be caused by various processes. Physical processes like drying or wetting can potentially change the texture of the product to a level that consumers reject it. Bread becomes hard because moisture evaporates; biscuits or potato chips become soft because they take up water from the atmosphere. Although the

Nanotechnology in the Agri-Food Sector: Implications for the Future, First Edition. Edited by Lynn J. Frewer, Willem Norde, Arnout Fischer, Frans Kampers.

nutritional quality is not necessarily affected, consumers tend to throw the product away or, at best, feed it to animals.

The influence of light can cause color changes that often are interpreted as quality deterioration, also causing the consumer to discard products that are still perfectly suitable for human consumption. Chemical processes can occur between food components, or between one or more components of the product and external substances like water or oxygen. Sometimes these processes are desirable – like the aging of wine or the ripening of fruits – but mostly they result in negative changes in taste and/or texture. Although they usually do not affect the safety of the product, these changes provide reasons for consumers to dispose of the product.

The most important processes for quality deterioration are of biological origin. Apart from the threats that rodents and insects constitute for the product, most food spoilage in industrialized countries originates at the microbial and fungal levels. For products that have been sterilized, the packaging has to prevent recontamination occurring, which necessitates the use of strong materials like metals for canned foods, thick plastics, and so on. Mildly processed or unprocessed foods still contain bacteria or fungi or the spores thereof. In those cases, it is important to prevent the rapid development of these organisms. Cooling/freezing or chemical additives like salt can be used in many cases, but for certain products "modified-atmosphere packaging" is more appropriate. In modified-atmosphere packaging the product is kept in an atmosphere in which a certain gas, necessary for the growth of the micro-organisms, usually oxygen, is absent. The concept relies on the packaging to maintain the modified atmosphere for as long as possible to extend the shelf-life of the product.

4.3 Physical Properties of Packaging Materials

In many food packaging applications, certain physical properties of the packaging materials are important. These physical properties enable the packaging concept to work and sometimes are the cause of the failure of the packaging to deliver. At this point, it is relevant to review the contributions that micro- and nanotechnologies can make to improve the physical properties of packaging materials.

4.3.1 Strength

In order that products are effectively contained, food packaging materials are required that have sufficient strength to withstand the pressure or the forces that the product exerts on the containment, or forces from outside that can occur under circumstances arising from normal use. The material from which a bottle is made should be strong enough to prevent the bottle from tearing, even when the pressure inside is raised under the influence of temperature or processes occurring within the product. For example, champagne bottles need to be extra-strong to withstand the pressure of the carbon dioxide that is generated in the champagne. Even the closure needs to be strengthened with iron wire to prevent the cork popping prematurely.

The strength of a material usually needs to be traded off against other properties like weight and transparency. Consumer preferences for convenience have resulted in food packaging being "multifunctional", insomuch as increasingly strong materials also need to fulfill other requirements like transparency, light weight, and so on. The rapid introduction and acceptance of polyethylene terephthalate (PET) bottles is an example of this trend. The PET bottle combines strength with reduced weight and reduced vulnerability to breakage. On the one hand, the PET bottle offers a lot of advantages, although, on the other, there is a disadvantage to PET bottles that will be discussed later in this chapter.

The ability to modify materials at the nanolevel to provide new functionality has already delivered one material that is extremely strong: carbon nanotubes. The remarkable properties of carbon nanotubes were one of the drivers for the "hype" associated with nanotechnology, when scientists, researchers, and developers started to realize that these remarkable properties were examples of what could be developed at the nanolevel. Carbon nanotubes per unit weight are much stronger than steel [2]. Indeed, multiwalled carbon nanotubes are the strongest material currently known to humanity. For this reason, they are used in the manufacture of sports equipment to improve the strength-to-weight ratio, and to enhance stiffness [3].

4.3.2
Barrier Properties

Packaging forms a barrier against contamination of the product from external elements. The properties of the packaging materials have to agree with the requirements of the packaging purpose under normal storage conditions. For example, if a product is sensitive to moisture, and usually stored in the open air, the packaging material should be water-tight. There is very little that nanotechnologies can add regarding improved waterproofing. However, when packaging is required to be gas-tight, nanotechnologies can make important contributions.

If a fizzy drink bottle is left open for some time, the carbon dioxide (CO_2) that was dissolved in the product evaporates and the properties of the drink are changed to such an extent that most people do not want to consume it, although other aspects of quality and taste are unaffected. Such products must therefore be packaged in containers that are impervious to CO_2 to prevent evaporation from occurring during storage. Glass, metal, and polymer materials like PET are suitable for this purpose.

In other cases, products are vulnerable to gases or vapours permeating the package. Potato chips (crisps) represent an example of a product where water diffusing through the polymer packaging material has, in the past, limited the shelf-life of these products. Prior to the advent of nanotechnology, the problem was solved through application of a very thin metal layer on top of the polymer. The disadvantage of this solution is that the consumer cannot see through the packaging in order to inspect the contents. Arguably this does not represent a very big problem for processed products like potato chips. However, when buying fresh produce, consumers prefer to see the product in order to be able to make a visual assessment of food quality.

In another example, certain products are vulnerable to oxidation. These products must be packaged in such a way that oxygen (O_2) is kept out of the package for as long as possible. Beer is an example of such a product. Unfortunately, because the O_2 molecule is smaller than the CO_2 molecule, it is much more difficult to keep oxygen from diffusing in through the packaging material than to keep CO_2 from diffusing out. Oxygen can diffuse through PET material and, before the use of nanotechnologies in packaging, it was not possible to package beer in a PET bottle and to maintain its quality for sufficiently long.

In the application of nanocomposites, nanoparticles are used to enhance the barrier properties of these materials. The advantage of nanocomposites is that the particles used to achieve the required functionality are too small to scatter visible light, which enables the development of clear transparent food packaging materials. The application of nanotechnologies is also a good way to improve the properties of more environmentally friendly biopolymers [4]. However, these materials are less suitable for packaging purposes because usually they are not transparent and can degrade over time.

A nanocomposite typically is a polymer matrix in which nanoparticles have been embedded to improve existing barrier properties. These nanoparticles can be of natural origin. For instance, several manufacturers use a special kind of clay, montmorillonite, or other silicates as an additive to different kinds of polymer materials.[1)] These clays are mined in Africa and typically are made up of larger particles of stacked platelets. In a special process, these particles are exfoliated, which results in individual clay platelets that are relatively large in two dimensions, but have a thickness in the order of nanometers (Figure 4.1). Gases cannot penetrate the platelets. By adding them to the polymer, a material is formed that forces the gas molecules to diffuse around the platelets, substantially elongating the path of the molecules (see Figure 4.1) and therefore increasing the time needed for the molecule to pass through the wall of the container. The platelets also improve the mechanical properties of the material such as the tensile strength and the elasticity.

In some cases "oxygen scavengers" are added to the polymer matrix. These react with any oxygen molecules that do manage to diffuse into the material.

In the case of modified-atmosphere packaging concepts, the normal atmosphere is replaced by one or more gases that are inert to the product. These prevent or

1) Imperm and other products by Nanocor, http://www.nanocor.com; Aegis by Honeywell, http://www51.honeywell.com/sm/aegis/products.html; Cloisite and Nanofil from Southern Clay Products Inc., http://www.nanoclay.com; Durethan from Bayer, http://www.research.bayer.com/edition_15/15_polyamides.pdfx.

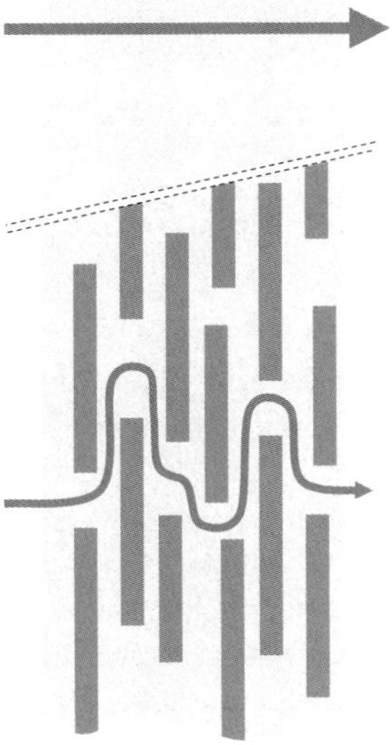

Figure 4.1 Clay platelets (rectangles) force gas molecules to follow a tortuous path, thus improving the barrier properties of the material.

reduce quality deterioration processes. Such applications can also benefit from the improved barrier properties of nanocomposite materials because they slow down the deterioration of the modified atmosphere caused by diffusion of atmospheric gases.

Nanocomposites or nanostructured materials can also be used as a film on top of other packaging materials to form multilayered materials in which the properties of the different layers combine to achieve the required overall specifications of the packaging material.

For example, recently, a new method of structuring polymer material into stacked layers of 20 nm thickness has been reported [5]. Polyethylene oxide is forced to crystallize in thin lamellae, or layers, which increase the gas permeability by two orders of magnitude, maintaining the modified atmosphere even longer.

4.3.3
Light Absorption

Certain food products are vulnerable to light irradiation. The products may change color, which potentially affects their appeal to the consumer. Chemical reactions, triggered by photons, can reduce the quality of the product. This is one of the reasons why beer used to be packaged in brown or green bottles. As is known from the use of certain nanoparticles in sunscreens, titanium oxide and zinc oxide nanoparticles are very effective in absorbing ultraviolet (UV) light. In sunscreen applications, the nano-sized particles are small enough not to scatter visible light, thus providing a clear fluid or cream that does not leave a white film on the skin, but still blocks the dangerous high-energy part of the solar spectrum.

These properties are also used in food packaging materials to provide a concept that is transparent, enabling consumer inspection of the contents, but which also locks out the UV radiation that can cause deterioration of certain quality aspects of the product.[2] In this application, the nanoparticles, mostly metal oxides, are embedded in the polymer matrix of the packaging material. As a consequence of their small size, they do not interfere with visible light, resulting in a clear package.

4.3.4 Structuring of Interior Surfaces

In specific cases, the product to be packaged is sticky and adheres to the inside of the package. For example, removing custard from packaging can be a time-consuming undertaking. Micro- and nanotechnology has been used to structure the surface of packaging materials in such a way that it mimics the water- and dirt-repellent effect of the lotus leaf [6, 7], which facilitates removal of the product from the package. With the lotus effect,[3] a micro- and a nanostructure are used to create a surface that is very hydrophobic [8], causing even very sticky substances to slide from the surface. The lotus effect can be implemented as a coating.[4] At the present time, these coatings are mostly used in non-packaging applications.[5] However, the potential for packaging material application is obvious.

It is also possible to modify the structure of the food packaging material to give it "self-cleaning properties".[6] In other words, a reusable packaging can be reused without extensive cleaning, and chances of contamination after reuse are reduced. At the moment, self-cleaning materials are in the development stage, but they will ultimately be applied in food storage containers.

4.4 Antimicrobial Functionality

Micro-organisms are usually responsible for the spoilage of food products. People have traditionally applied high-temperature processing (pasteurization and sterilization) or chemical treatment (salt, sugar, alcohol, smoke, etc.) to kill the organisms that are always present on or in foods to prevent or slow down the spoilage processes. Packaging was frequently required to prevent recontamination of the product after heat or chemical treatment. There is, however, a trend toward consumer preferences for the application of mild preservation techniques and the wish for fresh or minimally processed and preservative-free products. In order to

2) Light Stabilizer 210 by DuPont, http://www2.dupont.com/Titanium_Technologies/en_US/products/dls_210/dls_210_landing.html.
3) See http://www.lotus-effect.com.
4) Lotusan, http://www.stocorp.com/allweb.nsf/lotusanpage.
5) Mincor by BASF, http://www.basf.com/group/corporate/en/news-and-media-relations/science-around-us/mincor/index.
6) Lightmotif, http://www.lightmotif.nl.

achieve this, the need to slow down the development of micro-organisms near the product has increased. There are several options to achieve this, and nanotechnology can provide some of them [9].

It has long been known that silver has antimicrobial properties. Alexander the Great is reputed to have used large silver containers to ensure his personal supply of fresh water. Although the exact mechanism of these properties is not exactly known – experts dispute whether silver ions or metallic silver is the active species – it has been established that increased surface area of the silver enhances the antimicrobial activity [10]. Consequently, making silver particles smaller improves the antimicrobial properties, in the end resulting in silver nanoparticles as effective agents to contain microbial growth.

The antimicrobial effect of nano-silver is being exploited in several areas of application, including bandages for dressing wounds, and in textiles to inactivate the odor-producing bacteria on feet and in armpits. Nano-silver can also be used in food packaging. Food containers with nano-silver can be obtained commercially. Manufacturers claim substantially improved quality of food products even after extended storage. These containers are reusable, but the same effects can be achieved when nano-silver is incorporated into disposable food packaging materials.

Other antimicrobial concepts rely on the interaction of the microbe with cations in a polymer layer on top of the food packaging material, or a nanostructured agent. Based on delivery technology, a system has been developed that releases an antimicrobial chemical when the presence of a bacterium is detected through the chemicals it distributes in its direct environment.[7] Although applications in food packaging of this specific system are not foreseen in the near future, there is potential for comparable developments to be applied specifically for this purpose. An advantage of such a development is that they can be made more specific to the types of bacteria causing deterioration in certain high-cost food products such as meat and fish.

4.5 Visual Indicators

One of the reasons that so much good food is thrown away in the industrialized world is consumer reliance on "sell-by dates" to safeguard the quality of food products. Unfortunately, sell-by dates are based on unsophisticated models of quality deterioration. Manufacturers, afraid of image damage if too many products have below-standard quality when they reach the consumer, build in a large safety margin. The result is that perfectly good products cannot be sold because of sell-by date expiration. Similarly, consumers may reject these products and discard them unnecessarily. Even under these conservative conditions, in exceptional circumstances (e.g., elevated storage temperatures), a product may deteriorate to unac-

7) BioSwitch by TNO, http://www.tno.nl/content.cfm?context=markten&content=product&laag1=195&laag2=327&item_id=1126.

Figure 4.2 The OnVu system to check that a product has not been stored at elevated temperatures for too long.

ceptable levels before the sell-by date has expired. It would therefore be preferable to have some means to directly determine the quality of a food product in the package. Nanotechnology can provide applications that meet this need.

4.5.1 Quality Assessment

A concept that is based on the same principles as the sell-by date, but is more sophisticated, insomuch as environmental conditions are taken into account, indicates the product of temperature and time.[8] The indicator changes color more quickly if storage temperatures are higher (see Figure 4.2). This represents an improvement over the sell-by date system, in particular for fresh products. Although it is probably possible to tailor the change in color of the indicator to the specific spoilage behavior of the product under consideration–which would definitely increase the accuracy of the system–this could also reduce the economic viability of the concept for most food products.

However, an indicator still only represents an indirect measure of the quality of a food product. The main problem of these systems is that a certain storage period at a specific temperature might be perfectly all right for some products, while it may cause others to be totally spoiled. An accurate assessment of quality through the area below the storage temperature versus time graph requires intricate models of the product and models of the spoilage behavior. This, of course, can be incorporated into the system by calibrating it to different food products or classes of food product, but again that would make the system more expensive as it would require tailoring the indicators to different products. A better way to detect spoilage is through food safety indication.

8) OnVu, http://www.onvu.com.

4.5.2
Food Safety Indication

The food industry is very conscience of quality, and also needs to ensure that optimal hygiene standards are applied in order to improve food safety. Even in industrialized countries, there is still substantial room for improvement. In developing countries, many deaths can be attributed to poor water and food quality. In the industrialized countries, the economic damage caused by hospitalization from food poisoning is substantial [11], as is the economic cost of a food recall or loss of consumer confidence in a brand. A useful innovation in food packaging would be a method of "warning" consumers about food products that are not fit for consumption.

If food safety is compromised, the cause is usually micro-organisms such as bacteria or fungi that develop in and on the food product. Certain types of micro-organisms, if present in sufficient quantities, can cause health problems by themselves, whereas others produce toxins. In fresh products like meat and vegetables, bacteria and/or spores of fungi will be present from the harvest stage onward. After harvest, they start to multiply. The traditional method to deal with this problem was to treat the product in such a way that these micro-organisms were killed. As mentioned before, heat, chemicals (salt/sugar) or smoke have all proven to be effective in killing organisms and preserving the product. Modified atmospheres, chilling, and freezing slow down or virtually stop the development of the organisms. Often, food packaging is required to maintain the sterile status of the product.

Modern consumers want their food to be not only fresh, but also convenient. This means that consumers prefer the application of mild conservation methods, if any – but then they buy precut and pre-prepared vegetables, fruits, and meats in order to maximize convenience. From the perspective of food quality and safety, these two types of consumer preferences represent a bad combination. Little to no conservation leaves micro-organisms alive, and the "wounds" inflicted on food by cutting provide them with a substrate on which to feast.

Food quality deterioration and spoilage processes produce different characteristic chemical by-products. If these molecules are small enough, they will be volatile, and can be detected in the atmosphere surrounding the product. With chemical detection, such as "Toxin Guard technology",[9] suitable molecules are deposited on the inside of the packaging. When they react with certain characteristic volatiles, a color change signals the presence of these substances and warns the consumer that certain organisms have developed on the food to such an extent that consumption of the product is no longer safe.

4.5.3
Product Properties

The production of by-products from ripening processes may represent an important signaling mechanism between plants and fruits. For example, ethene

9) See http://www.toxinalert.com.

(ethylene) is a plant hormone the production of which speeds up ripening in fruits. The amount of ethene in the package of a fruit is a measure of the state of ripeness of the fruit. Of course, ripeness does not necessarily represent a deterioration in quality, but can be a highly desirable aspect of the food product. The principle of ethene detection is employed in the "Ripesense" system[10] in which a color change in a suitably prepared dot on the inside of the package indicates the ripeness of the product in question.

4.6
Information and Communication Technology

Although visual indicators can be helpful in providing information to the individual consumer, they are less suitable for integration into logistical systems that can add value in the product chains. In our highly automated society, effective monitoring of safety, quality or product characteristics could be delivered to computer and information and communication technology (ICT) systems located remotely to enable automated electronic control of the logistic process. Storage systems could monitor certain product characteristics directly, and decisions could be made on how to optimize the value of the product. For example, products that approach quality or ripeness limits could be taken out of storage and sold in nearby markets; products that have sufficient shelf-life remaining could be shipped to more distant markets, where they bring more money. Systems such as these are dependent on effective communication between the sensors and the outside world.

4.6.1
Sensors

In packaged products, the amounts of by-products of deterioration can be substantial, and can be detected with suitable electronic devices. These devices are currently in development and often mimic the operation of a human nose. Different receptors, much like the ones that also are situated in the nose, are placed on a semiconductor device in such a way that, when a molecule of interest gets close enough, it "docks" onto the receptor, causing charges to shift in the receptor molecule. These charge shifts can influence conductance in the semiconductor material and, therefore, can result in an electronic signal that can be interpreted digitally. The presence of more molecules results in more docking events, and therefore increases the signal. These receptors usually are not very specific. By using more than one type of receptor, the device generates a pattern that will be able to specifically detect certain processes. This is also how the nose works: the human nose holds about 350 different receptors, and the brain has learned to interpret the signal pattern that results from food deterioration. Receptors can be

10) See http://www.ripesense.com.

developed that are more specific to the volatiles involved in the food quality deterioration processes. This would make the detection simpler yet more accurate.

At the moment, these systems are still in the development stage, rely on silicon-based microelectronics to do the sensing, the data analysis, and communication, and are very expensive. They are unlikely to be used in food packaging applications in the near future. However, progress is being made in other areas of nanotechnology that will result in "printable electronics" with which the electronic circuitry necessary to measure the parameters, analyze the signals, and communicate the outcome to external computer systems can be printed with conductive inks in combination with polymer components. This technology could be sufficiently mature, advanced, and cheap within 15–20 years. If the electronic nose can be developed using printable electronics, the application will be very cheap, and will certainly be adopted in order to assure the quality of food products.

Electronics require power to operate. The usual solution for this is batteries. Unfortunately, the combination of a food product and batteries is not very attractive. Moreover, batteries possibly run out of power before the product is out of storage. Alternatives are to scavenge power from external sources like the Sun, temperature differences, movement, etc. They all have their own drawbacks. However, radiofrequency identification technology may not only solve the power problem, but also provide the necessary communication channel to transmit sensor data to the outside world.

4.6.2
Radiofrequency Identification Technology

Being able to extract an electronic signal from a packaged product in itself is not sufficient. The signal still has to be communicated to computers in the outside world. To this end, radiofrequency identification (RFID) technology has been developed, which can identify individual objects without requiring a line of sight. RFID technology can be applied as an electronic version of the barcode, with the difference that it can be read without opening the box. Furthermore, RFID can be used to identify animals [12], for electronic access systems, to identify tools for professional workshops, and so forth.

Radiofrequency identification technology consists of two elements: the transponder attached to the object to be identified, and a reader that transmits an electromagnetic field to read the transponder. The transponder does not contain a power source of its own. It uses the electromagnetic field from the reader to temporarily power its electronics, and to communicate the predefined code back to the reader. A drawback of the technology is that the reading distance is dependent on the specifics of the electromagnetic field, and the range is typically in the order of 1 m. Regulations in most countries do not allow higher fields and/or different frequency bands that would allow larger reading distances.

Although developed for identification purposes, RFID technology is also used in combination with sensors. If measurements are also required when there is not a reader present, then there must be some sort of power source in the transponder.

For this purpose, limited storage of power to allow measurements in between reads has been demonstrated.

Radiofrequency identification is a high-frequency technology, and, as such, at the moment can only be implemented by application of silicon technology, which makes it expensive and less suitable for incorporation in food packaging materials. However, recent advances in other areas of nanotechnology have resulted in the first implementation of RFID transponders in polymer electronics.[11] Although these can already be produced very cheaply, it is generally expected that these will be used in food products when the electronics can be printed directly on the package.

4.7 Discussion

Although applications of nanotechnology in food packaging are less controversial than those where the nanotechnology is in the food product itself, and is thus consumed by the consumer, there are still some aspects that need careful consideration before large-scale introduction of some applications is warranted. Since the benefits of some of the systems are not equally distributed along the value chains, some stakeholders in the chain may be unenthusiastic about implementing them.

On paper, food quality indicators, for instance, seem a very good idea from a consumer perspective. The retail sector, however, is not that enthusiastic. Although sell-by dates also have this problem, if more detailed information on freshness is provided, retailers fear customers scavenging the shelves for the freshest products and the supermarket being left with more products that cannot be sold any more. Thus quality labels may not reduce the amount of food wastage, but could easily increase it. In addition, in systems such as these, the costs and the benefits are usually not spread evenly along the food chain. Very often they will increase the costs for those stakeholders involved at the earlier stages of the food chain, but the benefits will be accrued by stakeholders at the end of the chain.

4.7.1 Health Risks

Food is something that, following consumption, enters the body and cannot easily be removed if something is wrong with it. People are conscious about what they eat and prefer food to be natural and fresh (see also Chapters 12 and 14 in this volume). The food packaging materials that are used to maintain the quality of the food are usually not consumed. However, consumer concerns about contact between nanotechnology applied to packaging and foods may be an important issue. The

11) See http://www.polyic.com.

general public believes that nanoparticles, one of the more commonly known forms of nanotechnology, cannot be seen, can easily migrate from one matrix to another, and can even cross barriers in the body that cannot be crossed by non-nanoscale particles. Nanoparticles in packaging materials could therefore migrate to the food, be ingested, get into the body, and end up in parts of the human body where they could result in health problems. In the case of nano-silver, this scenario could be realistic. Nano-silver particles can get out of the packaging materials matrix and get into the food product. If they have antimicrobial properties in the packaging material, they will also be capable of damaging cells in the body.

Arguably, the health risks of the applications discussed in this chapter are very small. The amount of nanostructured materials used in applications like sensors and indicators is very limited. Even if some of the nanostructures can transfer to foodstuffs, consumer exposure will be very small. There are more nanoparticles involved in improved barrier properties and antimicrobial layers, but these nanomaterials are embedded in the matrix of the packaging material. The amount of nanoparticles released is also very low. However, this does need to be verified for each of the applications to be developed and brought to market. It can therefore safely be concluded that the health risks involved in applications of nanotechnologies in food packaging are less than those associated with the risks of contamination by nanoparticles, from wear of the machines, in conventional processes that are used to prepare the products.

4.7.2
Environmental Risks

One of the aspects of food packaging is that the materials used are usually discarded after consumption of the food. This means that, at least in some part, they will end up in the environment. If they include nanostructured materials, these will also end up the environment. At the moment, it is largely unclear what the effects of nanoparticles in the environment will be. Research into this problem lags behind research into health effects.

Nanoclays, which are basically natural materials, are nanostructured materials used to improve the barrier properties of packaging. They are embedded in the matrix of the polymer. When they are freed, for instance when the packaging material is incinerated, they will be no more harmful than other clays that are deposited by rivers and the sea. This is not true of silver nanoparticles. If they exist in the environment as individual particles with a large surface-to-volume ratio, they will be as effective in killing micro-organisms in the environment as they were in the initial application in food packaging. These particles could pose serious problems for wastewater treatment plants that rely on micro-organisms to break down certain chemical components in the wastewater. There could also be a negative impact on ecological systems and biodiversity.

If small amounts of nanoparticles were to get into the environment, it could also be argued that the amounts of free nanoparticles would be small, and exposure would therefore be limited, reducing the risk. However, persistent free nanoparticles – particles that do not dissolve and are not broken down by physical, chemical or biological processes – that entered the environment could accumulate in certain compartments and remain there for a long time. Moreover, it has been seen in the past that certain chemicals can accumulate in organisms that are high up in the food chain. This same effect can also play a role in the uptake of free nanoparticles. Before large-scale application of persistent nanoparticles in food packaging applications, more research is necessary to characterize these effects.

4.7.3
Consumer and Societal Acceptance

There are benefits that are likely to be achieved from the application of nanotechnologies to food packaging. Whether or not these benefits will be realized largely depends on the acceptance of the technology and its applications by individual consumers and society as a whole (see also Chapter 14 in this volume). The consumer will consider each application in the context of benefits to be gained for themselves in relation to the perceived personal (or personally relevant) risks that accompany the application. Societal concerns will focus on risks for specific population groups, future generations or environmental impacts. Perceived risks and negative effects may include ethical and psychological impacts. For instance, in the case of applications of radiofrequency identification technology, privacy may be an important issue. In order for nano-packaging technology to be successfully introduced and commercialized, the benefits for individual consumers, the environment, and society as a whole must be assessed. At the same time, research should be conducted to enable possible negative effects (e.g., risks to human and environmental health or negative socio-economic effects) to be assessed and communicated in an objective and honest way. Both the consumer and society need to feel that they are in control of these kinds of application before they will accept their large-scale implementation.

References

1 Brody, A.L. (2003) *Food Technol.*, **57**, 52.
2 Yu, M.-F., *et al.* (2000) *Science*, **287**, 637.
3 Woodrow Wilson International Center for Scholars (2010) Project on Emerging Nanotechnologies, see Consumer Products: An inventory of nanotechnology-based consumer products currently on the market, http://www.nanotechproject.org/inventories/consumer/ (accessed 3 November 2010).
4 Rhim, J.W., and Ng, P.K.W. (2007) *Crit. Rev. Food Sci. Nutr.*, **47**, 411.
5 Wang, H., *et al.* (2009) *Science*, **323**, 757.
6 Barthlott, W., and Ehler, N. (1977) *Tropische subtropische Pflanzenwelt*, **19**, 367.
7 Barthlott, W., and Wollenweber, W. (1981) *Tropische subtropische Pflanzenwelt*, **32**, 7.
8 Lai, S.C.S. (2003) Mimicking Nature: Physical Basis and Artificial Synthesis of the Lotus-Effect, University of Leiden, http://members.ziggo.nl/scslai/lotus.pdf (accessed 3 November 2010).
9 Cha, D.S., and Chinnan, M.S. (2004) *Crit. Rev. Food Sci. Nutr.*, **44**, 223.

10 Wijnhoven, S.W.P., *et al.* (2009) *Nanotoxicology*, **3**, 109–138.

11 WHO (2002) WHO Global Strategy for Food Safety: Safer Food for Better Health, World Health Organization, Geneva, Switzerland, http://www.who.int/entity/foodsafety/publications/general/en/strategy_en.pdf (accessed 3 November 2010).

12 Kampers, F.W.H., Rossing, W., and Eradus, W.J. (1999) The ISO standard for radiofrequency identification of animals. *Comput. Electron. Agric.*, **24**, 27–43.

5
Using Nanoparticles in Agricultural and Food Diagnostics

Geertruida A. Posthuma-Trumpie and Aart van Amerongen

5.1
Introduction

There is growing interest in the safety of agricultural raw materials and of food and feed products. During growth, the production process, and the storage of food, sophisticated, low-cost, and rapid tests are increasingly being used. Safety items include the presence of pathogenic micro-organisms [1–4] or the toxins they produce during storage of the raw ingredients [5–14]. The presence of pesticides [15–20], anabolic steroids [21], antibiotics [22–25], or adulterating substances [26–28] is also a matter of concern. New regulation on food labeling requires notification with respect to the (possible) presence of allergenic substances to inform the allergic consumer of potential hazards [29–31]. The presence of the phrase "may contain" on the label is no longer sufficient.

Here we will present biosensors that use nanoparticles as detection labels. Some of these sensors can be applied on-site, needing a minimum amount of resources and training. Biosensor formats that will be discussed include lateral flow (immuno)assays, nucleic acid lateral flow (immuno)assays, flow-through (immuno) assays, antibody microarrays, and the surface plasmon resonance biosensor. The use of nanoparticles during sample pre-treatment will be mentioned and future prospects will be discussed.

5.2
Biosensors

According to the International Union of Pure and Applied Chemistry (IUPAC), a biosensor is "a self-contained integrated device, which is capable of providing specific quantitative or semi-quantitative analytical information using a biological recognition element (biochemical receptor) which is retained in direct spatial contact with a transduction element" [32]. Biosensors are being developed that recognize the micro-organism or analyte with high specificity, sensitivity, and efficiency [33].

Nanotechnology in the Agri-Food Sector: Implications for the Future, First Edition. Edited by Lynn J. Frewer, Willem Norde, Arnout Fischer, Frans Kampers.

A special kind of biosensor is an immunosensor that uses the highly specific interaction of antibody and antigen, often with high sensitivity and speed. Immunosensors are among the most often used sensors for toxin and microbial detection in agriculture, food, and feed [34]. The biosensor is constructed on a carrier material onto which the capturing element is immobilized, a transduction element is positioned, and often a connection is made to a device for reading the response and archiving purposes. A set-up for sample pre-treatment and automated delivery can be included. The workhorse of immunosensors is the antibody, which has to be highly specific, sensitive, and efficient. Although very important, this aspect is not the focus of this chapter. We will only mention here that a specific, sensitive, and efficient antibody has to be available. Especially when running rapid tests, the affinity between analyte and capture agent should be high, with short interaction kinetics to allow a relevant number of complexes to be formed within the timeframe of the test.

5.3 Transduction Principles

The transduction technology ultimately gives the result of the test. Although a wide variety of transduction principles are available, the most popular are of an electrochemical [35] or optical nature, although magnetic [36] and piezoelectric transduction [37] principles are gaining more attention. For rapid assays, it is preferable to develop a test where the results can be interpreted by visual examination. To this end, specific antibodies or secondary antibodies are labeled with colored nanoparticles. Today, the most used nanoparticles are based on colloidal gold with a diameter of 40 nm [5, 7, 10–14, 17–19, 21, 25, 38–53], giving a red color. Nanoparticles that are used less often include colloidal carbon [2, 54–57] (black), colored latex [15] (several colors), fluorescent silica particles [58], or dye-encapsulated liposomes [5, 31] (several colors or fluorescent).

Nanoparticles with magnetic properties [12, 36, 59–62], quantum dots having fluorescent properties [63] (several colors), and nanoparticles with up-converting phosphors have been developed as well [64–66]. However, no applications in the agricultural disciplines are known. Magnetic nanoparticles are, apart from signal-generating labels, also used in sample pre-treatment and washing [36, 59]. Although with colored nanoparticles the result is visible, it may be necessary to digitize the results for later evaluation. To that end, specialized readers are available, but a flatbed scanner and image analysis software are often sufficient [2, 4, 54, 56, 57, 67]. For fluorescent and magnetic particles, a dedicated reader is obligatory.

Coupling of the requested biological compound (for example, antibody or (strept)avidin to the nanoparticles) basically can be done using one of two strategies: coupling by adsorption [54] or by covalent interaction by means of a chemical reaction [61, 68].

5.4
Examples of Biosensors in Which Nanoparticles Are Being Used

5.4.1
Lateral Flow (Immuno)assay

One of the most popular immunochemical methods is the lateral flow (immuno) assay, well known from the pregnancy test. The test does not require trained personnel or expensive equipment, and its result is often a visual "yes" or "no" with a certain cut-off value. Usually the results are obtained within 30 minutes. Depending on the analyte, several formats may be developed. For high-molecular-weight analytes, the sandwich format is applicable. On a nitrocellulose strip with dimensions of, for example, 5 cm × 0.5 cm, a transverse stripe of a solution of the specific antibody at an appropriate concentration (100–1000 $\mu g\,ml^{-1}$) in a, preferably, low-salt buffer is sprayed at an appropriate distance, for example, 1.5 cm, from the origin, called the test line. A second line can be included at some distance from the test line, called the control line. This line may contain an antibody against the species of the labeled antibody to provide a test control.

A sample application pad and a conjugate release pad are mounted on one end of the strip and an adsorbance pad is on the opposite end. The conjugate release pad contains the nanoparticles labeled with antibody and used for the evaluation (see Section 5.3). This antibody can be the same as or different from the sprayed antibody. Often, one antibody is a polyclonal and the other is a monoclonal, recognizing different epitopes of the analyte. The strip can be mounted in a device for easier handling. The strip is dried and can be stored in a sealed aluminum pouch with desiccant for later use. Such a strip can be stored for a prolonged time without refrigeration. Running the test is possible by simple addition of a fixed volume, for example, 10–100 µl of (an extract of) the food or feed sample on the sample pad. An appropriate running buffer, for example, 100 mM borate buffer (pH 8.8) with 1–2% bovine serum albumin (BSA) or any other blocking compound and 0.05% Tween 20 can be added to make the volume up to100 µl to run the test when necessary. Using this format, a positive result is obtained when the analyte is present. The response at the control line has to be positive in all cases, to ensure a proper performance of the test. The test format is called lateral flow (immuno) assay (LFIA) or immunochromatography (ICG) in sandwich format [69]; a scheme is presented in Figure 5.1.

When the analyte is of low molecular weight, such as pesticides or antibiotics, the test has to be formatted in another way (inhibition format). A conjugate of an analog of the analyte to a carrier protein has to be sprayed at the test line. It is advisable to use a carrier protein other than the protein that has been used to produce the antibody. For example, if the antibody-inducing antigen used a conjugate to keyhole limpet hemocyanin (KLH), then the protein of the conjugate to be sprayed may be BSA instead.

Again, the conjugate release pad contains the label and a specific antibody that recognizes the analyte. The strips in this format can be stored and the test can be

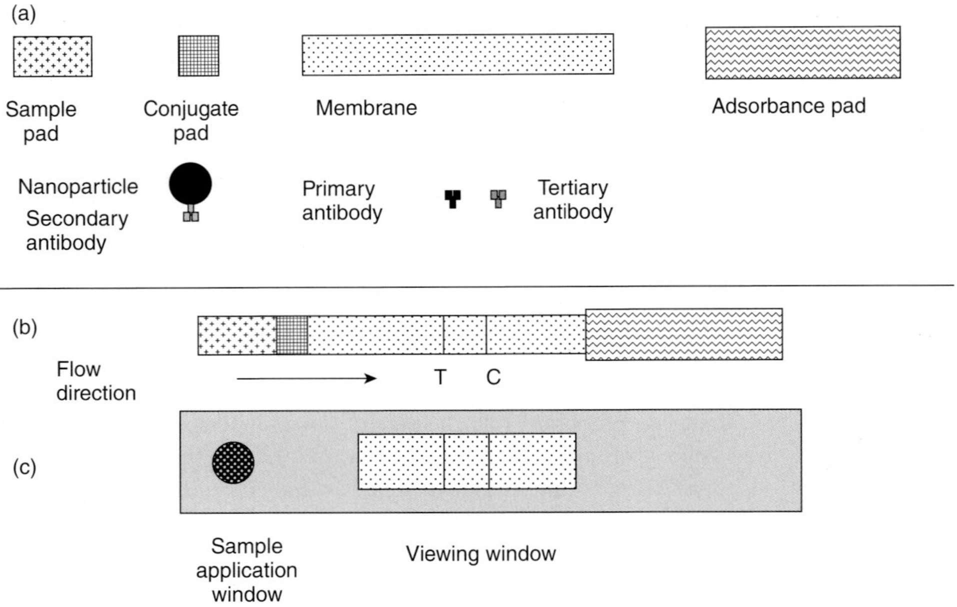

Figure 5.1 Scheme of a lateral flow immunoassay test strip and device in sandwich format: (a) parts of the test, (b) ready-made test strip (test line T; control line C), (c) test strip in device (not to scale).

processed in the same way as outlined above. However, the presence of a colored response at the test line now indicates the absence of the analyte, and the absence of a response indicates the presence of the analyte above a certain threshold. One can say that in this case the technician who performs the test needs more knowledge. The test format is called LFIA or ICG in inhibition format [69]. Several attempts are being made to reverse the response in this layout (presence of analyte yields a positive, colored response) using anti-idiotype antibodies (antibodies against antibodies) [70], or anti-complex antibody fragments produced in an expression system [71]. A scheme of this principle and layout is presented in Figure 5.2.

5.4.2
Nucleic Acid Lateral Flow (Immuno)assay

Another quite different format has to be developed when a specific and sensitive antibody cannot be generated. This is especially true when the absence of pathogenic micro-organisms has to be proven. To design such a test, a species-specific deoxyribonucleic acid (DNA) or ribonucleic acid (RNA) sequence has to be amplified using one of the currently available amplification procedures such as the polymerase chain reaction (PCR) [2, 38, 54, 72–77], loop-mediated isothermal amplification (LAMP) [43, 78–81], and the nucleic acid sequence-based amplification (NASBA) [68, 82].

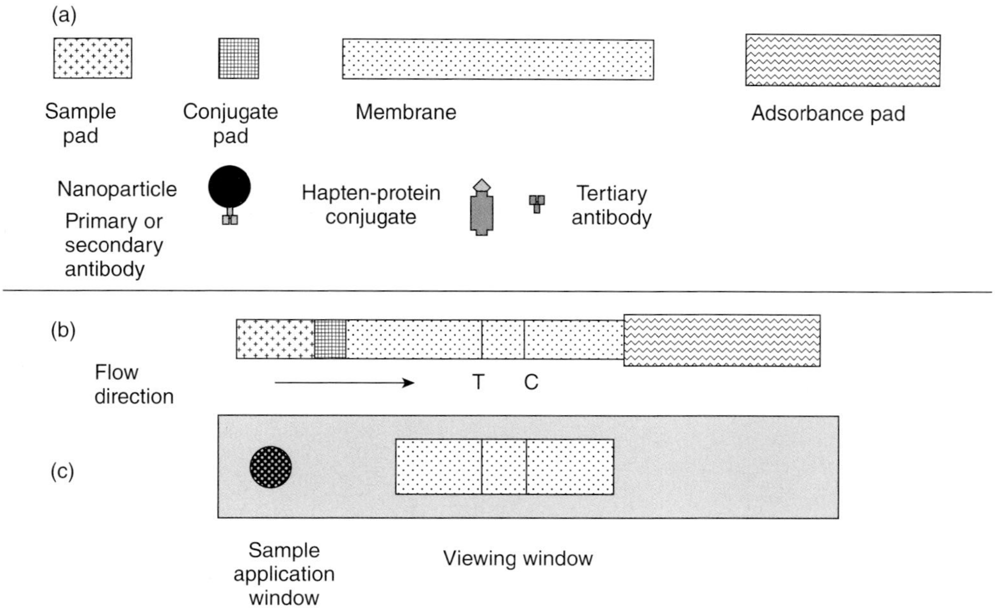

Figure 5.2 Scheme of a lateral flow immunoassay test strip and device in inhibition format: (a) parts of the test, (b) ready-made test strip, (c) test strip in device (not to scale).

As an example, a strategy of PCR-amplified species-specific templates is illustrated in Figure 5.3. A set of two primers is used, of which the forward primer usually contains a discriminating tag and the reverse primer has a biotin. Such primers can be ordered from primer suppliers according to the sequence required. Antibody against the discriminating tag is sprayed at the test line and a conjugate of the nanoparticles to (strept)avidin is used for visualization of the signal. Discriminating tags used include Texas Red (TxR), fluorescein isothiocyanate (FITC), cyanine 5 (Cy5), digoxigenin (DIG), and dinitrophenyl phosphate (DNP). Antibodies to these tags are commercially available and are sprayed at the test line. The control line contains biotin-labeled immunoglobulin G (IgG), also widely available. The response at the control line has to be positive in all cases, to ensure a proper performance of the test. This test format is called nucleic acid lateral flow (immuno) assay (NALFIA) [2, 4, 57, 59], and the response is positively correlated to the amount of amplicon.

5.4.3
Flow-Through (Immuno)assays

Apart from the above-mentioned format where the sample flows laterally through the membrane, it is also possible to design a format with a vertical sample flow, that is, through the membrane. In this case, several spots may be applied onto the membrane, of which one spot is the test control and the other(s) contain(s) the

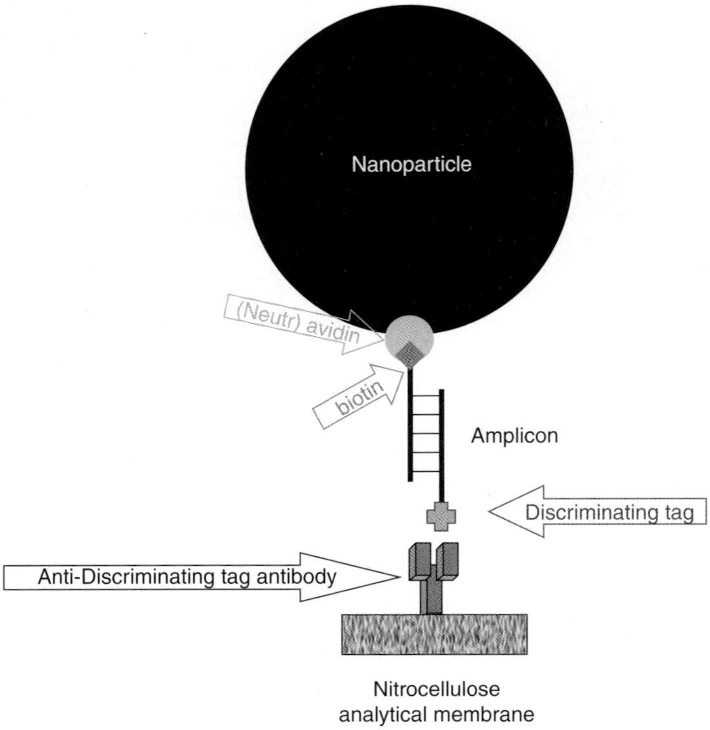

Figure 5.3 Scheme of a nucleic acid lateral flow test principle.

capturing agent(s) [18, 19]. A scheme of this format and test principle is shown in Figure 5.4.

5.4.4
Antibody Microarrays

Protein chips using antibodies as recognition elements are evolving very rapidly [83]. They are not yet as popular as the above-mentioned techniques and formats; this is mainly due to the paucity of suitable antibodies, lack of affinity, cross-reactivity, and loss of functionality upon binding. Antibody microarrays are used for multi-analyte testing [84–86]. Briefly, antibodies raised against the analytes of interest are spotted in an ordered way on a carrier chip, often in a microscope slide format [87]. It is also possible to spot a microarray in the wells of a multi-well plate. Often gold nanoparticles are used for detection upon binding of analyte to antibody. However, we recently showed that carbon nanoparticles are well suited to this task; see Figure 5.5 for a typical layout and principle.

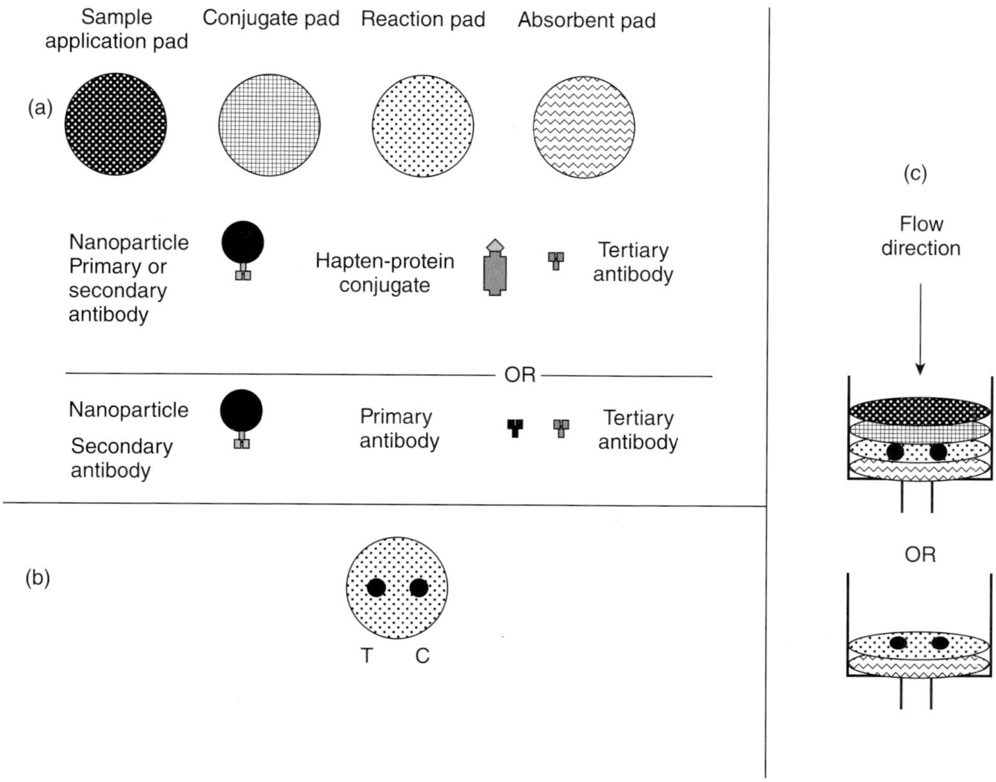

Figure 5.4 Scheme of a flow-through immunoassay, in sandwich or inhibition format: (a) parts of the test, (b) ready-made test pad, (c) test pad in device (not to scale).

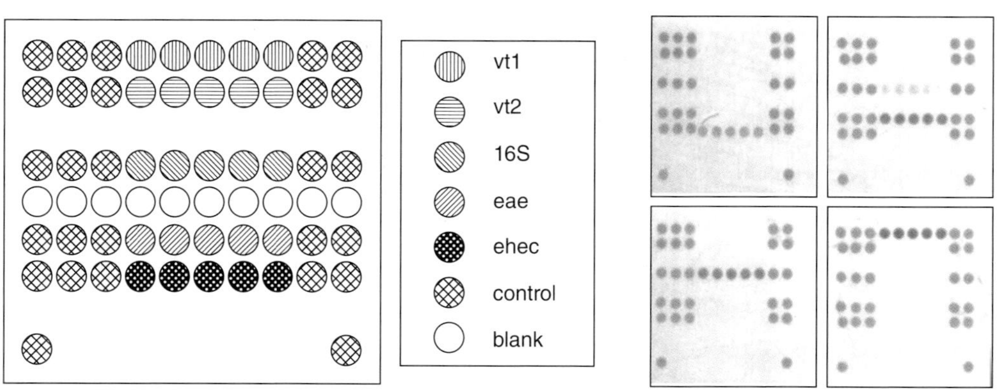

Figure 5.5 Layout of an antibody microarray and a typical set-up of a test.

5.4.5 Surface Plasmon Resonance Spectroscopy

Surface plasmon resonance spectroscopy is a technique where antibody–antigen interaction is visualized using the change of the refractive index upon binding. The technology relies on the generation of plasmons, quasi-particles resulting from the quantization of plasma oscillations that can be compared to photons for light waves. Optical evanescent waves are commonly found during total internal reflection. When a plasmon interacts with a molecule, characteristics depending on the molecular mass are changed and can be measured. Mainly the angle of reflection changes due to the interaction with coated molecules. A biosensor based on surface plasmon resonance is advertised as a label-free technique. However, sometimes nanoparticles are used to increase its sensitivity [88, 89].

5.5 Future Prospects

Preferably the whole assay covers automated sample-taking until the read-out of the results without any additional handling by the user. To that end, an integrated, so-called "lab-on-a-chip" layout may be used in which all necessary items, including sample pre-treatment, are combined in one, often disposable, housing. However, no applications have been presented for food or feed components, but the principles of those presented in the medical profession [90] use nanoparticles doped with up-converting phosphors. This format will be transferable very well to food and feed safety issues.

References

1 Baeumner, A.J. (2004) Nanosensors identify pathogens in food. *Food Technol.*, **58**, 51–55.

2 Blažková, M., Koets, M., Rauch, P., and van Amerongen, A. (2009) Development of a nucleic acid lateral flow immunoassay for simultaneous detection of *Listeria* spp. and *Listeria monocytogenes* in food. *Eur. Food Res. Technol.*, **229**, 867–874.

3 Nauta, M.J., van der Wal, F.J., Putirulan, F.F., Post, J., van de Kassteele, J., and Bolder, N.M. (2009) Evaluation of the "testing and scheduling" strategy for control of *Campylobacter* in broiler meat in The Netherlands. *Int. J. Food Microbiol.*, **134**, 216.

4 van Amerongen, A., and Koets, M. (2005) Simple and rapid bacterial protein and DNA diagnostic methods based on signal generation with colloidal carbon particles, in *Rapid Methods for Biological and Chemical Contaminants in Food and Feed* (eds A. van Amerongen, D. Barug and M. Lauwaars), Wageningen Academic Publishers, Wageningen, The Netherlands, pp. 105–126.

5 Delmulle, B.S., De Saeger, S.M.D.G., Sibanda, L., Barna-Vetro, I., and Van Peteghem, C.H. (2005) Development of an immunoassay-based lateral flow dipstick for the rapid detection of aflatoxin B1 in pig feed. *J. Agric. Food Chem.*, **53**, 3364–3368.

6 Ho, J.-A.A., and Wauchope, R.D. (2002) A strip liposome immunoassay for aflatoxin B1. *Anal. Chem.*, **74**, 1493–1496.

7 Kolosova, A., De Saeger, S., Sibanda, L., Verheijen, R., and Van Peteghem, C. (2007) Development of a colloidal gold-based lateral-flow immunoassay for the rapid simultaneous detection of zearalenone and deoxynivalenol. *Anal. Bioanal. Chem.*, **389**, 2103–2107.

8 Krska, R., and Molinelli, A. (2009) Rapid test strips for analysis of mycotoxins in food and feed. *Anal. Bioanal. Chem.*, **393**, 67–71.

9 Lindner, P., Molz, R., Yacoub-George, E., Durkop, A., and Wolf, H. (2004) Development of a highly sensitive inhibition immunoassay for microcystin-LR. *Anal. Chim. Acta*, **521**, 37–44.

10 Molinelli, A., Grossalber, K., Führer, M., Baumgartner, S., Sulyok, M., and Krska, R. (2008) Development of qualitative and semiquantitative immunoassay-based rapid strip tests for the detection of T-2 toxin in wheat and oat. *J. Agric. Food Chem.*, **56**, 2589–2594.

11 Molinelli, A., Grossalber, K., and Krska, R. (2009) A rapid lateral flow test for the determination of total type B fumonisins in maize. *Anal. Bioanal. Chem.*, **395**, 1309–1316.

12 Tang, D., Sauceda, J.C., Lin, Z., Ott, S., Basova, E., Goryacheva, I., Biselli, S., Lin, J., Niessner, R., and Knopp, D. (2009) Magnetic nanogold microspheres-based lateral-flow immunodipstick for rapid detection of aflatoxin B2 in food. *Biosens. Bioelectron.*, **25**, 514–518.

13 Tippkötter, N., Stückmann, H., Kroll, S., Winkelmann, G., Noack, U., Scheper, T., and Ulber, R. (2009) A semi-quantitative dipstick assay for microcystin. *Anal. Bioanal. Chem.*, **394**, 863–869.

14 Wang, S., Quan, Y., Lee, N., and Kennedy, I.R. (2006) Rapid determination of fumonisin B1 in food samples by enzyme-linked immunosorbent assay and colloidal gold immunoassay. *J. Agric. Food Chem.*, **54**, 2491–2495.

15 Campbell, K., Fodey, T., Flint, J., Danks, C., Danaher, M., O'Keeffe, M., Kennedy, D.G., and Elliott, C. (2007) Development and validation of a lateral flow device for the detection of nicarbazin contamination in poultry feeds. *J. Agric. Food Chem.*, **55**, 2497–2503.

16 Gabaldón, J.A., Cascales, J.M., Morias, S., Maquieira, A., and Puchades, R. (2003) Determination of atrazine and carbaryl pesticide residues in vegetable samples using a multianalyte dipstick immunoassay format. *Food Addit. Contam.*, **20**, 707–715.

17 Kaur, J., Singh, K.V., Boro, R., Thampi, K.R., Raje, M., Varshney, G.C., and Suri, C.R. (2007) Immunochromatographic dipstick assay format using gold nanoparticles labeled protein–hapten conjugate for the detection of atrazine. *Environ. Sci. Technol.*, **41**, 5028–5036.

18 Wang, S., Zhang, C., Wang, J., and Zhang, Y. (2005) Development of colloidal gold-based flow-through and lateral-flow immunoassays for the rapid detection of the insecticide carbaryl. *Anal. Chim. Acta*, **546**, 161–166.

19 Zhang, C., Zhang, Y., and Wang, S. (2006) Development of multianalyte flow-through and lateral-flow assays using gold particles and horseradish peroxidase as tracers for the rapid determination of carbaryl and endosulfan in agricultural products. *J. Agric. Food Chem.*, **54**, 2502–2507.

20 Zhou, P., Lu, Y., Zhu, J., Hong, J., Li, B., Zhou, J., Gong, D., and Montoya, A. (2004) Nanocolloidal gold-based immunoassay for the detection of the *N*-methylcarbamate pesticide carbofuran. *J. Agric. Food Chem.*, **52**, 4355–4359.

21 Huo, T., Peng, C., Xu, C., and Liu, L. (2006) Development of colloidal gold-based immunochromatographic assay for the rapid detection of medroxyprogesterone acetate residues. *Food Agric. Immunol.*, **17**, 183–190.

22 Kandimalla, V.B., Kandimalla, N., Hruska, K., and Franek, M. (2007) Detection of sulfamethazine in water, milk and pig manure by dipstick immunoassay. *Vet. Med. (Czech)*, **52**, 445–450.

23 Li, X., Zhang, G., Liu, Q., Feng, C., Wang, X., Yang, Y., Xiao, Z., Yang, J., Xing, G., Zhao, D., Cai, S., and Chen, H. (2009) Development of immunoassays for the detection of sulfamethazine in swine urine. *Food Addit. Contam. A*, **26**, 314–325.

24 Watanabe, H., Satake, A., Kido, Y., and Tsuji, A. (2002) Monoclonal-based

enzyme-linked immunosorbent assay and immunochromatographic assay for enrofloxacin in biological matrices. *Analyst*, **127**, 98–103.

25 Zhao, Y., Zhang, G., Liu, Q., Teng, M., Yang, J., and Wang, J. (2008) Development of a lateral flow colloidal gold immunoassay strip for the rapid detection of enrofloxacin residues. *J. Agric. Food Chem.*, **56**, 12138–12142.

26 Klein, F., Lupo, T., Pielack, D., Mozola, M., Pinero, D., Coates, S., Thiex, N., Holst, C., and Drouillard, J. (2005) Validation study of a lateral-flow immunoassay for detection of ruminant by-product material in animal feeds and feed ingredients: performance-tested method SM 010405. *J. AOAC Int.*, **88**, 1583–1592.

27 Martín-Hernández, C., Muñoz, M., Daury, C. Weymuth, H., Kemmers-Voncken, A.E.M., Corbatón, V., Toribio, T., and Bremer, M.G.E.G. (2009) Immunochromatographic lateral-flow test strip for the rapid detection of added bovine rennet whey in milk and milk powder. *Int. Dairy J.*, **19**, 205–208.

28 Newgard, J.R., Rouse, G.C., and McVicker, J.K. (2002) Novel method for detecting bovine immunoglobulin G in dried porcine plasma as an indicator of bovine plasma contamination. *J. Agric. Food Chem.*, **50**, 3094–3097.

29 Röder, M., Vieths, S., and Holzhauser, T. (2009) Commercial lateral flow devices for rapid detection of peanut (*Arachis hypogaea*) and hazelnut (*Corylus avellana*) cross-contamination in the industrial production of cookies. *Anal. Bioanal. Chem.*, **395**, 103–109.

30 Schubert-Ullrich, P., Rudolf, J., Ansari, P., Galler, B., Führer, M., Molinelli, A., and Baumgartner, S. (2009) Commercialized rapid immunoanalytical tests for determination of allergenic food proteins: an overview. *Anal. Bioanal. Chem.*, **395**, 69–81.

31 Wen, H.-W., Borejsza-Wysocki, W., DeCory, T.R., and Durst, R.A. (2005) Development of a competitive liposome-based lateral flow assay for the rapid detection of the allergenic peanut protein Ara h1. *Anal. Bioanal. Chem.*, **382**, 1217–1226.

32 Thévenot, D.R., Toth, K., Durst, R.A., and Wilson, G.S. (1999) Electrochemical biosensors: recommended definitions and classification. *Pure Appl. Chem.*, **71**, 2333–2348.

33 Baeumner, A.J. (2003) Biosensors for environmental pollutants and food contaminants. *Anal. Bioanal. Chem.*, **377**, 434–445.

34 Nayak, M., Kotian, A., Marathe, S., and Chakravortty, D. (2009) Detection of microorganisms using biosensors – a smarter way towards detection techniques. *Biosens. Bioelectron.*, **25**, 661–667.

35 Wang, J. (2005) Nanomaterial-based electrochemical biosensors. *Analyst*, **130**, 421–426.

36 Bruls, D.M., Evers, T.H., Kahlman, J.A.H., van Lankvelt, P.J.W., Ovsyanko, M., Pelssers, E.G.M., Schleipen, J.J.H.B., de Theije, F.K., Verschuren, C.A., van der Wijk, T., van Zon, J.B.A., Dittmer, W.U., Immink, A.H.J., Nieuwenhuis, J.H., and Prins, M.W.J. (2009) Rapid integrated biosensor for multiplexed immunoassays based on actuated magnetic nanoparticles. *Lab Chip*, **9**, 3504–3510.

37 Ross, S. (2008) Developing an optically stimulated piezofilm immunoassay. *IVD Technol.*, **14**, 42.

38 Aveyard, J., Mehrabi, M., Cossins, A., Braven, H., and Wilson, R. (2007) One step visual detection of PCR products with gold nanoparticles and a nucleic acid lateral flow (NALF) device. *Chem. Commun.*, 4251–4253.

39 Cheng, Q.-Y., Meng, X.-L., Xu, J.-P., Lu, W., and Wang, J. (2007) Development of lateral-flow immunoassay for WSSV with polyclonal antibodies raised against recombinant VP (19+28) fusion protein. *Virol. Sinica*, **22**, 61–67.

40 Chiao, D.-J., Wey, J.-J., Shyu, R.-H., and Tang, S.-S. (2008) Monoclonal antibody-based lateral flow assay for detection of botulinum neurotoxin type A. *Hybridoma*, **27**, 31–35.

41 Drygin, Y., Blintsov, A., Osipov, A., Grigorenko, V., Andreeva, I., Uskov, A., Varitsev, Y., Anisimov, B., Novikov, V., and Atabekov, J. (2009) High-sensitivity express immunochromatographic method for detection of plant infection by tobacco

mosaic virus. *Biochemistry (Moscow)*, **74**, 986–993.

42 Kolosova, A.Y., Sibanda, L., Dumoulin, F., Lewis, J., Duveiller, E., Van Peteghem, C.,
and De Saeger, S. (2008) Lateral-flow colloidal gold-based immunoassay for the rapid detection of deoxynivalenol with two indicator ranges. *Anal. Chim. Acta*, **616**, 235–244.

43 Nimitphak, T., Kiatpathomchai, W., and Flegel, T.W. (2008) Shrimp hepatopancreatic parvovirus detection by combining loop-mediated isothermal amplification with a lateral flow dipstick. *J. Virol. Methods*, **154**, 56–60.

44 Oku, Y., Kamiya, K., Kamiya, H., Shibahara, Y., Ii, T., and Uesaka, Y. (2001) Development of oligonucleotide lateral-flow immunoassay for multi-parameter detection. *J. Immunol. Methods*, **258**, 73–84.

45 Shyu, R.-H., Shyu, H.-F., Liu, H.-W., and Tang, S.-S. (2002) Colloidal gold-based immunochromatographic assay for detection of ricin. *Toxicon*, **40**, 255–258.

46 Verheijen, R., Osswald, I.K., Dietrich, R., and Haasnoot, W. (2000) Development of a one step strip test for the detection of (dihydro)streptomycin residues in raw milk. *Food Agric. Immunol.*, **12**, 31–40.

47 Verheijen, R., Stouten, P., Cazemier, G., and Haasnoot, W. (1998) Development of a one step strip test for the detection of sulfadimidine residues. *Analyst*, **123**, 2437–2441.

48 Wang, X., Li, K., Shi, D., Xiong, N., Jin, X., Yi, J., and Bi, D. (2007) Development of an immunochromatographic lateral-flow test strip for rapid detection of sulfonamides in eggs and chicken muscles. *J. Agric. Food Chem.*, **55**, 2072–2078.

49 Weetall, H.H., and Rogers, K.R. (2002) A simple assay for 2,4-dichlorophenoxyacetic acid using coated test-strips. *Anal. Lett.*, **35**, 1341–1348.

50 Zhang, G., Wang, X., Zhi, A., Bao, Y., Yang, Y., Qu, M., Luo, J., Li, Q., Guo, J., Wang, Z., Yang, J., Xing, G., Chai, S., Shi, T., and Liu, Q. (2008) Development of a lateral flow immunoassay strip for screening of sulfamonomethoxine residues. *Food Addit. Contam.*, **25**, 413–423.

51 Zhang, G.P., Wang, X.N., Yang, J.F., Yang, Y.Y., Xing, G.X., Li, Q.M., Zhao, D., Chai, S.J., and Guo, J.Q. (2006) Development of an immunochromatographic lateral flow test strip for detection of [beta]-adrenergic agonist clenbuterol residues. *J. Immunol. Methods*, **312**, 27–33.

52 Zhang, M.-Z., Wang, M.-Z., Chen, Z.-L., Fang, J.-H., Fang, M.-M., Liu, J., and Yu, X.-P. (2009) Development of a colloidal gold-based lateral-flow immunoassay for the rapid simultaneous detection of clenbuterol and ractopamine in swine urine. *Anal. Bioanal. Chem.*, **395**, 2591–2599.

53 Zhao, J., He, S.-P., Liu, W., Deng, A.-X., Nan, T.-G., Wang, B.-M., Zhai, Z.-X., and Li, Z.-H. (2006) Development of a lateral flow dipstick immunoassay for the rapid detection of glycyrrhizic acid. *Food Agric. Immunol.*, **17**, 173–181.

54 Aldus, C.F., van Amerongen, A., Ariens, R.M.C., Peck, M.W., Wichers, J.H., and Wyatt, G.M. (2003) Principles of some novel rapid dipstick methods for detection and characterization of verotoxigenic *Escherichia coli*. *J. Appl. Microbiol.*, **95**, 380–389.

55 Bogdanovic, J., Koets, M., Sander, I., Wouters, I., Meijster, T., Heederik, D., van Amerongen, A., and Doekes, G. (2006) Rapid detection of fungal [alpha]-amylase in the work environment with a lateral flow immunoassay. *J. Allergy Clin. Immun.*, **118**, 1157–1163.

56 Koets, M., Sander, I., Bogdanovic, J., Doekes, G., and van Amerongen, A. (2006) A rapid lateral flow immunoassay for the detection of fungal alpha-amylase at the workplace. *J. Environ. Monit.*, **8**, 942–946.

57 Capps, K.L., McLaughlin, E.M., Murray, A.W.A., Aldus, C.F., Wyatt, G.M., Peck, M.W., van Amerongen, A., Ariëns, R.M.C., Wichers, J.H., Baylis, C.L., Wareing, D.R.A., and Bolton, F.J. (2004) Validation of three rapid screening methods for detection of verotoxin-producing *Escherichia coli* in foods:

interlaboratory study. *J. AOAC Int.*, **87**, 68–77.
58 Xia, X., Xu, Y., Zhao, X., and Li, Q. (2009) Lateral flow immunoassay using europium chelate-loaded silica nanoparticles as labels. *Clin. Chem.*, **55**, 179–182.
59 Koets, M., van der Wijk, T., van Eemeren, J.T.W.M., van Amerongen, A., and Prins, M.W.J. (2008) Rapid DNA multi-analyte immunoassay on a magneto-resistance biosensor. *Biosens. Bioelectron.*, **24**, 1893–1898.
60 Taton, K., Johnson, D., Guire, P., Lange, E., and Tondra, M. (2009) Lateral flow immunoassay using magnetoresistive sensors. *J. Magn. Magn. Mater.*, **321**, 1679–1682.
61 Wang, Y., Xu, H., Wei, M., Gu, H., Xu, Q., and Zhu, W. (2009) Study of superparamagnetic nanoparticles as labels in the quantitative lateral flow immunoassay. *Mater. Sci. Eng. C*, **29**, 714–718.
62 Xu, Q., Xu, H., Gu, H., Li, J., Wang, Y., and Wei, M. (2009) Development of lateral flow immunoassay system based on superparamagnetic nanobeads as labels for rapid quantitative detection of cardiac troponin I. *Mater. Sci. Eng. C*, **29**, 702–707.
63 Lingerfelt, B.M., Mattoussi, H., Goldman, E.R., Mauro, J.M., and Anderson, G.P. (2003) Preparation of quantum dot–biotin conjugates and their use in immunochromatography assays. *Anal. Chem.*, **75**, 4043–4049.
64 Corstjens, P., Zuiderwijk, M., Brink, A., Li, S., Feindt, H., Niedbala, R.S., and Tanke, H. (2001) Use of up-converting phosphor reporters in lateral-flow assays to detect specific nucleic acid sequences: a rapid, sensitive DNA test to identify human papillomavirus type 16 infection. *Clin. Chem.*, **47**, 1885–1893.
65 Li, L., Zhou, L., Yu, Y., Zhu, Z., Lin, C., Lu, C., and Yang, R. (2009) Development of up-converting phosphor technology-based lateral-flow assay for rapidly quantitative detection of hepatitis B surface antibody. *Diagn. Microbiol. Infect. Dis.*, **63**, 165–172.
66 Yan, Z., Zhou, L., Zhao, Y., Wang, J., Huang, L., Hu, K., Liu, H., Wang, H., Guo, Z., Song, Y., Huang, H., and Yang, R. (2006) Rapid quantitative detection of *Yersinia pestis* by lateral-flow immunoassay and up-converting phosphor technology-based biosensor. *Sens. Actuators B Chem.*, **119**, 656–663.
67 Lönnberg, M., and Carlsson, J. (2001) Quantitative detection in the attomole range for immunochromatographic tests by means of a flatbed scanner. *Anal. Biochem.*, **293**, 224–231.
68 Edwards, K., Curtis, K., Sailor, J., and Baeumner, A. (2008) Universal liposomes: preparation and usage for the detection of mRNA. *Anal. Bioanal. Chem.*, **391**, 1689–1702.
69 Posthuma-Trumpie, G.A., Korf, J., and van Amerongen, A. (2009) Lateral flow (immuno)assay: its strengths, weaknesses, opportunities and threats. A literature survey. *Anal. Bioanal. Chem.*, **393**, 569–582.
70 Barnard, G., Osher, J., Lichter, S., Gayer, B., De Boever, J., Limor, R., Ayalon, D., and Kohen, F. (1995) The measurement of progesterone in serum by a non-competitive idiometric assay. *Steroids*, **60**, 824–829.
71 Winger, L.A., Dessi, J.L., and Self, C.H. (1996) Enhanced specificity for small molecules in a convenient format which removes a limitation of competitive immunoassay. *J. Immunol. Methods*, **199**, 185–191.
72 D'Agostino, M., Wagner, M., Vazquez-Boland, J.A., Kuchta, T., Karpiskova, R., Hoorfar, J., Novella, S., Scortti, M., Ellison, J., Murray, A., Fernandes, I., Kuhn, M., Pazlarova, J., Heuvelink, A., and Cook, N. (2004) A validated PCR-based method to detect *Listeria monocytogenes* using raw milk as a food model – towards an international standard. *J. Food Prot.*, **67**, 1646–1655.
73 Lane, C.R., Hobden, E., Walker, L., Barton, V.C., Inman, A.J., Hughes, K.J.D., Swan, H., Colyer, A., and Barker, I. (2007) Evaluation of a rapid diagnostic field test kit for identification of *Phytophthora* species, including *P. ramorum* and *P. kernoviae* at the point of inspection. *Plant. Pathol.*, **56**, 828–835.
74 Litos, I.K., Ioannou, P.C., Christopoulos, T.K., Traeger-Synodinos, J., and

Kanavakis, E. (2009) Multianalyte, dipstick-type, nanoparticle-based DNA biosensor for visual genotyping of single-nucleotide polymorphisms. *Biosens. Bioelectron.*, **24**, 3135–3139.

75 Mens, P.F., van Amerongen, A., Sawa, P., Kager, P.A., and Schallig, H.D.F.H. (2008) Molecular diagnosis of malaria in the field: development of a novel 1-step nucleic acid lateral flow immunoassay for the detection of all 4 human *Plasmodium* spp. and its evaluation in Mbita, Kenya. *Diagn. Microbiol. Infect. Dis.*, **61**, 421–427.

76 Sithigorngul, W., Rukpratanporn, S., Sittidilokratna, N., Pecharaburanin, N., Longyant, S., Chaivisuthangkura, P., and Sithigorngul, P. (2007) A convenient immunochromatographic test strip for rapid diagnosis of yellow head virus infection in shrimp. *J. Virol. Methods*, **140**, 193–199.

77 Thornton, C.R., Groenhof, A.C., Forrest, R., and Lamotte, R. (2004) A one-step, immunochromatographic lateral flow device specific to *Rhizoctonia solani* and certain related species, and its use to detect and quantify *R. solani* in soil. *Phytopathology*, **94**, 280–288.

78 Tomlinson, J.A., Dickinson, M.J., and Boonham, N. (2010) Rapid detection of *Phytophthora ramorum* and *P. kernoviae* by two-minute DNA extraction followed by isothermal amplification and amplicon detection by generic lateral flow device. *Phytopathology.*, **100**, 143–149.

79 Soliman, H., and El-Matbouli, M. (2010) Loop mediated isothermal amplification combined with nucleic acid lateral flow strip for diagnosis of cyprinid herpes virus-3. *Mol. Cell. Probes*, **24**, 38–43.

80 Puthawibool, T., Senapin, S., Kiatpathomchai, W., and Flegel, T.W. (2009) Detection of shrimp infectious myonecrosis virus by reverse transcription loop-mediated isothermal amplification combined with a lateral flow dipstick. *J. Virol. Methods*, **156**, 27–31.

81 Jaroenram, W., Kiatpathomchai, W., and Flegel, T.W. (2009) Rapid and sensitive detection of white spot syndrome virus by loop-mediated isothermal amplification combined with a lateral flow dipstick. *Mol. Cell. Probes*, **23**, 65–70.

82 Lo, W.-Y., and Baeumner, A.J. (2007) RNA internal standard synthesis by nucleic acid sequence-based amplification for competitive quantitative amplification reactions. *Anal. Chem.*, **79**, 1548–1554.

83 Hartmann, M., Roeraade, J., Stoll, D., Templin, M.F., and Joos, T.O. (2009) Protein microarrays for diagnostic assays. *Anal. Bioanal. Chem.*, **393**, 1407–1416.

84 Han, A., Dufva, M., Belleville, E., and Christensen, C.B.V. (2003) Detection of analyte binding to microarrays using gold nanoparticle labels and a desktop scanner. *Lab Chip*, **3**, 329–332.

85 Järås, K., Tajudin, A.A., Ressine, A., Soukka, T., Marko-Varga, G., Malm, J., Laurell, T., and Lilja, H. (2008) ENSAM: europium nanoparticles for signal enhancement of antibody microarrays on nanoporous silicon. *J. Proteome Res.*, **7**, 1308–1314.

86 Kim, D., Daniel, W.L., and Mirkin, C.A. (2009) Microarray-based multiplexed scanometric immunoassay for protein cancer markers using gold nanoparticle probes. *Anal. Chem.*, **81**, 9183–9187.

87 Kusnezow, W., and Hoheisel, J.D. (2003) Solid supports for microarray immunoassays. *J. Mol. Recognit.*, **16**, 165–176.

88 Ling, J., Li, Y.F., and Huang, C.Z. (2008) A label-free visual immunoassay on solid support with silver nanoparticles as plasmon resonance scattering indicator. *Anal. Biochem.*, **383**, 168–173.

89 Huang, L., Reekmans, G., Saerens, D., Friedt, J.-M., Frederix, F., Francis, L., Muyldermans, S., Campitelli, A., and Hoof, C.V. (2005) Prostate-specific antigen immunosensing based on mixed self-assembled monolayers, camel antibodies and colloidal gold enhanced sandwich assays. *Biosens. Bioelectron.*, **21**, 483.

90 Wang, J., Chen, Z., Corstjens, P.L.A.M., Mauk, M.G., and Bau, H.H. (2006) A disposable microfluidic cassette for DNA amplification and detection. *Lab Chip*, **6**, 46–53.

Part Three
Food Applications

6
Nano-Functionalized Techniques in Crop and Livestock Production: Improving Food Productivity, Traceability, and Safety

Niall O'Brien and Enda Cummins

6.1
Introduction

Safe and efficient food production relies on the detection and monitoring of food characteristics, environment, processing, and contaminants at all stages in the food production chain. Nanotechnology not only has the potential to optimize existing and create new monitoring techniques at early food production stages, but also may be employed in nano-functionalized agricultural products such as fertilizers, pesticides, and herbicides to maximize productivity. The integration of nanotechnological techniques in primary food production (i.e., preprocessing) is explored in this chapter.

Crop and animal diseases threaten productivity from the very beginning of the food chain. In countries with agriculture-driven economies, food safety and market protection are high on the list of priorities, especially in relation to the livestock industry. In the recent past, bovine spongiform encephalopathy (BSE) in the UK, dioxin contamination in Belgium and Ireland, and swine flu in the USA have all damaged consumer confidence in the safety of the food we eat and resulted in significant loss of global markets. Measures to improve food traceability and to monitor farm management practices, including administration of medicines, fertilizers, and pesticides, are becoming increasingly important issues, with more stringent regulation in industrialized countries including the USA and member states of the European Union [1, 2]. Preventative application of these treatments or untargeted general application in response to detection is inefficient and may contribute to the evolution of resistant strains of viruses, pests, and pathogens [3–5]. Nanotechnology may aid in tackling these issues on two levels: nano-based sensors and nano-functionalized sample pre-treatment, extraction, and amplified diagnostic techniques may aid in early detection of pathogens and spoilage organisms, and allow targeted responses; while nano-functionalized medicines, vaccines, and pesticides may

Nanotechnology in the Agri-Food Sector: Implications for the Future, First Edition. Edited by Lynn J. Frewer, Willem Norde, Arnout Fischer, Frans Kampers.

allow automatic responses to disease by triggered release in response to specific markers.

As well as adopting and implementing practices and technologies that optimize productivity along the food production chain, the food production industry must also consider practices and technologies that ensure safety and transparency, as required in Europe by EU Directive 178/2002. Current tools employed in the food industry to enable traceability include alphanumeric codes, barcode labels, and radiofrequency identification (RFID) tags, all of which may be optimized in terms of their scope of application, specificity, and security through nanoscaling and nano-functionalization. RFID tags employed in the food industry typically monitor temperature, humidity, and light exposure, though nano-functionalized "lab-on-a-chip" developments – for example, handheld deoxyribonucleic acid (DNA) sensors – may lead to tags that monitor pathogen and/or chemical exposure throughout the food production chain as part of an integrated monitoring network.

The scope of integration of nanotechnology into an early-stage agri-food sensor network may be seen in Figure 6.1. The nano-functionalized sensing systems, data analysis and management, and subsequent nano-functionalized responses highlighted in this figure may be used in isolation to enhance current agricultural practices or in tandem as part of a fully nano-functionalized agricultural management system. The information gathered from the sensing devices for use in optimizing agricultural practices may also be employed in parallel food traceability

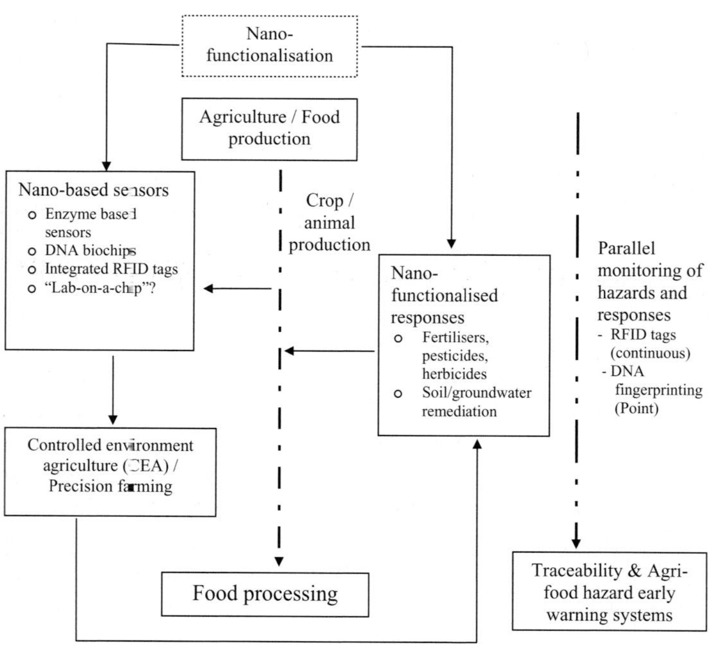

Figure 6.1 Scope of agri-food nanosensor network.

and/or safety systems. The devices and products highlighted in this figure are those with current application, or potential application in the near future (0–10 years), though, as further nano-specific characteristics of materials are discovered, and new methods to control materials on the nanoscale are developed, further agricultural applications are envisaged.

The Woodrow Wilson International Center for Scholars, a non-partisan institution engaged in the study of national and world affairs, compiled a database of largely US government-funded, application-oriented, research and development projects with food and agriculture applications [6]. Projects concerned with the agri-ecosystem, pre-harvest and post-harvest sectors include pathogen and contaminant detection, animal and crop identity preservation and tracking, smart treatment and delivery of veterinary medicines and pesticides, smart systems integration, nano-functionalized devices for molecular and cell biology, and environment and agriculture waste management. Some of these projects may have commercial application in the next five years, while many others have potential application in the next 15 years. Table 6.1 highlights basic, applied, and developmental projects concerned with the application of nanotechnology in primary food production from the Woodrow Wilson Agrifood database, EU research projects, and the literature. Research into the direct application of nanotechnology into agriculture is set to increase in the future, with research funding in Europe, the USA, Asia, and other developing countries increasingly geared toward nanotechnology sectors such as portable measurement devices, food quality and food safety assurance, nanosensors for optimization of bioprocesses, water treatment technologies, and the development of sustainable products and markets through bio-production of green chemicals and materials from agricultural waste [16].

6.2
Sensors

Monitoring of agricultural environments and activities is an integral part of controlled-environment agriculture (CEA) and precision farming. Much of the data on which agricultural management systems are built comes from relatively infrequent point-source measurements, a necessity considering labor and analytical costs. Continuous monitoring of agricultural systems may become a reality with nano-functionalized sensor networks based on bioassays and wireless communication technology.

There has been much development into the application of nanomaterials into the design of enzymatic and DNA-based biosensors in the field of environmental monitoring and protection, while nanotechnology has many applications in communications technology, ranging from printed electronics to power supplies. Biosensors may be applied in the monitoring of specific analytes, such as pesticides, herbicides, phenols, endocrine disruptors, and surfactants, or for monitoring whole biological effects, such as toxicity or estrogenicity [17]. These sensors provide analytical information relating to the analyte or biological effect by converting a

Table 6.1 Basic, applied, and developmental nano-based primary agri-food projects.

Project/study	Nanomaterial	Function	Time to commercialization	References
Remediation of PCB-contaminated soils using iron nanoparticles	Fe_2O_3	Bioremediation – contaminant sequestering and destruction	Current	[7]
Large-scale application of nanotechnology for wood protection	$CuCO_3$	Surface coating – wood preservative	Current	[8]
Pesticides: nano-emulsions and triggered release nano-encapsulates	Lipid/natural polymer	Nanoscaling and coating – increased efficacy, water solubility, and triggered (local) release	Current	[9]
Ultra-sensitive detection of mutated papillary thyroid carcinoma DNA using square wave stripping voltammetry method and amplified gold nanoparticle biomarkers	Au/Ag	Sample labeling – signal transduction and amplification	0–5 years	[10]
Poly-silicon nanowire field-effect transistor for ultra-sensitive and label-free detection of pathogenic avian influenza DNA	SiO_2	Silicon nanowire field-effect transistor – specific and ultra-sensitive detection of virus DNA	0–5 years	[11]
Veterinary medicines: triggered release nano-encapsulates	Lipid/natural polymer	Nanoscaling and coating – novel administration routes: pulmonary, optical, and topical; improved dose uniformity, bioavailability, and drug stability	0–5 years	[12]
Microfluidics for livestock breeding	N/A	DNA fingerprinting – marker-assisted breeding	0–5 years	[13]

Development of nanosensors for the detection of quality parameters along the food chain	N/A	Nano-functionalized immunoassays – detection of pathogenic micro-organisms, mycotoxins, and drug residues	0–5 years	[14]
Nanotechnologies for bio-inspired polysaccharides: biological decoys designed as knowledge-based, multifunctional biomaterials	Chitosan	Nanoscale materials – drug/gene delivery	0–5 years	[14]
Handheld gene analyzer based on dye-doped nanoparticles	SiO_2	Sample labeling – signal transduction and amplification	5–10 years	[6]
Optochemical nanosensor PEBBLEs: photonic explorers for bioanalysis with biologically localized embedding	Polyacrylamide/sol–gel	Nanosized sensors encapsulating an analyte-specific dye and a reference dye within a biologically inert matrix – detection of chemical/virus specific analytes	5–10 years	[15]
Detection of foot-and-mouth disease virus by anchoring transitions of liquid crystals	Au	Nanoscale topography – specific detection of virus DNA	5–10 years	[6]
Adhesion-specific nanoparticles for removal of *Campylobacter jejuni* from poultry	N/A	Bioactive nanoparticles – bind to surfaces of *Campylobacter* and excreted	5–10 years	[6]
Engineering advanced polymeric surfaces for smart systems in biomedicine, biology, material science and nanotechnology: a cross-disciplinary approach of biology, chemistry, and physics	Elastin-like polymers	Nano-functionalized polymeric surfaces – drug/gene delivery, nanobiotechnology, lab-on-a-chip systems	5–10 years	[14]
Development of nano-based passive sensors for RF/wireless sensing systems	C	Carbon nanotube resonator – gaseous detection and signal transduction	10–15 years	[6]

biological event into a detectable signal by means of a transducer or processor. It is this transduction element/event that may be optimized or amplified through nano-functionalization, resulting in more efficient diagnostic tools for agricultural management and veterinary applications.

A future application of these nano-functionalized sensing and diagnostic methods is their integration into "lab-on-a-chip" (LOC) technologies. The ambition of these sensing systems is the development of a rapid (real-time) and partly reusable LOC biosensor for bacterial, pest, and viral detection in agricultural samples. In order to achieve the desired attributes of a LOC biosensor, such as near real-time parallel detection of multiple analytes, high sensitivity and specificity, reagentless operation, and minimal cost, nano-functionalization of sensing methods will be essential.

6.3 Enzyme Biosensors and Diagnostics

Enzymes are biomolecules that increase the rate of chemical reactions. Enzymes are very specific, catalyzing only one specific chemical reaction or, at most, a few related reactions. It is on this principle that enzyme-based biosensors operate, measuring the inhibition of a specific enzyme resulting from the presence of a target analyte or the catalytic transformation of a target analyte by a specific enzyme. The majority of transduction elements associated with enzyme-based biosensors have an electrochemical basis [17]. Electrochemical transduction elements convert the interactions between electro-active species produced or consumed by the actions of biological elements (e.g., enzymes) into a primary signal. The relatively large surface area of nanomaterials has led to enhanced catalytic properties in many materials, especially metals. Combining this catalytic action with the potential for control over the size and structure of these materials has led to the incorporation of metal nanoparticles such as gold, copper, and platinum into electrodes for the improved detection of analytes [18, 19].

A challenge in the development of enzyme-based biosensors is establishing an electronic communication between the active site of the enzyme and the electrode. In many sensor systems, a co-substrate or mediator is required to allow transduction of the biological event, owing to the presence of an electrically insulating protein shell surrounding most relevant enzymes. The need for co-substrates or mediators may be removed through the incorporation of "molecular wires" in the form of carbon nanotubes (CNTs). These nanotubes may allow electrical communication between the electrode and redox proteins [20]. Reducing the need for sample preparation and mediators in sensors, and the amplification and optimization of electrochemical detection signals (resulting in lower limits of detection), through nano-functionalization, may result in the development of autonomous nano-based sensors for use in detecting chemical contamination, pest infestation, or the development of disease in crops. Gold and magnetic nanoparticles functionalized with the appropriate chemistry for enzyme immobilization have been investigated for use in biosensor technologies for the detection and characterization of micro-organisms such as *E. coli*, *Salmonella*, and *Staphylococcus* [6].

Future applications envisioned for enzyme- or antibody-immobilized nanoparticles developed as part of this project included real-time biosensor platforms for the detection and diagnosis of diseases for human and animal health. Further applications of nanoscale materials include the use of electron-beam lithography of polymeric materials to produce features on the nanoscale for virus diagnostics [6]. Nanostructured gold films with topography matched to that of the size of smaller viruses, such as foot-and-mouth disease virus (FMDV), may be functionalized with a single chain antibody specific for FMDV in such a way that liquid crystals will uniformly anchor on these surfaces. This technology may aid in early intervention in animal disease monitoring, allowing preventative and containment measures to be taken.

6.4 DNA-Based Biosensors and Diagnostics

Deoxyribonucleic acid (DNA) is the largest biomolecule and stores biological information on living organisms. The structure of DNA is very sensitive to the influence of environmental pollutants, such as heavy metals, polychlorinated biphenyls (PCBs), and polyaromatic hydrocarbons (PAHs) [17]. Based on this known carcinogenesis and mutagenesis, DNA-based biosensors have been developed for rapid testing of potential pollutants for mutagenic and carcinogenic activity. DNA sequencing is the determination of the precise sequence of nucleotide bases, adenine, guanine, cytosine, and thymine, in a molecule of DNA. DNA chips, consisting of DNA immobilized on a substrate, are designed to explore a biological sample for genetic information [21]. This information may be read from the chip through optical, optoelectronic, electrochemical, and magnetic methods. Many of these detection methods require sample processing, consisting of radioactive or fluorescent tagging, which limits the application of this technology in remote sensor operations of LOC technologies.

The large relative surface area and biocompatibility of many nanomaterials has led to their incorporation into biosensors for the immobilization of biomolecules, which lends itself to labeling, sensor substrate, and electrode enhancement applications. Nano-functionalization of substrates, tracers, and electrodes has improved the sensitivity of bioelectronic assays by several orders of magnitude. DNA detection platforms incorporating nanoparticle labeling (e.g., gold nanoparticles enhanced with silver), and photodiode and optical illumination to detect and examine DNA sequences, with the potential for integration into circuit technologies, have been developed [10, 22, 23]. Silica, dye-doped nanoparticles have been demonstrated to produce 10 000- to 100 000-fold increases in detectable signal when sequencing a DNA sample when compared to commonly used fluorescent tagging [6]. The combination of integrated circuit technology and electrochemical DNA immobilization techniques may allow the development of portable DNA nanoarray systems with electronic read-out. This is expected to result in rapid (10 minutes), low-cost ($5–10 per test) detection and/or diagnosis of environmental pathogens in water, animals, and crops and feed [6].

The potential for nanoscale pores to be used as biosensors has been investigated, as they are an established tool for analyzing the structure and composition of DNA [24]. The nanopores are used to measure the binding of enzymes to their DNA substrates. In this technique, a DNA molecule bound to an enzyme is drawn into the nanopore (consisting of a protein pore inserted into a thin membrane) by an applied voltage, and the force applied to the DNA molecule by this voltage allows the measurement of the enzyme–DNA interactions. From these enzyme–DNA interactions, DNA templates may be identified by this nanopore device. These nanopore devices combined with digital logic hardware may allow real-time discrimination and control of single molecules and detection of enzymes that bind or modify DNA. These nanopore-based sensors may have eventual application in rapid DNA fingerprinting of crops or disease and pathogen diagnostics.

DNA-based molecular electronics or "nanoelectronics" has been of interest to researchers for over a decade. The ability for DNA to be coated with metals – for example, gold nanoparticles – allows the possibility of conductive wires with the self-assembly characteristics of DNA [21]. From these DNA wires, a three-dimensional structure may be constructed for use in biosensors, biotransistors, diodes, and, potentially, nano-engines. As discussed in the section on enzyme biosensors (Section 6.3), carbon nanotubes (CNTs) have ideal electronic properties and dimensions for use in nanoelectronics. The unique electronic properties of these molecular electronics, incorporating DNA scaffolding and nanomaterials, have resulted in the proposal for a DNA–single-electron transistor (DNA-SET) for single-molecule DNA sequence analysis. A single-electron transistor consists of two tunnel junctions sharing one common electrode with a low self-capacitance, the floating island, the electrical potential of which can be tuned by a third electrode, the gate. In a DNA-SET, the space between the floating island and the gate is known as a "nano-eye" and the passage of any analyte molecule through the nano-eye alters the potential of the floating island by virtue of the charge and permittivity of the analyte molecule. This change in the potential of the floating island produces a change in the SET current. A silicon nanowire field-effect transistor has been demonstrated to achieve specific and ultra-sensitive detection of high pathogenic strain virus DNA of avian influenza [11]. The nano-functionalized sensor device developed in this study was capable of high uniformity and reproducibility, high yield, and excellent scalability and manufacturability using commercially available procedures, leading to its potential commercial application in the near future (within five years).

The future of DNA-based biosensors lies with reagentless technology, technology that relies on minimal processing or labeling of samples. The incorporation of transduction technologies into integrated circuits, such as electrical and electrochemical approaches, may lead to diagnostic tools for use in veterinary medicine, such as handheld DNA chips with electronic read-out and direct approach-based, generic DNA sensors [11, 21]. Advances in the speed, cost, and ease of obtaining genetic information on animals and crops due to nano-functionalization of DNA-based sensors and diagnostic tools will have application in marker-assisted breeding of livestock and selection of crops with improved yield, nutrient profiles, and uptake. The quick and cost-efficient detection and selection of feedstuffs with preferable nutrient profiles through DNA-based sensor technology will have application in livestock production and food processing.

6.5 Radiofrequency Identification (RFID)

The future of agricultural management lies in the allying of enzyme and DNA-based nanosensor technologies with RFID data storage and tracking systems, especially in the area of livestock systems. The nano-functionalized sensing methods described earlier have the potential to become part of a remote management system, once integrated with appropriate data management systems. Information on the early detection of viruses in livestock, through implants, and on the bacterial, contaminant, and pest exposure in crops and feed, through localized sensors, may be relayed and pre-emptive, targeted action taken.

Passive RFID tags have been available for application in livestock for many years. These implanted tags log the location of an animal every time it passes an RFID transceiver, normally located at drinking and feeding areas, and at various points in the transportation chain. Active RFID tags are finding more application, where the animal's location is automatically logged whether it passes a transceiver or not, over a range of 100–200 m. These tags may also log each individual animal's temperature, which is stored in a database along with manually logged information on inoculations and antibiotic treatments. Raised temperature in animals may be a symptom of an infectious disease, and the logged data on location and temperature allow determination of contact between potentially infected animals [25]. The advent of nano-functionalized sensor technology, such as enzyme and DNA sensors, and its integration into RFID systems may allow the automatic logging of inoculation and antibiotic information through tracers associated with their absorption or ingestion. Other than that of animal temperature, the active monitoring of further disease tracers may also be possible through nano-functionalized sensor technologies.

The carbon nanotube resonator is a potential wireless remote-sensing technology not incorporating enzyme- or DNA-based sensing [6]. The proposed sensor consists of an electromagnetic resonator integrated with vertically aligned carbon nanotubes used as a chemical transducer. The carbon nanotube resonator sensor is a passive sensing device and requires no power consumption. The microwave carbon nanotube resonator sensor exhibits changes in resonant frequency when exposed to gases, with these frequencies read by a remote transceiver. The passive nature and zero power consumption of this proposed device are elements that may lead to its potential application in crop/feed monitoring and animal welfare.

6.6 Integrated Nanosensor Networks: Detection and Response

Containing and controlling potential hazards in the food chain requires data on a vast range of environmental and product characteristics to be collected and stored at all points. In the area of food production, systems are in place to monitor chemical and pathogenic contaminants on a point-source basis, while environmental factors are being monitored through the use of RFID tagging. The integration of nano-functionalized sensors and traceability systems for early-stage agricultural activities, that is, livestock and crop cultivation, into agri-food management and early-warning hazard systems must be considered. Through DNA libraries and datasets on the enzymatic actions of pathogens, chemical contaminants, and pests, enzyme- and DNA-based sensor technologies may be combined with communication systems to develop targeted response actions on the farm, state, or country-wide scale. Secondary benefits, not related directly to the product but affecting overall productivity, may be gleaned from autonomous nano-functionalized sensor systems, such as improved water and fuel economy.

Responses to a detected hazard may also be nano-functionalized in order to optimize their effectiveness, timeliness, and efficiency. Fertilizers such as phosphates, nitrates, and urea are often added to agricultural land to optimize conditions for growing certain crops, though their usage is often strictly regulated and monitored due to adverse environmental effects that may arise through leaching to water sources. This leaching of applied fertilizer also leads to a lower fraction of applied fertilizer being available to the target crop, resulting in reduced productivity, as the fertilizer application limits are strictly controlled. The answer to this problem of fertilizer availability may be in the form of nanoparticle coated fertilizers. Because of the high surface area associated with nanoparticles, this high surface tension will hold nanoparticle coated fertilizers more strongly to the plant than conventional fertilizers [26]. This nano-coating may be a sulfur-based coating, where the rate of dissolution, and so availability to the plant, may be determined through its thickness, porosity, surface area, and chemical phase. It has also been suggested that, with genetic engineering, fertilizers consisting of nutrient solutions with precise dissolution behaviors may be tailored to specific crops [26].

The advent of biofuels and the unstable global oil situation has led to competition for agricultural land. In this age of food security, sustainable agriculture, increasing populations, and shifting diets, land previously unsuitable for agricultural practices may need to be re-examined. Land may be unsuitable for agricultural practices for a number of reasons: naturally occurring high concentrations of elements such as arsenic, heavy metals, and so on, or contamination through industrial practices, for example, Superfund sites in the USA. The remediation or optimization of these environments for agricultural practices has traditionally been an expensive operation, though with the advent of nanoscale catalysts this may become a more viable option. The remediation of contaminated land may be performed *in situ*, where contaminants are treated in place through the injection of catalytic material to sequester or degrade contaminants, or *ex situ*, where the contaminants are removed from the soil and treated in above-ground reactors, known as "pump and treat" technologies.

Nanomaterials may be employed in more efficient catalysts for *ex situ* treatment of contaminants, but it is the application of nanoparticles for treatment technologies *in situ* that is of more interest, as this has the potential to be more cost effective. The high relative surface area and tunability of surface properties of nanoparticles result in unique catalytic properties and mobility behaviors in soils and aquatic systems. These mobility characteristics aid in the delivery of the catalytic material to the contaminants, and so reduce the concentrations of catalytic material traditionally used in technologies *in situ* to ensure distribution in soil systems. Other advantages noted for nanoscale catalysts are the more rapid degradation of contaminants compared to the traditional micrometer-scale catalysts used for soil and groundwater remediation, the degradation of contaminants that do not react with micrometer-scale particles of similar material at any detectable rate, and the production of more favorable by-products as a result of contaminant degradation [27]. While the mobility characteristics of nanomaterials are the driving force behind their application in remediation technologies, they also lie at the center of some concern over their environmental fate and impact. While the usage of nanoscale catalysts for soil and groundwater remediation is becoming more widespread in the USA, there has been a voluntary precautionary suspension on the usage of nanoscale catalysts by industry in UK until the environmental fate of these materials has been satisfactorily considered [28].

The application of pesticides and herbicides to crops and of veterinary medicines to livestock may be pre-emptive actions or cover-all responses to the detection of contamination or disease. The application of pesticides or herbicides to crops is associated with many of the same problems as those faced for the application of fertilizers. For example, leaching or drift to water sources may have adverse environmental effects and reduces the dose available to tackle the problem. Increasing the applied concentrations of pesticides and herbicides may fall foul of regulatory limits, trends for more organically grown foodstuffs, and result in the evolution of resistant strains of bacterial contaminants. In the area of livestock health and productivity, viruses and parasitic infections are often immunized against through pre-emptive dosing, though in other cases, once a certain hazard is detected, whole stocks are treated.

The use of nano-functionalized sensors for the early detection of the onset of an infection or contamination and the subsequent targeted response or isolation has been considered in this chapter. These sensor technologies may work in tandem with nano-functionalized responses in the form of nanoparticle coated pesticides, herbicides, and veterinary medicines, triggered to release upon the detection of tracers associated with certain viruses, pathogens or bacterial infections. Studies with porous hollow silica nanoparticles as the controlled delivery system for water-soluble pesticides found that the resulting quick initial release burst of pesticide after application may satisfy the immediate treatment need, while the sustained release at the later stages will achieve a continuous long-time treatment [29]. Nanostructured medicines will improve the rate of adsorption upon ingestion by animals, reducing the quantities required and the subsequent environmental contamination.

The potential integration of nano-functionalized sensor technology with that of RFID technology has been discussed in this chapter. The vast quantities of data that may be obtained from such integrated technology may have application in agri-food early-warning hazard systems. Countrywide and global-scale early-warning systems for emerging food-borne hazards may be reactive or predictive systems [30]. Reactive surveillance systems track and forecast emerging food-borne safety risks through the collection, integration, analysis, and interpretation of data, disseminating this information through reports, advisories, and warnings. End-point-based systems record the occurrence of diseases or intoxication as caused by pathogens and toxicants present in foods, from which a report detailing this outbreak is filed.

Nano-enhanced diagnostic tools will allow sensitive, rapid, and low-cost analysis of human and/or food samples. Data gained from nanosensor-enhanced RFID tagging of livestock or crop/feed loads may aid in tracing this disease or intoxication back through the early-stage agricultural chain. Hazard-oriented systems measure the presence of a pathogen, toxicant, or other hazardous agent present in food and feed. This also includes the occurrence of diseases in livestock that may be potentially transmitted to humans through consumption of derived edible products or through animal–human contacts. Nanosensors and nano-enhanced diagnostic tools will also allow the improved and active detection of pathogens, toxicants, or other hazardous agents at the early stages of the food chain, preventing the development of animal diseases and food-borne hazards.

While endpoint- and hazard-based early warning systems for threats to public health are reasonably well developed and in operation, predictive early-warning systems for food-borne hazards require more development. Endpoint- and hazard-based early-warning systems command greater resources than predictive modeling, as health agencies are expected to deal with current food-borne hazards rather than those which may (or may not) emerge in the future. Predictive models employ data on environmental and animal/crop characteristics to predict possible disease and pathogen outbreaks. Investment in predictive modeling systems has the potential to reduce human fatalities, suffering, and economic loss. A great deal of data is required for this modeling, however, which may be provided by the autonomous nano-functionalized agriculture sensor networks described in this chapter. Automatic data collection is an attractive option compared to the current costly and labor-intensive sample collection and analysis systems. The full potential of food-borne hazard warning systems, whether reactive or predictive in nature, may be realized through nano-functionalization of sensing and diagnostic methods.

6.7 Conclusions

The commercial application of nanotechnology into early-stage food-chain activities in the short term, that is, within five years, is likely to be dominated by the enhancement of current sensor and diagnostic technologies in terms of sample pre-treatment and detection limit amplification and specificity. The characteristics of nanomaterials, such as large relative surface area and surface reactivity, will allow their application in biosensor labeling, sensor substrate, and electrode enhancement. This nano-functionalization will lead to signal enhancement and improved sensor sensitivity. Nanoparticle-enhanced transduction elements in biosensors will result in reagentless sensors, paving the way for autonomous, reusable operation. Improved sensitivity and reagentless operation will also improve diagnostic speed and reduce analytical costs.

In the medium term, that is, 5–10 years, this nano-functionalized sensor technology may be incorporated into "lab-on-a-chip" technologies, where several target analytes are detected in parallel in real time with minimal processing, and the production of handheld DNA-based sensors with electronic read-out. The integration of "lab-on-a-chip" sensor technologies with RFID tagging and data communication will allow the development of nanosensor networks, where agricultural practices are managed on a previously unimaginable scale, that is, single animal or field. The management of agricultural practices as a result of the improved data collection and communication may also be on the nanoscale, through the use of nano-functionalized fertilizers, pesticides, and herbicides with slow release characteristics or triggered release in response to a specific hazard. Nano-functionalized medicines and inoculations will allow better absorption and triggered release in animals. Traceability in food systems and human health hazard warning schemes may also be improved as a result of nano-functionalized sensor, tagging, and diagnostic technologies. Improved speed and reduced cost in sample processing will allow potential hazards to be detected more efficiently, resulting in more timely public warning and disease prevention schemes to be put in place. Nano-enhanced RFID tagging will allow the improved tracing of potential hazards through the early stages of the food chain.

In order for industrialized and developing countries to fulfill the aim of self-sustainability and to improve food productivity, the application of nanotechnology at many levels, ranging from nano-structured fertilizers to autonomous nanosensor networks, will be essential over the next 20 years.

References

1 European Commission (1991) Council Directive 91/676/EEC of 12 December 1991 concerning the protection of waters against pollution caused by nitrates from agricultural sources. *Official J. Eur. Communities*, **L375**, 1–8.

2 USEPA (2007) Major Existing EPA Laws and Programs that Could Affect Agricultural Producers, Environmental Protection Agency, Agricultural Counselor, Office of the Administrator, National Service Center for Environmental Publications, Cincinnati.

3 Foil, L.D., Coleman, P., Eisler, M., Fragoso-Sanchez, H., Garcia-Vazquez, Z., Guerrero, F.D., Jonsson, N.N., Langstaff, I.G., Li, A.Y., Machila, N., Miller, R.J., Morton, J., Pruett, J.H., and Torri, S. (2004) Factors that influence the prevalence of acaricide resistance and tick-borne diseases. *Vet. Parasitol.*, **125**, 163–181.

4 Srinivas, R., Udikeri, S.S., Jayalakshmi, S.K., and Sreeramulu, K. (2004) Identification of factors responsible for insecticide resistance in *Helicoverpa armigera*. *Compar. Biochem. Physiol. C*, **137**, 261–269.

5 Zhua, Y.C., Snodgrassa, G.L., and Chen, M.S. (2004) Enhanced esterase gene expression and activity in a malathion resistant strain of the tarnished plant bug, *Lygus lineolaris*. *Insect. Biochem. Mol. Biol.*, **34**, 1175–1186.

6 Kuzma, J., and VerHage, P. (2006) Nanotechnology in Agriculture and Food Production: Anticipated Applications, Project on Emerging Nanotechnologies, Woodrow Wilson International Center for Scholars, Washington, DC.

7 Varanasi, P., Fullana, A., and Sidhu, S. (2007) Remediation of PCB contaminated soils using iron nano-particles. *Chemosphere*, **66**, 1031–1038.

8 Evans, P., Matsunaga, H., and Kiguchi, M. (2008) Large-scale application of nanotechnology for wood protection. *Nat. Nanotechnol.*, **3**, 577.

9 Bouwmeester, H., Dekkers, S., Noordam, M.Y., Hagens, W.I., Bulder, A.S., de Heer, C., ten Voorde, S.E.C., Wijnhoven, S.W.P., Marvin, H.J.P., and Sips, A.J.A.M. (2009) Review of health safety aspects of nanotechnologies in food production. *Regul. Toxicol. Pharmacol.*, **53**, 52–62.

10 Liao, K., Cheng, J., Li, C., Liu, R., and Huang, H. (2009) Ultra-sensitive detection of mutated papillary thyroid carcinoma DNA using square wave stripping voltammetry method and amplified gold nanoparticle biomarkers. *Biosens. Bioelectron.*, **24**, 1899–1904.

11 Lin, C., Hung, C., Hsiao, C., Lin, H., Ko, F., and Yang, Y. (2009) Poly-silicon nanowire field-effect transistor for ultrasensitive and label-free detection of pathogenic avian influenza DNA. *Biosens. Bioelectron.*, **24**, 3019–3024.

12 Martinez, M., Rathbone, M., Burgess, D., and Huynh, M. (2008) *In vitro* and *in vivo* considerations associated with parenteral sustained release products: a review based upon information presented and points expressed at the 2007 Controlled Release Society Annual Meeting. *J. Controlled Release*, **129**, 79–87.

13 Makkar, H.P.S. (2008) A review of the use of isotopic and nuclear techniques in animal production. *Anim. Feed Sci. Technol.*, **140**, 418–443.

14 EU (2007) Food Quality and Safety in Europe: Project Catalogue, EUR 22393, Office for Official Publications of the European Communities, Luxembourg.

15 Buck, S.M., Koo, Y.L., Park, E., Xu, H., Philbert, M.A., Brasuel, M.A., and Kopelman, R. (2004) Optochemical nanosensor PEBBLEs: photonic explorers for bioanalysis with biologically localized embedding. *Curr. Opin. Chem. Biol.*, **8**, 540–546.

16 EU (2009) FP7 Cooperation Work Programme: Food, Agriculture and Fisheries, and Biotechnology, C(2008)4598, European Commission, Luxembourg.

17 Farre, M., Kantiani, L., Perez, S., and Barcelo, D. (2009) Sensors and biosensors in support of EU Directives. *Trends Anal. Chem.*, **28**, 171–185.

18 Kim, G., Shim, J., Kang, M., and Moon, S. (2008) Optimized coverage of gold nanoparticles at tyrosinase electrode for measurement of a pesticide in various water samples. *J. Hazard. Mater.*, **156**, 141–147.

19 Male, K.B., Hrapovic, S., Liu, Y., Wang, D., and Luong, J.H.T. (2004) Electrochemical detection of carbohydrates using copper nanoparticles and carbon nanotubes. *Anal. Chim. Acta*, **516**, 35–41.

20 Yu, X., Chattopadhyay, D., Galeska, I., Papadimitrakopoulos, F., and Rusling, J.F. (2003) Peroxidase activity of enzymes bound to the ends of single-wall carbon nanotube forest electrodes. *Electrochem. Commun.*, **5**, 408–411.

21 Khanna, V.K. (2007) Existing and emerging detection technologies for DNA (deoxyribonucleic acid) finger printing, sequencing, bio- and analytical chips: a multidisciplinary development unifying molecular biology, chemical and electronics engineering. *Biotechnol. Adv.*, **25**, 85–98.

22 Li, J., Xu, C., Zhang, Z., Wang, Y., Peng, H., Lu, Z., and Chan, M. (2005) A

DNA-detection platform with integrated photodiodes on a silicon chip. *Sens. Actuators B*, **106**, 378–382.

23 Cai, H., Wang, Y., He, P., and Fang, Y. (2002) Electrochemical detection of DNA hybridization based on silver-enhanced gold nanoparticle label. *Anal. Chim. Acta*, **469**, 165–172.

24 Benner, S., Chen, R.J.A., Wilson, N.A., Abu-Shumays, R., Hurt, N., Lierberman, K.R., Deamer, D.W., Dunbar, W.B., and Akeson, M. (2007) Sequence-specific detection of individual DNA polymerase complexes in real time using a nanopore. *Nat. Nanotechnol.*, **2**, 718–724.

25 Eigenberg, R.A., Brown-Brandl, T.M., and Nienaber, J.A. (2008) Sensors for dynamic physiological measurements. *Comput. Electron. Agric.*, **62**, 41–47.

26 Wilson, M.A., Tran, N.H., Milev, A.S.K., Kannangara, G.S., Volk, H., and Lu, G.Q.M. (2008) Nanomaterials in soils. *Geoderma*, **146**, 291–302.

27 Tratnyek, P.G., and Johnson, R.L. (2006) Nanotechnologies for environmental cleanup. *Nano Today*, **2**, 44–48.

28 Klaine, S.J., Alvarez, P.J.J., Bately, G.E., Fernandes, T.F., Handy, R.D., Lyon, D.L., Mahendra, S., McLaughlin, M.J., and Lead, J.R. (2008) Nanomaterials in the environment: behaviour, fate, bioavailability and effects. *Environ. Toxicol. Chem.*, **27**, 1825–1851.

29 Liu, F., Wen, L., Li, Z., Yu, W., Sun, H., and Chen, J. (2006) Porous hollow silica nanoparticles as controlled delivery system for water-soluble pesticide. *Mater. Res. Bull.*, **41**, 2268–2275.

30 Marvin, H.J.P., Kleter, G.A., Prandini, A., Dekkers, S., and Bolton, D.J. (2009) Early identification systems for emerging foodborne hazards. *Food Chem. Toxicol.*, **47**, 915–926.

7
Nanotechnologies for Improving Food Quality, Safety, and Security

Douglas K.R. Robinson and Mark Morrison

7.1
Introduction

Nanotechnologies promise to enable improvements in many sectors. One key promising area is in the diagnostics and sensing area, where a variety of nanotechnology platforms are being developed for application to land management, process diagnostics, quality control, and authenticity.

This chapter gives an overview of some of the key areas of activity in nanotechnologies for improving food quality, safety, and security as identified through the research activities of the European Commission-funded ObservatoryNANO project. It provides an indication of the variety of developments in this area, and is broken down into three application subdomains:

a) agricultural production;
b) food processing;
c) packaging and distribution.

Owing to its breadth, this chapter provides a glimpse of the various activities, and provides references for more detailed information.

7.2
Improving Quality, Safety, and Security of Agricultural Production

Global agriculture today faces several issues: maximizing land use in different environments, sustainable use of resources (in particular, fresh water), and ensuring that practices do not have an adverse impact on the environment (e.g., accumulation of pesticides and fertilizers). At the same time, there are opportunities for agriculture to expand into new areas, for example, the utility of what would previously have been regarded as agricultural waste, which can now be used for industrial processes. An area to which nanotechnology promises to contribute is that of "precision farming". This is the use of the Global Positioning System

Nanotechnology in the Agri-Food Sector: Implications for the Future, First Edition. Edited by Lynn J. Frewer, Willem Norde, Arnout Fischer, Frans Kampers.

(GPS), geographic information systems (GISs), and networks of sensors and actuators throughout an agricultural area that measure and report on (and in some cases respond to) a number of different environmental, crop, and pest variables. These effectively support the farmer by providing data that allow the farmer to make informed choices for irrigation, fertilization, pest control, and even harvest. Although costly, this is becoming largely offset by the rising cost of food, the need for higher quality, and increasing legislation.

In agricultural production, sensors and diagnostic devices allow farmers to closely monitor environmental conditions, plant and animal health, and growth. As part of precision farming, they can facilitate targeted and early intervention, thus increasing productivity and decreasing the use of agrochemicals (e.g., antibiotics, pesticides, and nutrients).

Sensors and diagnostic devices can be used to measure a number of important variables for agriculture:

- physiological status of crops (such as growth rates, nutritional levels, crop maturity, disease status);
- physiological status of livestock (such as body temperature, respiration rate, blood biochemistry, disease status);
- presence and identification of pests or pathogens;
- environmental variables (ambient temperature, levels of water and nutrients in soil).

They are also an essential element in the measuring of the environmental impact of the agricultural process itself, in particular the levels of pesticides and fertilizers in soil and run-off.

A variety of different sensor and diagnostic systems based on nanotechnologies have potential application in the agricultural industry – these are described in the following sections and are summarized in Table 7.1.

Biosensors utilize biomolecules to detect targets. However, the format of biosensors varies from free molecules to those bound to a substrate such as nanoparticles, nanowires, nanotubes, and thin films. The interaction of the target with the biosensor can be measured either directly or indirectly via changes in color, fluorescence, and electrical potential. When placed in arrays, multiple biomolecules are fixed to a substrate, allowing many things to be measured simultaneously.

A variety of sensors incorporating single biomolecular species are being researched, including acetylcholinesterase (AChE) [1–3], glucose oxidase [4, 5], glucose dehydrogenase [6], and tyrosinase [7, 8]. For the agricultural and environmental industries, it is the development of technologies based on AChE and tyrosinase that have drawn greatest interest. AChE is an enzyme involved in nerve signaling in many different species, and is inhibited by organophosphate and carbamate pesticides, as well as heavy metals. This inhibition can be measured by the failure of AChE to catalyze the conversion of substrate, or an analog, that would normally result in a pH decrease. This pH decrease can be measured electrochemically, or by using a dye molecule that is sensitive to pH changes and exhibits a

Table 7.1 Different sensor systems that could be used in agricultural production.

Technology	Description	Principal agents detected	Maturity
Unimolecular sensors	Biomolecules enclosed by or attached to nanostructured materials such as liposomes, nanoparticles or carbon nanotubes. Detection is measured by electrochemical or optical read-out	Pesticides, gases	Basic and applied research
Bioarrays	Biomolecules conjugated to substrates. Read-out by chemical or electronic means	Different chemical species and microbes	Some mature, but application in the field still at the applied stage
Solid-state sensors	Thin film or nanowire sensors. Read-out by electronic means	Gases	Early stage
Optical and spectrographic sensors	CCD, lasers and spectrometers	Plant growth, presence of various chemical species	Some mature, but application in the field still at the applied stage
Sensor networks	Individual sensor nodes that can be dispersed throughout an area, measure local variables, and report to a central processing unit	Potentially all desired variables	Microsystems technology is mature. Nanotechnology developments still at basic and applied research level

change in color or fluorescence. Tyrosinase can catalyze the oxidation of phenolic compounds, which are present in many industrial wastewaters and are also used as pesticides.

Unimolecular species can be either attached to an electrode or encapsulated in some form of matrix or other capsule. The nanostructured materials include liposomes [9], self-assembled monolayers [10], carbon nanotubes [11, 12], and nanoparticles [5, 7, 8]. Each provides increased sensitivity through the greater surface area of the nanoparticle, allowing either more of the biomolecule to be present, or greater access to the analyte.

In the case of electronic detection, the nanostructured material coated with biomolecule can either form the electrode itself or be used to coat the electrode (e.g., nanoparticles and self-assembled monolayers). While electrode-based systems offer the convenience of an electrical read-out, encapsulation affords greater stability (e.g., AChE is stable for at least 50 days at 4 °C when enclosed within liposomes) [13]. Whichever approach is used, sensitivity is better than that required to detect the minimum legal safe limits. While AChE sensors do not show specificity toward individual pesticides, they are cheap to manufacture and useful for an overall measurement, and therefore could be a tool for rapid assessment in the field, with follow-up as required in an analytical lab. However, to date, there have been no field trials using such sensors with non-purified samples, a point that will need to be addressed.

Bioarrays link several different biomolecules to a substrate in such a way that each is individually addressable. Arrays have been made using a number of different biomolecules, but have tended to concentrate on proteins (or parts of proteins), such as antibodies and enzymes, or deoxyribonucleic acid (DNA) and ribonucleic acid (RNA). These are quite mature technologies, having been developed and marketed by a number of companies for use in basic research and diagnostic sciences (including forensics and medicine). To date, they have largely been based on microsystem technologies and been used in laboratory settings, for measuring analyte concentrations in semi-purified (e.g., filtered and buffered) samples. Bioarrays have the capability to measure and quantify simultaneously many different analytes (in some cases thousands). Such arrays are a mature technology manufactured by a number of different companies and used in fields as diverse as clinical diagnostics, environmental monitoring, and bioscience research.

Nanotechnology is beginning to have an impact on bioarrays. The advantages that it brings to such systems are: further miniaturization, allowing more variables to be measured; greater sensitivity, thus requiring less sample material; faster detection rates, allowing read-out in real time; and novel detection methodologies (e.g., electronic, colorimetric, fluorometric, and mass changes).

There are a number of different formats for bioarrays, including: planar forms, with biomolecules directly attached to the flat array surface; cantilevers, with biomolecules attached to a number of individual micro-sized levers (which resemble diving boards); and biomolecules attached to nanowires or nanotubes, which in turn are attached to a planar surface, with each attachment point being a unique electronic address.

Array technologies can be used at different stages of the food chain, such as: detecting the presence of pathogens in livestock or crops; measuring the levels of toxins or nutrients in soils; and monitoring the quality of processed food.

Cantilever arrays are perhaps one of the most interesting, as they detect the presence of specific target molecules in a mixed environment based on mass displacement of the cantilever when the target binds a reporter molecule attached to the cantilever [14]. In addition, they can operate in gas or liquid phases, giving rise to the electronic nose and tongue. These topics will be discussed in greater detail in the section on food processing (Section 7.3).

One interesting example comes from the European Automated Water Analyzer Computer Supported System (AWACCS) project, which produced and field-tested a device that was capable of providing information (by means of an integrated optical biochip) on up to 32 different analytes from water from a variety of different sources with only a pre-filtration step required. The detection limits obtained were at levels below the EU recommended safe limits, and the chip could be reused up to 500 times [15, 16].

Solid-state sensors have at their core a conducting or semiconducting material, which can bind target molecules and record this as a change in an electronic property (conductance, capacitance, or resistance). Materials that have been used include oxides of tin, indium, aluminum, zinc, and many others, as well as composites of these materials [17].

Solid-state sensors are primarily used to detect gases, for example, nitrogen oxides, oxygen, carbon monoxide, and carbon dioxide. Although largely used for the monitoring of combustion processes (e.g., automotive) and environmental pollution, they have potential applications in agriculture, where the levels of the above gases give good indication of the growth status of the plant being monitored.

Optical and spectrographic sensor microsystems already exist that can monitor and analyze plant growth using cameras and sophisticated software. However, these are expensive and can only provide trend analysis, with little real-time data. A useful system would monitor overall plant health and inform the farmer if growth is retarded, if physiological changes are evident, but not yet manifest on a visual level, and the best time to harvest.

Hyperspectral sensing (imaging spectrometry) was originally developed for mining and geology. It measures reflected radiance as hundreds to thousands of contiguous spectral bands. From this, it is possible to distinguish between different mineral types. However, it has more recently been applied to vegetation and can be used to determine plant coverage and growth rates. Largely this technology depends on aerial or satellite imaging, but portable ground-based sensors can also be used [18].

Sensor networks for crops combine the different sensor technologies described above with a means to communicate this to a central processing unit. Ideally, they would be distributed over the area to be monitored (e.g., a field) and provide sufficient real-time data for the farmer to be able to monitor any localized changes in the environment, crop or livestock. Such systems exist and are already in use for high-value crops, for example, in vineyards; however, they are expensive and relatively bulky (being based on microsystems). Nanotechnology advances promise a real impact through decreased size, cost, durability, and longevity due to advances not just in sensor technologies, but in energy supply and durability of materials.

By monitoring variables such as body temperature, heart and respiration rate, and eating and drinking frequency, farmers can evaluate livestock, compare with previous statistics, and make informed decisions in case of deviation. Of all of these parameters, respiration rate has been shown to be a good indicator of animal stress [19]. However, conventional systems are ill-suited to the task of automated recording of such physiological data. Where monitoring livestock is not possible, measuring local weather conditions can be used to infer impacts on respiration rate.

Sensor systems for agricultural production based on nanotechnologies are still largely at the fundamental research level. Progressing to application status will require demonstration of an ability to operate in the field, as well as to scale up to manufacturing levels. It is most likely that such systems will be based on solid-state materials, as these are most mature and have demonstrated application in other sectors. In this area, solid-state sensors and bioarrays are the most developed. These are, however, largely used in other application fields, such as bioscience research and development, environmental monitoring, and internal combustion engine management systems.

7.3
Improving Quality, Safety, and Security in Food Processing

The food processing industry transforms raw materials from agricultural and fishery industries, and manufactures foodstuffs for consumers. In the main, food processing is driven by a number of factors, such as increasing output and decreasing waste.[1)] Quality control is also an important aspect, from the delivery and storage of raw materials, through the various processing stages, to the packaged foodstuffs that leave the factory for the retailer.

Quality control in the food processing industry ideally requires in-line monitoring so that contamination can be identified as quickly as possible, and production halted. This is a highly repetitive process, with sampling points at various stages of production. While nanotechnology-enabled sensors have the potential to deliver on this in terms of sensitivity and real-time response, these are not sufficiently mature.

One of the major considerations for the detection of chemical and biological contaminants that will need to be addressed, before such systems become commonplace, is the pre-treatment of samples to remove interfering components. In this regard, a pre-filtration or lysis step, followed by an enriching system (such as functionalized magnetic nanoparticles[2)]), could be used to remove other materials from the food sample prior to assaying for a specific contaminant. These steps are actively researched in the area of integrated microfluidics, although, for applications, it still remains a promise to be delivered.

Novel coatings for food processing equipment, and filters for handling of waste and purification of low-concentration bioactive components, are expected to bring efficiency savings and improved products. However, coatings have to demonstrate longevity, lower biofilm production, and compatibility with different foodstuffs before they are adopted by the processing industry.

1) This is true of Western industrialized agri-food production, but is not the case for other areas around the globe and for the, currently, niche area of organic food production, which are driven by other values, such a minimum artificial pesticide use.

2) A label-free, microfluidics, and interdigitated array microelectrode-based impedance biosensor in combination with nanoparticles immunoseparation for detection of *Escherichia coli* O157:H7 in food samples.

Ultrafiltration is an established technology for the separation of different food products and waste; nanofiltration, in contrast, has largely been employed within laboratory settings. Nanofiltration offers the opportunity to discriminate between lower-molecular-weight components, through engineering of the different membrane layers. However, much fundamental research still needs to be undertaken before the fluid mechanics of these systems are fully understood and therefore controlled.

From the moment food enters the food processing chain it is monitored for a number of different aspects to ensure that it meets specific standards (and continues to do so). The terminology for this process is hazard analysis and critical control point (HACCP), which emphasizes rapid and thorough analysis of raw foodstuffs before entering the processing chain and in-line assessment, to ensure the quality of the final product. Key aspects are the detection and quantification of: agrochemicals (such as pesticides, fertilizers, antibiotics); other chemical contaminants (e.g., heavy metals); pathogens (in particular, bacteria and fungi); and overall food quality (measured by variables such as visual appearance, freshness).

Detection of chemical contaminants, such as pesticides, heavy metals, and antibiotics, within food products is performed by gas or liquid chromatography followed by mass spectrometry, after extraction of a sample by suitable means. This is the industry standard, used by analytical agencies and departments worldwide, and developments as far as these systems are concerned focus mainly on decreasing the turnaround time while retaining sensitivity [20].

Nanotechnology applications in this area have the potential for greater sensitivity and real-time detection, with a lower sampling level; however, they are very much at the level of basic research. Platforms include unimolecular sensors, sensor arrays, and solid-state systems (outlined earlier in this chapter). In contrast to other application areas (such as field monitoring), such systems do not need to be portable and can operate on standardized samples. However, they must detect accurately and quantify multiple analytes, if they are to compete with (and replace) the industry standard. Promising candidates includes electrochemical detection of various important chemical species (hydrazine, sulfite, nitrite) using composite electrodes containing gold nanoparticles [21] and immunodetection using cantilever arrays conjugated to antibodies against specific pesticides [22].

According to the European Food Safety Authority (EFSA), in the 23 Member States during 2005 there were a total of 5311 food-borne outbreaks, involving 47 251 people and resulting in 5330 hospitalizations and 24 deaths, the majority caused by *Salmonella* and *Campylobacter* [23]. Many of these illnesses are caused by bacterial enterotoxins, which are not easily removed from food, as they are often stable at the temperatures used in normal cooking. To combat this, it is critical to be able to detect food spoilage through bacterial, fungal or viral contamination at each stage in the food processing industry.

This is a major market, with an estimated 558 million tests performed each year, worth 1.45 billion euros. More than 90% are performed by service laboratories;

however, the use of rapid test kits is increasing. Some 70% of all tests are for *Salmonella* and *Listeria* [20].

Currently, such methods are largely based on classical immunoassays – for example, enzyme-linked immunosorbent assay (ELISA) – or DNA assays – for example, polymerase chain reaction (PCR) – which require some sample preparation and have a turnaround of a day or two. Although this is much quicker than other techniques (such as isolation and cultivation of microbes), there is still scope for greater sensitivity and faster detection times. The key drivers are lower detection limits, real-time detection, higher throughput, and discrimination between different species.

Most systems detect microbial components, rather than intact cells. Protein detection systems are favored, as this increases the probability that the intact microbe is present and also screens for the presence of important bacterial enterotoxins and fungal mycotoxins (which can be present in the absence of viable microbes, and are responsible for significant illnesses).

In general, such systems must be able to detect the presence of 10–100 infectious particles per milliliter. There are various biosensor platforms in development that are based on nanostructured materials. While there are large amounts of research on the development of electronic platforms (principally amperometric, but also voltametric and impedance), there are also efforts in the area of optical and mass-change detection. In each case the nanostructured material is decorated with biomolecules capable of interacting specifically with the target analyte. This interaction is transduced by the nanomaterial into a quantifiable signal.

1) **Electronic biosensors** These are based on protein-conjugated nanowires [24, 25] and carbon nanotubes (CNTs) [26]. They directly quantify the presence of specific analytes (e.g., proteins, nucleic acids, metabolites) that directly or indirectly indicate the presence of the microbe. Because the output is an electrical signal, such platforms have the potential to be linked to other devices, allowing data to be transmitted, shared and analyzed further. By virtue of the nanoscale dimensions, these demonstrate much faster electron transfer rates than microelectrodes, which manifests as higher sensitivity. CNTs have been combined with nanoparticles (e.g., gold or platinum nanoparticles or quantum dots) and polymer matrices to form composite materials with improved robustness and high porosity (facilitating entry of target biomolecules) [26]. Such composite electrodes exhibit even greater sensitivity.

2) **Optical biosensors** These have read-out by a number of different techniques, including surface plasmon resonance (SPR), fluorescence, and colorimetric changes, and are based on a number of biomolecule-conjugated platforms, including CNTs [27], silica [28], gold [29, 30], and latex [31] nanoparticles.

3) **Mass-change biosensors** These are based on cantilever arrays and piezoelectric devices [32, 33]. Binding of analyte to the conjugated biomolecule results in changes in the resonant frequency of the nanomaterial, which is directly proportional to the amount of target bound, and can be read by, for example, the deflection of a laser beam.

Most of these technologies are still at the level of basic research. However, Biophage Pharma Inc.,[3)] in collaboration with NRC-Biotechnology Research Institute, has developed electronic biosensors capable of discriminating between different bacteria (in a process termed Electric Cell–Substrate Impedance Sensing, or ECIS). This is now at the pre-commercialization stage and is expected to have applications for the detection of bacteria in water, food, and biological fluids.

While it is important to detect and identify contaminants, it is equally as important to manufacturers (particularly of high-value foods) to measure the quality of their produce: primarily color, smell, taste, and mouth-feel. Traditionally such quality control would have been performed by experienced individuals; however, this is not always appropriate, especially for high-volume foodstuffs. Developments over the last two decades based on semiconductor and polymer materials are going some way to automate the quality control procedures as far as taste and smell are concerned. These are commonly known as electronic tongues and noses [34]. The presence of specific chemicals within a sample (gas or liquid) can be quantified through changes in the electronic properties of the detector material as a result of binding that chemical species. By using different materials, or by doping the detection material, variable sensitivities to different chemicals can be engineered. These different detector materials are then arranged within the electronic nose or tongue, each constituting a separate electronic address. The detection profile (or fingerprint) from a sample can be used to determine the chemical composition and distinguish between different but related products.

Microtechnology-based systems have been developed. For example, commercially available electronic noses have been used to detect the presence of microbial contamination (indirectly, through the measurement of volatile metabolites) [35]. However, nanotechnology advances are expected to increase sensitivity and breadth of chemicals that can be measured, thereby giving greater discrimination between different chemical species over a wider range of concentrations. Recent work has demonstrated the potential for greater sensitivity, with electronic noses based on doped tin oxide thin films discriminating between two different red wines [36], and doped zinc oxide nanoparticulates discriminating between different vinegars [37, 38].

There are six events in food processing that together account for the greatest losses in productivity: breakdowns; set-up and adjustments resulting in downtime; small stops; reduced speeds; start-up rejects; and production rejects. In-line quality control monitoring as discussed above can help to resolve some of these issues, while advances in equipment coatings and new materials for waste management can help to keep production in full flow.

Coatings for food processing equipment must be non-hazardous to human health, should minimize biofilm formation (which can lead to food spoilage and contamination), and should be durable. Traditionally, such equipment was manufactured from stainless steel, as this is both durable and non-hazardous to human health. However, stainless steel is susceptible to pitting and scoring, which serve

3) See http://www.biophagepharma.net/index.php?option=com_content&task=view&id=54&Itemid=36&lang=en.

as focal points for microbial growth. As a result, such equipment requires regular cleaning and disinfection, which at the very least means some production downtime, and can often require partial dismantling to allow access to internal spaces. The areas that are most prone to biofouling are heat exchangers.

It has been known for a number of years that biofilms will grow in any nutrient-rich medium and will strongly adhere to a variety of different surfaces [39]. More recently, it has been shown that the nanoscale structure of a surface can control the adhesion of biomolecules and, by extension, microbes [40–42]. Biofilms are a major concern for the food processing industry, as bacteria within the biofilm are resistant to antibiotics and normal cleaning practices, but have the potential to "break off" and contaminate foodstuffs.

Nanotechnology-enabled processes can help to resolve the issues of durability and biofilm prevention. This can be achieved through application of a coating or through the direct nanostructuring of the surface layers of the material. Both act to decrease the material's surface free energy, thus decreasing the strength of microbial adherence. This can either help to prevent adherence in the first instance or increase cleaning efficiency [43]. An established material that is widely used is polytetrafluoroethene (PTFE, Teflon), which has a low surface free energy, but poor abrasion resistance.

For certain equipment parts, high durability is required. Coatings of diamond-like carbon (DLC), which are deposited by gas-phase processes, show high durability and minimal biofouling. They are used in many different industries: for example, personal care (e.g., razor blades), car engine parts, and the medical device industry (e.g., implants such as stents and catheters). In food processing, their applications are more likely in non-food-contact areas, as there is experimental evidence that DLC coatings do not withstand the repeated cleaning cycles necessary in the food processing industry [44].

Other promising research in this area includes electrodeless plating with nickel and PTFE to produce a nanostructured surface on stainless steel [45], and the use of polymer coatings with and without antimicrobial nanoparticulates on a variety of surfaces, but which do not require high wear resistance [46, 47].

There are some applications of nanostructured coatings within the food processing industry. SPX Process Equipment are applying DLC coatings in their Waukesha Cherry-Burrell pump range.[4] SuSoS AG manufacture nanostructured antimicrobial coatings using PTFE or polyethylene glycol (PEG) by a sol–gel process that have a lifespan of up to two years (personal communication to Kshitij Singh, 2008). Few Chemicals GmbH have developed a sol–gel coating using hybrid polymers, which provides easy-to-clean and anticorrosion properties on metals, and scratch resistance on glass, that has a lifespan up to five years (personal communication to Kshitij Singh, 2008). Sarastro GmbH produce antimicrobial, hygiene, and anti-fingerprinting coatings based on sol–gel technologies with lifespans up to several years (personal communication to Kshitij Singh, 2008).

4) See http://www.gowcb.com/products/pumps/PDF/ff-1104_compdiacoat_wcb.pdf.

Filtration is an important process for several different foodstuffs, including milk, oils, wine, and beer, as well as for purifying bioactive components that are present at low concentration. In addition, it is estimated that the food industry uses more water per unit mass of product than any other industry [48]; for example, the dairy industry produces between 0.2 and 10 liters of effluent per liter of processed milk [49].

As a result, filtration technologies are becoming increasingly important in the drive to minimize and recycle as much of this wastewater as possible. In the dairy industry, in particular, much of this waste also contains useful proteins (such as whey) but has a high mineral content. Nanofiltration technologies are seen as one solution to these issues. Nanofiltration systems employ multiple membrane layers where molecules and ions can be separated based on charge, size, and water solubility. Most employ ceramic and polymer layers. They have demonstrated ability to separate and concentrate useful components from waste [50–52]. However, one issue that still needs to be addressed is biofouling, which is estimated to be the biggest contributor to decreased filtration efficiency [53] and to the loss of desirable proteins and peptides (through retention in the membrane) [54].

Alongside the technical and manufacturing challenges, the other major issue is to ensure that all food contact materials (coatings, filters) and ingredients are safe for human health. Although there is no specific EU legislation governing the use of nanomaterials in food, it is likely, at present, that current legislation can be applied. For example, Article 14 Reg. (EC) 178 of 2002 states that "unsafe food" cannot be placed on the market. The Novel Food Regulation (EC) No. 258/97 includes all foodstuffs or ingredients that have not been consumed to a significant degree before 1997. This could potentially be adapted to encompass nanotechnology, and has in fact already been cited by the Finnish government to prevent the importation of a liposomal nutriceutical.[5)]

7.4
Improving Quality, Safety, and Security in Packaging and Distribution

Food packaging acts to enclose processed food in a stable environment and protect it from environmental changes (such as moisture, light, oxidation, and temperature), physical damage, and contamination by micro- and macro-organisms. In addition, it provides information to the consumer. Food packaging innovations have been covered elsewhere in this volume (cf. Chapter 4), but below we highlight some of the recent developments in antimicrobial and antimycotic packaging as particularly interesting for improving quality and safety. By doing so, it improves quality, extends the shelf-life of processed food, and allows the consumer to assess whether the product is suitable. Food packaging also provides important ancillary

5) Lypo-spheric™ Vitamin C claims to increase bioavailability of vitamin C through liposomal encapsulation, and was refused an import license in 2008 by Finland under the Novel Food Regulation.

functions: authentication of the foodstuff and product, and evidence of tampering or breach of package integrity, thus improving food security.

Active packaging is an area where nanotechnology is expected to have a large impact. Radiofrequency identification (RFID) tags, temperature, and gas sensors based on nanomaterials are in development, and in some cases these have already been commercialized.

In conjunction with an effective packaging system, improvements in identification of items and stock control ensure that delivery is efficient and that foodstuffs are maintained in the appropriate conditions throughout the supply chain. This includes RFID tags for logging the movement of stock at all stages of the supply chain, and other tags to provide covert or overt identification and authentication.

In addition to acting as a passive barrier, packaging can contribute to the control of microbial growth in foodstuffs that leads to spoiling. Most activities to combat this have centered around nanoparticulates of silver and zinc oxide, but there is also research into the antimicrobial effects of natural biological compounds [55].

Silver nanoparticles have been incorporated in a wide variety of consumer goods, including clothing, electrical goods, kitchenware, and wound dressings [56]. Nanoparticulate silver releases ions more efficiently than bulk metal, and it is the silver ions that have a bactericidal effect due to the inhibition of a wide variety of biological processes within bacteria [57]. As the levels of silver ions liberated are too low to have toxic effects in humans, it is likely that nanoparticulate silver will be included in further composite materials. However, there is some concern over the effects of large amounts of silver ions being discharged into the environment and accumulating in ecosystems, as silver ions are known to be toxic to aquatic life.

Zinc oxide exhibits antibacterial activity that increases with decreasing particle size [58]. This activity does not require the presence of ultraviolet (UV) light (unlike titanium dioxide) but is stimulated by visible light [59]. The exact mechanism(s) of action is (are) still unknown. Zinc oxide nanoparticles have been incorporated in a number of different polymers, including polypropylene [60]. In addition, zinc oxide effectively absorbs UV light, without re-emitting as heat, and therefore improves the stability of polymer composites.

Chitosan is a biopolymer derived from chitin (a polysaccharide constituent of crustacean shells). It has seen much interest in recent years as a material for the encapsulation of nutriceuticals. In addition to its utility as a packaging material, it also exhibits antimicrobial properties [61]. This has led a number of groups to investigate its incorporation into different composite materials, which could have applications in healthcare and food packaging, including using it as a "green" reagent to reduce and stabilize silver ions [62], in combination with clays such as rectorite, which could then be used in polymer composites [63, 64]. Companies such as Nanograde GmbH market polymer composites containing nanoparticles of silver and calcium phosphate that demonstrate microbicidal activity.

Smart packaging responds to its environment either to regulate an external effect or to produce a visual read-out of a change. It includes materials that can regulate the internal environment of packaged foodstuffs to maintain food quality (e.g., through the release or absorption of substances), sensors that provide an indication of the storage history of the product and whether it is still fresh, and materials that can repair minor damage (self-heal) [65, 66].

Regulating the internal environment of the packaging, at its simplest, is the control of the temperature of the foodstuff. Manufacturers of chilled or fresh foods want to ensure that their produce reaches the consumer in good condition. However, there are inevitable breaks in the cold chain, for example, due to transfer between different transport systems. If these occur in high ambient temperatures, food quality can quickly deteriorate.

Ideally, it would be useful to have a protective material that is cheap, recyclable or reusable, and does not add significantly to package weight or volume. Traditional insulating materials (such as polystyrene) are bulky and inappropriate for this use, as they would add significantly to transport costs. In contrast, nanostructured foams, which are considerably thinner than conventional materials for the same thermal properties, could be an alternative, if available at low enough cost (at present, these are used more for building insulation). An alternative system based on low-cost materials has been developed by researchers in New Zealand. This system, based on nanoporous calcium silicate, is loaded with a phase-change material (such as paraffin wax) that can mitigate the effects of an increase in external temperature over a short period of time (five hours), while having similar dimensions to bubble wrap [67].

Self-heating or self-cooling systems are an attractive option for consumers. Essentially the chemistry is simple. Exothermic reactions are used for self-heating (e.g., mixing water and calcium oxide), while evaporation of a refrigerant (e.g., water or carbon dioxide) is used for self-cooling. There are several examples of self-heating systems on the market, and at least one for self-cooling. It is unclear whether nanomaterials would offer significant improvements to self-heating efficiencies. However, they may provide increased efficiencies for self-cooling, and there is at least one patent, based on fullerenes, for this purpose [68].

In the longer term, completely different platforms such as combination thin-film photovoltaic and thermoelectric systems could be used (to harness solar power to drive the cooling effect of thermoelectric materials, in much the same way as solid-state coolers).

Gas scavenging or absorbing systems are also of interest for food packaging. There are several on the market using conventional technologies, such as the "Ageless" system from Mitsubishi Gas Chemical Co., which contains iron salts and vitamin C, and absorbs oxygen within a sealed package.[6] Multisorb Technologies Inc. has patented technology using oxidizable submicrometer particles for use as oxygen scavengers in packaging [69]. Research using nanostructured materials may offer enhancements by increasing the surface area of the active component (through nanoparticles, or loading of a nanoporous material such as silica, with active material). For example, preliminary work with polymer nanocomposites containing titanium dioxide shows that these exhibit similar oxygen scavenging properties, in the presence of UV, as conventional iron- and polymer-based materials [70].

6) See http://www.mgc.co.jp/eng/products/abc/ageless/index.html.

Other research themes have looked at the active release of compounds to help maintain food quality. Mostly these are based on conventional technologies to release preserving compounds such as carbon dioxide or ethanol. However, the last few years has seen the development of systems based on nanomaterials. Research patented from SouthWestern Research Institute provides a means for the release of antimicrobial agents (such as chlorine dioxide) inside packaging to inhibit microbial growth. This uses nanoscale capsules that release chlorine dioxide upon exposure to moisture [71] or nanoparticles of materials such as titanium dioxide to photocatalyze the production of such gases from inert reactants [72]. This research is now being developed by the Microactive Corporation.

Sensor technologies for packaging should provide a visible indicator to the supplier or consumer that foodstuffs are still fresh, or whether the packaging has been breached, kept at the appropriate temperatures throughout the supply chain, or spoiled. Key factors in their use are cost, robustness, and compatibility with different packaging materials.

The ability to detect the presence of oxygen within packages of, for example, fresh meat, at the earliest stage, would alert the consumer that the packaging has been compromised, even if there are no visual indications to suggest this. Such systems, for the purpose of food packaging, rely on changes in the color of dyes in the presence or absence of oxygen. One commercialized microtechnology product is "Ageless Eye",[7] which is pink in the absence of oxygen and blue in its presence. Advances using nanoparticles are expected to produce more sensitive systems that respond faster and produce stronger color changes. For example, researchers at the University of Strathclyde have produced a hydroxyethyl cellulose polymer film oxygen sensor, containing titanium dioxide nanoparticles and the blue dye, indigo-tetrasulfonate. Following incorporation in the packaging, the sensor is exposed to UV light, the dye is photo-bleached (a reaction catalyzed by the titanium dioxide) and remains so until exposed to atmospheric oxygen levels, when it rapidly (within three minutes) returns to a deep blue color (even in the dark) [73].

Time–temperature indicators (TTIs) allow suppliers to confirm that processed foods requiring refrigeration have been kept at the appropriate temperatures throughout the supply chain. They fall into two categories: one relies on the migration of a dye through a porous material, which is temperature- and time-dependent; the other makes use of a chemical reaction (initiated when the label is applied to the packaging), which results in a color change, the rate of which is temperature-dependent. These have limitations in that they require multiple components (dyes, reactants, porous layers), which can affect accuracy under some circumstances, and so a single-component system would be an improvement. Timestrip plc has developed a colloidal gold-based system (Timestrip) [74], which is red in color at temperatures above freezing. Freezing leads to the irreversible agglomeration of the gold nanoparticles, resulting in a clear solution, a useful indicator to detect the accidental freezing of chilled goods.

7) See http://www.mgc.co.jp/eng/products/abc/ageless/eye.html.

Radiofrequency identification (RFID) tags have been in use for a number of years, but only for high-value items such as clothing and electronics. They consist of two modules, one to process and store information, the second (an antenna) to transmit and receive information. A second device, the reader, is used to obtain information from the tag, and, depending on the radio frequency used, this can be at distance of several tens of meters.

RFID tags for the packaging industry are passive, that is, they have no associated power source, and gain energy to transmit information from the incoming radio waves from the reader. Their utility is that multiple items can be monitored at every stage in the supply chain without the need for line of sight, therefore increasing the speed and efficiency of distribution. This is a critical factor in modern supply chains, where large amounts of raw materials may be coming from different global regions to be processed at one site, then distributed to consumers (in many different global regions). Eventually RFID tags are expected to replace barcodes [75].

RFID tags at present are largely based on silicon semiconductor technologies. However, recent research could change this, allowing cheaper and easier production on a number of different materials.

Printable electronics (using conducting polymers, such as pentacene and oligothiophene, and metallic inks, including copper, silver, and gold nanoparticles) are being developed by a number of institutes and companies around the globe [75, 76]. While at present most are based on desktop inkjet printing, other forms more suited to high production levels (as already used in the printing industry) could be developed.

With regards to printable electronics and RFID tags, there are several companies developing and marketing these technologies. Companies such as Cima NanoTech and Novacentrix manufacture copper and silver nanoparticle-based inks. These can be formulated in aqueous or organic suspensions and printed onto a variety of substrates. Other active players include Du Pont, HP, Samsung, and Hitachi.

In addition to printed systems, some research groups are exploring the use of carbon nanotubes as antenna [77, 78]. However, this technology is not as highly developed as conductive inks based on metal nanoparticles.

Interestingly, there is some research into combining RFID tags with chemical sensing functions. One group has produced a prototype for ethylene (ethene) sensing (for fruit ripeness) [79], while another has demonstrated the potential of this technology by constructing a moisture sensor [80]. While these are both microelectronic systems, the potential for nanotechnology to enhance such systems is clear.

For ensuring authenticity of a product, a number of different systems are being developed, including nanoscale barcodes, quantum dots, and magnetic nanoparticles. Whether these are likely to be used widely within food packaging is unclear, and will be dependent on cost per unit and ease of use. It is more likely that RFID tags will serve a dual purpose of tracking and authenticating items. For a full description of anti-counterfeit and authentication technologies, please see the security section of the ObservatoryNANO online resource for Anti-counterfeiting[8)] and Authentication[9)].

7.5 Wrapping Up

This chapter has given a broad overview of some of the areas of nanotechnology that are actively being developed for improving food quality, safety, and security. It is a quickly growing field, and there are indications, especially from industry associations and consortia, that food quality, safety and security are becoming key focus areas in technology development for the agrifood sector. What role nanotechnology will play is not yet clear – mostly nanotechnology remains in the research laboratory. However, in the form of sensors, advanced coatings, and active packaging, there are unique advantages that nanostructured innovations can provide, and we anticipate a considerable growth of activity in these areas.

References

1 Andreescu, S., and Marty, J.L. (2006) Twenty years research in cholinesterase biosensors: from basic research to practical applications. *Biomol. Eng.*, **23**, 1–15.

2 Liu, G.D., and Lin, Y.H. (2006) Biosensor based on self-assembling acetylcholinesterase on carbon nanotubes for flow injection/amperometric detection of organophosphate pesticides and nerve agents. *Anal. Chem.*, **78**, 835–843.

3 Vamvakaki, V., and Chaniotakis, N.A. (2007) Pesticide detection with a liposome-based nano-biosensor. *Biosens. Bioelectron.*, **22**, 2848–2853.

4 Lin, Y.H., Lu, F., Tu, Y., and Ren, Z.F. (2004) Glucose biosensors based on carbon nanotube nanoelectrode ensembles. *Nano Lett.*, **4**, 191–195.

5 Hrapovic, S., Liu, Y.L., Male, K.B., and Luong, J.H.T. (2004) Electrochemical biosensing platforms using platinum nanoparticles and carbon nanotubes. *Anal. Chem.*, **76**, 1083–1088.

6 Zhang, M.G., Smith, A., and Gorski, W. (2004) Carbon nanotube–chitosan system for electrochemical sensing based on dehydrogenase enzymes. *Anal. Chem.*, **76**, 5045–5050.

7 Sanz, V.C., Mena, M.L., Gonzalez-Cortes, A., Yanez-Sedeno, P., and Pingarron, J.M. (2005) Development of a tyrosinase biosensor based on gold nanoparticles-modified glassy carbon electrodes – Application to the measurement of a bioelectrochemical polyphenols index in wines. *Anal. Chim. Acta*, **528**, 1–8.

8 Li, Y.F., Liu, Z.M., Liu, Y.L., Yang, Y.H., Shen, G.L., and Yu, R.Q. (2006) A mediator-free phenol biosensor based on immobilizing tyrosinase to ZnO nanoparticles. *Anal. Biochem.*, **349**, 33–40.

9 Viswanathan, S., Wu, L.C., Huang, M.R., and Ho, J.A.A. (2006) Electrochemical immunosensor for cholera toxin using liposomes and poly(3,4-ethylenedioxythiophene)-coated carbon nanotubes. *Anal. Chem.*, **78**, 1115–1121.

10 Yonzon, C.R., Jeoung, E., Zou, S.L., Schatz, G.C., Mrksich, M., and Van Duyne, R.P. (2004) A comparative

8) See http://www.observatorynano.eu/project/document/2331/.
9) See http://www.observatorynano.eu/project/document/2338/.

analysis of localized and propagating surface plasmon resonance sensors: the binding of concanavalin A to a monosaccharide functionalized self-assembled monolayer. *J. Am. Chem. Soc.*, **126**, 12669–12676.

11 Joshi, K.A., Tang, J., Haddon, R., Wang, J., Chen, W., and Mulchandani, A. (2005) A disposable biosensor for organophosphorus nerve agents based on carbon nanotubes modified thick film strip electrode. *Electroanalysis*, **17**, 54–58.

12 Chen, H.D., Zuo, X.L., Su, S., Tang, Z.Z., Wu, A.B., Song, S.P., Zhang, D.B., and Fan, C.H. (2008) An electrochemical sensor for pesticide assays based on carbon nanotube-enhanced acetycholinesterase activity. *Analyst*, **133**, 1182–1186.

13 Vamvakaki, V., and Chaniotakis, N.A. (2007) Pesticide detection with a liposome-based nano-biosensor. *Biosens. Bioelectron.*, **22**, 2848–2853.

14 McKendry, R., Zhang, J.Y., Arntz, Y., Strunz, T., Hegner, M., Lang, H.P., Baller, M.K., Certa, U., Meyer, E., Guntherodt, H.J., and Gerber, C. (2002) Multiple label-free biodetection and quantitative DNA-binding assays on a nanomechanical cantilever array. *Proc. Natl. Acad. Sci. USA*, **99**, 9783–9788.

15 Tschmelak, J., Proll, G., and Gauglitz, G. (2005) Optical biosensor for pharmaceuticals, antibiotics, hormones, endocrine disrupting chemicals and pesticides in water: assay optimization process for estrone as example. *Talanta*, **65**, 313–323.

16 Rickerby, D.G., and Morrison, M. (2007) Nanotechnology and the environment: a European perspective. *Sci. Technol. Adv. Mater.*, **8**, 19–24.

17 Eranna, G., Joshi, B.C., Runthala, D.P., and Gupta, R.P. (2004) Oxide materials for development of integrated gas sensors – a comprehensive review. *Crit. Rev. Solid State Mater. Sci.*, **29**, 111–188.

18 Ye, X.J., Sakai, K., Okamoto, H., and Garciano, L.O. (2008) A ground-based hyperspectral imaging system for characterizing vegetation spectral features. *Comput. Electron. Agric.*, **63**, 13–21.

19 Eigenberg, R.A., Brown-Brandl, T.M., and Nienaber, J.A. (2008) Sensors for dynamic physiological measurements. *Comput. Electron. Agric.*, **62**, 41–47.

20 GoodFood Project (2004) Food Safety and Quality Monitoring with Microsystems, EU funded under FP6-IST-1-508744-IP, http://www.goodfood-project.org (accessed 4 November 2010).

21 Maduraiveeran, G., and Ramaraj, R. (2007) A facile electrochemical sensor designed from gold nanoparticles embedded in three-dimensional sol–gel network for concurrent detection of toxic chemicals. *Electrochem. Commun.*, **9**, 2051–2055.

22 Suri, C.R., Kaur, J., Gandhi, S., and Shekhawat, G.S. (2008) Label-free ultra-sensitive detection of atrazine based on nanomechanics. *Nanotechnology*, **19**, 6.

23 EFSA (2007) The Community Summary Report on Trends and Sources of Zoonoses, Zoonotic Agents, Antimicrobial Resistance and Foodborne Outbreaks in the European Union in 2005, European Food Safety Authority, Parma, Italy, http://www.efsa.europa.eu/en/scdocs/doc/zoonosesreport2005.pdf (accessed 4 November 2010).

24 Yang, L., Chakrabartty, S., and Alocilja, E.C. (2007) Fundamental building blocks for molecular biowire based forward error-correcting biosensors. *Nanotechnology*, **18**, 42.

25 Mishra, N.N., Maki, W.C., Cameron, E., Nelson, R., Winterrowd, P., Rastogi, S.K., Filanoski, B., and Maki, G.K. (2008) Ultra-sensitive detection of bacterial toxin with silicon nanowire transistor. *Lab Chip*, **8**, 868–871.

26 Pumera, M., Sanchez, S., Ichinose, I., and Tang, J. (2007) Electrochemical nanobiosensors. *Sens. Actuators B Chem.*, **123**, 1195–1205.

27 Yang, M., Kostov, Y., and Rasooly, A. (2008) Carbon nanotubes based optical immunodetection of staphylococcal enterotoxin B (SEB) in food. *Int. J. Food Microbiol.*, **30**, 78–83.

28 Zhao, X.J., Hilliard, L.R., Mechery, S.J., Wang, Y.P., Bagwe, R.P., Jin, S.G., and Tan, W.H. (2004) A rapid bioassay for single bacterial cell quantitation using

bioconjugated nanoparticles. *Proc. Natl. Acad. Sci. USA*, **101**, 15027–15032.

29 Huang, S.H. (2007) Gold nanoparticle-based immunochromatographic assay for the detection of *Staphylococcus aureus*. *Sens. Actuators B Chem.*, **127**, 335–340.

30 Arora, K., Chand, S., and Malhotra, B.D. (2006) Recent developments in bio-molecular electronics techniques for food pathogens. *Anal. Chim. Acta*, **568**, 259–274.

31 Jaakohuhta, S., Harma, H., Tuomola, M., and Lovgren, T. (2007) Sensitive *Listeria* spp. immunoassay based on europium(III) nanoparticulate labels using time-resolved fluorescence. *Int. J. Food Microbiol.*, **114**, 288–294.

32 McKendry, R., Zhang, J.Y., Arntz, Y., Strunz, T., Hegner, M., Lang, H.P., Baller, M.K., Certa, U., Meyer, E., Guntherodt, H.J., and Gerber, C. (2002) Multiple label-free biodetection and quantitative DNA-binding assays on a nanomechanical cantilever array. *Proc. Natl. Acad. Sci. USA*, **99**, 9783–9788.

33 Chen, S.H., Wu, V.C.H., Chuang, Y.C., and Lin, C.S. (2008) Using oligonucleotide-functionalized Au nanoparticles to rapidly detect foodborne pathogens on a piezoelectric biosensor. *J. Microbiol. Methods*, **73**, 7–17.

34 Deisingh, A.K., Stone, D.C., and Thompson, M. (2004) Applications of electronic noses and tongues in food analysis. *Int. J. Food Sci. Technol.*, **39**, 587–604.

35 Magan, N., Pavlou, A., and Chrysanthakis, I. (2001) Milk-sense: a volatile sensing system recognises spoilage bacteria and yeasts in milk. *Sens. Actuators B Chem.*, **72**, 28–34.

36 Lozano, J., Arroyo, T., Santos, J.P., Cabellos, J.M., and Horrillo, M.C. (2008) Electronic nose for wine ageing detection. *Sens. Actuators B Chem.*, **133**, 180–186.

37 Zhang, Q.Y., Zhang, S.P., Xie, C.S., Zeng, D.W., Fan, C.Q., Li, D.F., and Bai, Z.K. (2006) Characterization of Chinese vinegars by electronic nose. *Sens. Actuators B Chem.*, **119**, 538–546.

38 Zhang, Q.Y., Zhang, S.P., Me, C.S., Fan, C.Q., and Bai, Z.K. (2008) "Sensory analysis" of Chinese vinegars using an electronic nose. *Sens. Actuators B Chem.*, **128**, 586–593.

39 Costerton, J.W., Lewandowski, Z., Caldwell, D.E., Korber, D.R., and Lappinscott, H.M. (1995) Microbial biofilms. *Annu. Rev. Microbiol.*, **49**, 711–745.

40 Emerson, R.J., and Camesano, T.A. (2004) Nanoscale investigation of pathogenic microbial adhesion to a biomaterial. *Appl. Environ. Microbiol.*, **70**, 6012–6022.

41 Verran, J., and Boyd, R.D. (2001) The relationship between substratum surface roughness and microbiological and organic soiling: a review. *Biofouling*, **17**, 59–71.

42 Diaz, C., Schilardi, P.L., Salvarezza, R.C., and De Mele, M.F.L. (2007) Nano/microscale order affects the early stages of biofilm formation on metal surfaces. *Langmuir*, **23**, 11206–11210.

43 Rosmaninho, R., Santos, O., Nylander, T., Paulsson, M., Beuf, M., Benezech, T., Yiantsios, S., Andritsos, N., Karabelas, A., Rizzo, G., Muller-Steinhagen, H., and Melo, L.F. (2007) Modified stainless steel surfaces targeted to reduce fouling – evaluation of fouling by milk components. *J. Food Eng.*, **80**, 1176–1187.

44 Saikhwan, P., Geddert, T., Augustin, W., Scholl, S., Paterson, W.R., and Wilson, D.I. (2006) Effect of surface treatment on cleaning of a model food soil. *Surface Coatings Technol.*, **201**, 943–951.

45 Zhao, Q., and Liu, Y. (2006) Modification of stainless steel surfaces by electroless Ni–P and small amount of PTFE to minimize bacterial adhesion. *J. Food Eng.*, **72**, 266–272.

46 Krishnan, S., Weinman, C.J., and Ober, C.K. (2008) Advances in polymers for anti-biofouling surfaces. *J. Mater. Chem.*, **18**, 3405–3413.

47 Lengke, M.F., Fleet, M.E., and Southam, G. (2007) Biosynthesis of silver nanoparticles by filamentous cyanobacteria from a silver(I) nitrate complex. *Langmuir*, **23**, 2694–2699.

48 Mavrov, V., and Belieres, E. (2000) Reduction of water consumption and wastewater quantities in the food industry by water recycling using

membrane processes. *Desalination*, **131**, 75–86.

49 Vourch, M., Balannec, B., Chaufer, B., and Dorange, G. (2008) Treatment of dairy industry wastewater by reverse osmosis for water reuse. *Desalination*, **219**, 190–202.

50 Sarrade, S.J., Rios, G.M., and Carles, M. (1998) Supercritical CO_2 extraction coupled with nanofiltration separation – applications to natural products. *Sep. Purif. Technol.*, **14**, 19–25.

51 Atra, R., Vatai, G., Bekassy-Molnar, E., and Balint, A. (2005) Investigation of ultra- and nanofiltration for utilization of whey protein and lactose. *J. Food Eng.*, **67**, 325–332.

52 Cuartas-Uribe, B., Alcaina-Miranda, M.I., Soriano-Costa, E., and Bes-Pia, A. (2007) Comparison of the behavior of two nanofiltration membranes for sweet whey demineralization. *J. Dairy Sci.*, **90**, 1094–1101.

53 Herzberg, M., and Elimelech, M. (2007) Biofouling of reverse osmosis membranes: role of biofilm-enhanced osmotic pressure. *J. Membr. Sci.*, **295**, 11–20.

54 Cudennec, B., Ravallec-Ple, R., Courois, E., and Fouchereau-Peron, M. (2008) Peptides from fish and crustacean by-products hydrolysates stimulate cholecystokinin release in STC-1 cells. *Food Chem.*, **111**, 970–975.

55 Cha, D.S., and Chinnan, M.S. (2004) Biopolymer-based antimicrobial packaging: a review. *Crit. Rev. Food Sci. Nutr.*, **44**, 223–237.

56 Woodrow Wilson International Center for Scholars (2010) Project on Emerging Nanotechnologies, see Consumer Products: An inventory of nanotechnology-based consumer products currently on the market, http://www.nanotechproject.org/inventories/consumer/ (accessed 3 November 2010).

57 Sondi, I., and Salopek-Sondi, B. (2004) Silver nanoparticles as antimicrobial agent: a case study on *E. coli* as a model for Gram-negative bacteria. *J. Colloid Interf. Sci.*, **275**, 177–182.

58 Yamamoto, O. (2001) Influence of particle size on the antibacterial activity of zinc oxide. *Int. J. Inorg. Mater.*, **3**, 643–646.

59 Jones, N., Ray, B., Ranjit, K.T., and Manna, A.C. (2008) Antibacterial activity of ZnO nanoparticle suspensions on a broad spectrum of microorganisms. *FEMS Microbiol. Lett.*, **279**, 71–76.

60 Chandramouleeswaran, S., Mhaske, S.T., Kathe, A.A., Varadarajan, P.V., Prasad, V., and Vigneshwaran, N. (2007) Functional behaviour of polypropylene/ZnO-soluble starch nanocomposites. *Nanotechnology*, **18**, 8.

61 Qi, L.F., Xu, Z.R., Jiang, X., Hu, C.H., and Zou, X.F. (2004) Preparation and antibacterial activity of chitosan nanoparticles. *Carbohydr. Res.*, **339**, 2693–2700.

62 Sanpui, P., Murugadoss, A., Prasad, P.V.D., Ghosh, S.S., and Chattopadhyay, A. (2008) The antibacterial properties of a novel chitosan–Ag–nanoparticle composite. *Int. J. Food Microbiol.*, **124**, 142–146.

63 Wang, X.Y., Du, Y.M., Yang, H.H., Wang, X.H., Shi, X.W., and Hu, Y. (2006) Preparation, characterization and antimicrobial activity of chitosan/layered silicate nanocomposites. *Polymer*, **47**, 6738–6744.

64 Wang, X.Y., Du, Y.M., Luo, J.W., Lin, B.F., and Kennedy, J.F. (2007) Chitosan/organic rectorite nanocomposite films: structure, characteristic and drug delivery behaviour. *Carbohydr. Polym.*, **69**, 41–49.

65 Yam, K.L., Takhistov, P.T., and Miltz, J. (2005) Intelligent packaging: concepts and applications. *J. Food Sci.*, **70**, R1–R10.

66 Brody, A.L., Bugusu, B., Han, J.H., Sand, C.K., and McHugh, T.H. (2008) Innovative food packaging solutions. *J. Food Sci.*, **73**, R107–R116.

67 Johnston, J.H., Grindrod, J.E., Dodds, M., and Schimitschek, K. (2007) Composite nano-structured calcium silicate phase change materials for thermal buffering in food packaging, in 3rd International Conference on Advanced Materials and Nanotechnology, Elsevier Science, Wellington, New Zealand.

68 Anthony, M.M. (1999) Self-cooling beverage and food container using fullerene nanotubes, US Patent Appl. 5946930.

69 Solovyov, S.E. (2007) Dry-coated oxygen-scavenging particles and methods of making them, US Patent Appl. 2007/0020456.

70 Mills, A., Doyle, G., Peiro, A.M., and Durrant, J. (2006) Demonstration of a novel, flexible, photocatalytic oxygen-scavenging polymer film. *J. Photochem. Photobiol. A Chem.*, **177**, 328–331.

71 Wellinghoff, S.T., Kampa, J.J., Barenberg, S.A., and Gray, P.N. (1999) Sustained release, transparent biocidal compositions, US Patent Appl. 5922776.

72 Wellinghoff, S.T., Kampa, J.J., Lelah, M.D., Barenberg, S.A., Gray, P.N., and Dixon, H. (2007) Energy-activated compositions for controlled sustained release of a gas, US Patent Appl. 2008/0026029.

73 Mills, A., Tommons, C., Bailey, R.T., Tedford, M.C., and Crilly, P.J. (2008) UV-activated luminescence/colourimetric O_2 indicator. *Int. J. Photoenergy*, **2008**, 547301, doi: 10.1155/2008/547301.

74 Taylor, D.H., Prusik, T., Smith, D.E., and Baughman, R.H. (2007) Freeze indicators, flexible freeze indicators and manufacturing methods. Marketed as Timestrip. World patent number 2007148321,

75 Subramanian, V., Frechet, J.M.J., Chang, P.C., Huang, D.C., Lee, J.B., Molesa, S.E., Murphy, A.R., and Redinger, D.R. (2005) Progress toward development of all-printed RFID tags: materials, processes, and devices. *Proc. IEEE*, **93**, 1330–1338.

76 Tentzeris, M.M. (2008) Novel paper-based inkjet-printed antennas and wireless sensor modules, in IEEE International Conference on Microwaves, Communications, Antennas and Electronic Systems, Tel-Aviv, Israel, IEEE, New York.

77 Loh, K.J., Lynch, J.P., and Kotov, N.A. (2007) Passive wireless sensing using SWNT-based multifunctional thin film patches, in 13th International Symposium on Applied Electromagnetics and Mechanics, East Lansing, MI, IOS Press, Amsterdam, The Netherlands.

78 Demoustier, S., Minoux, E., Le Baillif, M., Charles, M., and Ziaei, A. (2008) Review of two microwave applications of carbon nanotubes: nano-antennas and nano-switches. *Comptes Rendus Physique*, **9**, 53–66.

79 Jedermann, R., Behrens, C., Westphal, D., and Lang, W. (2006) Applying autonomous sensor systems in logistics – combining sensor networks, RFIDs and software agents. *Sens. Actuators A Phys.*, **132**, 370–375.

80 Potyrailo, R.A., Mouquin, H., and Morris, W.G. (2008) Position-independent chemical quantitation with passive 13.56-MHz radio frequency identification (RFID) sensors. *Talanta*, **75**, 624–628.

8
Food Functionality and the Physics of Bionanotechnology: Some Examples and Challenges

Erik van der Linden

8.1
Introduction: How Are Foods and Bionanotechnology Related?

Foods can be defined as materials that are edible and that provide nutritional value. Naturally, foods are also supposed to be attractive from a sensory point of view, that is, they should trigger one or more of our senses (touch, sight, smell, taste, sound) in such a manner that we can actually eat the material and like it during consumption and afterwards, and also preferably wish to consume it on subsequent occasions.

In relating foods with bionanotechnology, the first part is to relate the term "foods" with the term "technology". Before phrasing a definition of food technology, it is important to note that foods can be either fresh or prepared, and that they are made available throughout the world in different manners, like, for example, in local markets, supermarkets, elderly homes, hospitals, schools, restaurants, and in the kitchen at home. This leads us to a definition of food technology as:
the area of application of knowledge that allows one to decide

- in the case of prepared foods, which ingredient has to be put in where, when and how, during the preparation of the food, in relation to ensuring desired functionalities during processing, storage, transport, consumption and digestion;
- in the case of fresh food (fruits, vegetables, ...), what treatments have to be utilized in relation to ensuring desired functionalities during processing, storage, transport, consumption and digestion.

Food technology is one of the areas that plays an important role in addressing future challenges in providing food around the world. The challenges encompass:

- affordable and sufficient availability for a growing world population;
- need for sustainable production taking into account optimization of energy and water supplies, ...);

Nanotechnology in the Agri-Food Sector: Implications for the Future, First Edition. Edited by Lynn J. Frewer, Willem Norde, Arnout Fischer, Frans Kampers.

- specific requirements on combined health-, taste-, flavor-, and texture-related functionalities for different regions and different population groups.

These challenges relate to different disciplines at the same time, and therefore can only be met by an integrative effort. The systematic approach to such an effort is however not clear cut. This is due first to the fact that it concerns different disciplines each with their own "language" and phenomena (sociology, nutrition, chemistry, physics, ...), and second to the fact that not all physical laws that describe phenomena emerging on one scale necessarily follow in a direct manner from laws that describe phenomena on the adjacent smaller scale [1].

The second issue of scale is an important issue for food technology, since all food material properties should also be related to the physical and chemical properties of the ingredients on a molecular scale in order to comply with the definition of food technology.

Examples of consumer-relevant food material properties are firmness, pourability, spreadability, color, thickness, crunchiness, crispiness, and so on. These properties refer to one or more of the five senses (smell, taste, sight, sound, touch). Examples of molecular-scale physical and physicochemical properties of ingredients deal with the shape and size of the molecule, chemical fine structure, and so on. The main types of molecules one deals with are proteins, carbohydrates, fat/oil molecules, water, and several molecules in gas form (e.g., air in foams), as these form the nutritional and material functional components of foods.

There exist two important aspects in formulating physical and physicochemical relationships between properties of molecules and food (material) properties.

a) There is a factor of a billion difference in length scale between the molecular scale and the macroscopic scale (nanometers to meters).
b) Foods are usually not homogeneous on a length scale of micrometers. Examples are mayonnaise, bread, beer foam, and margarine

For instance, consider mayonnaise in more detail. Observed under the microscope, it consists of droplets (of oil) embedded in an aqueous phase. By way of another example, margarine is seen to consist of water droplets embedded in a continuous oil phase, which is partially crystallized. The properties of mayonnaise, such as spreadability, depend on the deformability of the droplets, which is in part determined by the droplet–droplet interaction and in part by the deformability of the interface of the droplets. The properties of the various constituents of the microstructure, in this case the oil in the droplets, of course in turn depend on the molecular properties of the oil molecules, their interactions, pressure, and temperature. Also, the properties of the aqueous phase depend on the properties of the water in it, which in turn depend on the molecular properties of the water molecules, their molecular interactions, temperature, and pressure. It is the domain of droplet size that makes the description easier, since the product properties are a function of the properties of the droplets, in this case of mayonnaise.

In order to formulate the desired relationships for any technological question in a useful way, one needs to know about this intermediate, inhomogeneous domain (in the case of mayonnaise, the droplet regime). Of particular relevance are domain sizes between several nanometers (macromolecular and supramolecular assembly size) and a few hundred nanometers. Here the nanodomain enters into our problem. Using the definition of nanotechnology as the control of matter at dimensions between approximately 1 and 100 nm, one realizes that the physics and physical chemistry regarding this nanodomain act as a bridge and therefore facilitator for formulating physical and chemical relationships between properties on a molecular scale and a macroscopic scale. Ingredients can form a multitude of different structures on the nanoscale. One can, for example, have platelets [2], spheres like casein micelles in milk, long thin threads [3], and threads that forming a coil-like structure, such as those of many polysaccharides. But also, for example, more complicated topologies exist, such as the bicontinuous structures of surfactant-based cubic phases [4]. The macroscopic properties are dependent on how the structures exhibit interaction, and on the properties of the nanostructures themselves, which in turn depend on the interactions between the molecular constituents.

By considering the nanodomain as an important intermediate scale, one simplifies the formulation of the relationships between molecular properties and macroscopic food properties, a requirement for reaching the aim of food technology. Both the sensory and the nutritional and health functionalities are related to the behavior in the nanodomain, and in the end to the molecular properties of the constituents. The main questions are: What are the rules that describe how the nanostructures are being formed (relevant to making the food)? And what are the rules that describe how they are broken down (relevant to the sensory and nutritional aspects of the food)? One should realize that, because foods are inhomogeneous, on the one hand this complicates their description, while on the other it provides the wealth of different food sensory properties in conjunction with optimized nutritional and health functionalities that are so much appreciated and desired by the consumer.

One can conclude that food technology is facilitated by knowledge and associated technology in the nanodomain, that is, by the area of nanotechnology. Furthermore, since food constituents are mainly biomolecules (proteins, carbohydrates, fat/oil molecules, water molecules), it is clear from the above how food and bionanotechnology are related and how food bionanotechnology can be used to address part of the major challenges in food. In other words, the nanodomain provides opportunities for co-tailoring the sensory and nutritional and health functionalities of foods.

8.2
Physics and Structures in Food Bionanotechnology

For the rest of this chapter, we will focus on the physical aspects of food bionanotechnology, which belongs to the area of food physics. We will neglect the many ways that can change the chemical fine structure of the constituents, and the concomitant changes in physical, sensory, and nutritional properties.

In order to come up with a systematic approach regarding foods, one may put forward categorizations of foods. Many of such categorizations boil down to more or less the same idea, as, for example, formulated by McGee [5]. He considers dairy products, egg-based products, meat, fruits, vegetables and herbs, grains and nuts, bread and dough, sauces, confectionaries (chocolate, sweets, sugar), alcoholic beverages, drinks, and juices.

Though this categorization may be helpful from a food point of view, it does not help much to reach the technological goal formulated above. Instead, we should consider the nanodomain as the bridge between food properties and the ingredients, and consider the principal parameters by which one can change the properties of matter.

These principal parameters, relevant to the physics of foods, are temperature (denoting effectively interaction strength), concentration of ingredients, and pressure as related to, for example, flow fields, as suggested by Liu and Nagel in their jamming diagram [6]. Pressure may also originate by any other externally applied field for that matter, or simply hydrostatic pressure. An additional very important parameter in the many non-equilibrium occasions is time.

From the above, it is clear that the physics-based functionalities in sensory and nutritional aspects are all coupled to nanostructures within the food. So, the physics of these nanostructures also provides a bridge between the sensory and nutrition- and health-related functions of food. Therefore, we have chosen to organize by means of examples of different existing nanostructures, and to mention some of their functionalities in both sensory as well as nutritional space, where known. Examples are distinguished in terms of the morphology of the structures, that is, fibrillar structures, plate-like structures, and spherical structures, either in water or oil phase, and bicontinuous structures. There exist many more types of topologies but we decided to limit ourselves to only a few examples to make the point of how the physics of nanostructures is relevant to foods. The examples do not cover the entirety – nor even a large portion – of what is published in the literature, and are by no means meant as a review of any kind. Instead, the examples are only aimed at forming a small "sample platter" to "get a flavor" for what is possible in foods using nanotechnology with (edible) biomolecules. In this respect, we also devote one section (Section 8.7) to developments within the restaurant setting that make use of knowledge and technologies from nanoscale science. The examples in addition are also intended to experience a "nano-view" toward connecting the molecular world with foods, which is key to addressing the challenges ahead for food in the future.

8.3
Fibrillar Structures

In water, various ingredients are found to give rise to fibrillar structures. This accounts for both proteins (milk proteins, egg proteins, ...) and polysaccharides (xanthan, carrageenans). One of their sensory functionalities is that they form weight-efficient viscosifiers, and weight-efficient water gelators or structurants. This facilitates the formulation of products that are high in water content, but at the same time have a solid-like behavior, allowing increased formulation flexibility for producing satiation products with low amounts of fat.

8.3.1 Protein-Based Fibrils

Formation of protein-based fibrils has lately received considerable attention. The diameters of protein-based fibrils are in the range of one to a few nanometers and their length can reach in the order of 10–100 μm. On the one hand, the fibrils give rise to interesting functionality, yet on the other the structures inside these fibrils show similarities with beta amyloid fibrils, which are associated with amyloid fibers and disease states. Examples of fibrillar protein assemblies have been reported by several authors, as early as several decades ago by Joly's group [7–9]. It took until about two decades ago before the topic of assemblies of food proteins again received much wider attention. A more detailed overview of food-relevant fibril aggregation has recently appeared in an extensive review on protein gelation [10], and in short reviews addressing issues on fibrillar protein assembly [11, 12].

Parameters that influence fibril formation, including the level of branching, are pH and salt concentration, as for example β-lactoglobulin fibrillization within a region of experimental conditions [13]. Properties like the persistence length of these fibrils were discussed later [14]. For ovalbumin gels, experimental conditions that lead to fibril formation were, for example, examined by Weijers *et al.* [15].

Regarding fibril formation, most food proteins had been thought until recently to be intact in forming the fibrils. However, recently it was reported that, for the assembly of proteins like lysozyme [16], β-lactoglobulin [17], and α-lactoglobulin [18], chemical changes were observed in the proteins *before* fibrillization took place, and it was found for these cases that only certain peptide types are built into the fibril, and the existence of pre-aggregates of these peptides were hypothesized [16]. Hydrolysis is one of the reactions that take place during the fibrillization process. Interestingly, hydrolysis was utilized to induce gelation by Doucet and Foegeding, where the hydrolysis was induced by an enzyme [19]. In that case both gelation and hydrolysis took place at pH 8. Enzymatic hydrolysis was also induced in the case of β-lactoglobulin, but at pH 7, and electron microscopy reveals fibril formation only after the pH was changed to 2 [20]. Thus, the pH is important in the aggregation and thought to affect the interaction between the according peptides before fibrillization. These results show the possibility of an enzymatic route for fibrillization, with first enzymatically induced peptide formation, and fibril formation by pH adjustment.

Recently, fibril formation for the case of β-lactoglobulin was reported to exhibit a critical aggregation concentration [21] and to follow the rules of thermodynamic self-assembly, completely analogous to micellar assembly.

Early stages of fibrillization were recently addressed by Meersman and Dobson [22] and the results suggest that the initial assembly is caused by hydrophobicity versus electrostatic interactions, while the secondary state is characterized by reorientation and intra-β-sheet formation [22]. In line with this are results for different proteins forming fibrils as a function of solvent. For instance, ethanol, methanol, propan-2-ol, and 2,2,2-trifluoroethanol [23] as well as urea [24] have been used for fibrillization.

We conclude that the formation of fibrils based on proteins, with nano-sized thickness while having micrometer length, shows several intermediate stages, follows self-assembly characteristics, and is dependent on the specific peptides that are built into the fibrils. These peptides can be formed by enzymatic reactions or chemically induced hydrolysis.

8.3.2
Extremely Low-Weight Gels Using Fibrils

Extremely low-weight-fraction gels have been realized down to 0.07%. Long linear fibrils are formed for β-lactoglobulin at pH 2 and low ionic strength. These fibrils exhibit stability even when, after their formation, the pH is changed toward 7 or 8 [3, 25, 26]. If subsequently one adds $CaCl_2$, one obtains calcium bridges, which lead to gelation at much lower concentration than for the case without $CaCl_2$. According to this new multi-step cold gelation method, one can make gels down to 0.07% protein matter. This is an order of magnitude lower than using the conventional cold gelation method, or heat-induced gelation. The new technology makes use of ionic bridging between nano-thick fibrils, and in this way provides an efficient route for gelation [26].

8.3.3
Helix-Based Fibrils in Gelatin Gels

Gelatin gels form the basis of different foods, in particular gelled deserts. Recently, their elasticity in the low-concentration in combination with the higher-concentration regime was elucidated [27]. It turns out that the persistence length of the helices is responsible for the critical gel concentration, and for the strength of the gelatin gels in the low-concentration regime. In fact, it is entropy that accounts for the fact that a gel is formed and for its strength. In the higher-concentration regime, the nanoscale fluctuations of the helices start to influence one another more and more, leading to an additional important nano-range length scale, that is, the deflection length [28, 29]. The elasticity of the gels in this higher-concentration regime then depends on the number of contacts between chains per unit volume, which is determined by the deflection length [27]. The overall description allows the elasticity to be predicted as a function of concentration over the entire food-applicable regime.

An important point to realize with the gels of the previous sections is that these gels are in fact one-phase systems, that is, there are no parts that are phase-separated from any other, as is the case in, for example, carrageenan gels. Their strength cannot be simply described as in this section on gelatin.

8.3.4
Fibrils in Oil

Recently, mixtures of specific plant sterols and plant sterol esters have been shown to form nanotubes in edible oil [30]. This structuring allows one to solidify oils that do not show gelation at regular temperatures due to the fact that they contain many polyunsaturated fatty acids (PUFAs). The latter are known to have a health benefit, as opposed to saturated fatty acids (SAFAs) [31].

Self-assembling plant sterol (ester) systems have been investigated in the context of low-SAFA alternative hard-stocks of edible oils [32, 33]. Such systems would allow the formulation of firm emulsion gels, like margarine, without having to resort to fats containing relatively unhealthy saturated fatty acids [31].

The self-assembly rules of the various compounds as a function of temperature and oil composition are the subject of current investigations. Notably, also polysaccharides, specifically cellulose derivatives, are reported to give rise to structuring of oils, by a still unknown mechanism [34].

8.3.5
Fibril-Enforced Composite Structures

Apart from the fibrils in bulk solution leading to gelation, one may also use the fibrils, with their relatively long persistence length and long fibrillar length, for making fibril-enforced composite capsule materials [35]. Depending on the desired size of the capsule, one may have to adjust the fibril length accordingly in order to accommodate the surface load of the fibrils. Long (larger than the size of the capsule) rod-like fibrils will tend to stick out of the surface, forming a thicker and more penetrable layer, while short rod-like ones will be able to cover the entire surface, and form a thinner layer. Therefore, it is important to be able to adjust the size of the fibrils, for example, by shear treatment [36].

8.4
Plate-Like Structures

Heertje *et al.* [2] have reported on the use of liquid-crystalline phases in the structuring of food systems. They mention as a practical example a fat-free margarine that contains 30 mg monoglyceride per gram of product in a so-called coagel state. This coagel state is formed from the liquid-crystalline lamellar phase, via the so-called α-gel state. Thus, for this particular monoglyceride system, the α-gel state is not the final state. However, for other compounds, like the lactic acid ester of monoglyceride [2], the α-gel state is the final state.

Schematically, the structure of the α-gel state is given in Figure 8.1, which can be deduced from the photographs (figures 7 and 8) in reference [2]. The structure consists of a continuous phase of bilayers (platelets), which are stacked parallel to one another in regions with a specific domain size, and of droplets embedded in

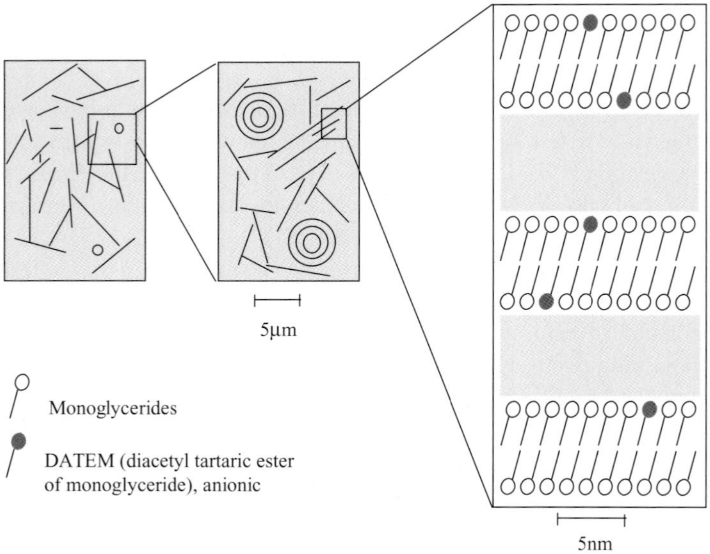

Figure 8.1 A schematic structure of a lamellar phase system. (Adapted from E. van der Linden [40].)

this continuous phase. The platelets are relatively large and a nanometer thick, and therefore provide a weight-efficient way for the structures to interact and form a gel. The droplets consist of concentrically stacked bilayers, like the layers in an onion, where the bilayers are separated from one another by water. For obvious reasons, these droplets are sometimes denoted by the term “onions”.

The rheological (i.e., functional) properties of a product consisting of such an α-gel state depend on parameters such as the number of droplets, their size, the concentration of bilayers, and whether bilayers have folded themselves around several droplets, thus forming entanglements. Optimal control of the structure would be when the bilayers are still in the liquid state, making them most deformable. This phase is also the phase encountered during processing of products. The first issue then is to investigate what type of structural transitions are possible in a liquid-crystalline phase, how to induce them, and whether there would be a key parameter identifiable. The parameter in the liquid-crystalline state in relation to shear effects turns out to be an effective surface tension of the onion phase. The relation between size, effective surface tension, and shear has been outlined before [37]. The subsequent relation between the effective surface tension and the rigidity of the platelets, and their interactions, was also laid down, and connected with the molecular structure and properties of the compounds building up the platelets [38].

The number of droplets versus bilayers also has a topology effect to it, as described in detail for a liquid-crystalline AOT (the surfactant sodium bis(2-ethylhexyl) sulfosuccinate) phase by van der Linden and Buytenhek [39]. In the end, the effects of flow (i.e., pressure) as well as type of molecules have been combined with the resulting structure by means of an effective surface tension of the lamellar phase. As such, this is again an example of how the understanding of the nanodomain leads to control of product properties. In this case how the nanodomain parameter "effective surface tension" connects to the structure and rheology of platelet-containing samples. The subject has also been reviewed in much more detail elsewhere [40].

8.5 Spherically Symmetric Structures

8.5.1 Protein Fractal Structures in Water

At suitable pH ranges, food proteins can form particulates, which are so-called fractal clusters. Their presence can lead to gelation once the cluster concentration is high enough. (For an extensive recent review on protein gelation and the role of fractality, see reference [10].) So, the structures that build up the gel are fractal clusters. The formation of particulates instead of fibrils, within a certain pH range (close to the protein's isoelectric point) has recently been proposed to be another generic feature of proteins, in addition to the feature of fibrillization [41]. Both features are suggested to be dependent not only on the specific amino acid sequence but also on how all molecular details together give rise to specific physicochemical properties on a molecular scale, such as hydrophobic spots, charge distribution, and dipole moments, for example. This has also been put forward recently [11], where a first attempt was made to explain the morphology of protein aggregates, ranging from fibrillar to particulate, on the basis of a balance between electrostatic and hydrophobic interactions. That analysis gives an explanation for why fibrils are expected at high charges, while particulates are expected at low charge and large hydrophobic interactions. The latter general feature of protein assembly was also put forward qualitatively in Krebs *et al.* [41].

8.5.2 Micelles

The so-called casein micelles present in milk are in fact an efficient way of agglomerating four different kinds of protein together. The exact structural features are not yet completely known. In sodium caseinate solutions the individual casein molecules interact with each other to form associate structures with a radius of approximately 10 nm. By slow acidification these structures can aggregate to eventually form a gel. This aggregation process has been studied by static light scattering and rheometry, and it was found that the adhesive hard-sphere model is a suitable model to describe the system [42]. In the case of sodium caseinate emulsions, it was found that diffusing wave spectroscopy is a very useful technique to measure the different structural transitions that take place within the system. It was found that two transitions took place in a system with an excess of stabilizer [43].

8.5.3
Spherically Symmetric Fractal Structures in Oil

So-called organogels have been reported by structuring liquid oil with mixtures of stearic acid and stearyl alcohol [44]. The observed macroscopic behavior of the organogels was related to the microscopic fractal spherically symmetric stacking of the crystals, and subsequently to the molecular ordering within the crystals.

8.6
Bicontinuous Structures in Protein–Polysaccharide Systems

Many foods contain both proteins as well as polysaccharides. Aqueous protein–polysaccharide mixtures often exhibit phase separation, even at low concentrations of one or both of the components. These mixtures, thermodynamically, lead to separation into usually two phases in equilibrium with one another, one phase being highly concentrated in protein and low in concentration of polysaccharide, and the other phase being the reverse [45, 46]. If the mixture is such that the volume of one of the two final phases is low (order of 10–30%), one may disperse this phase back into the other (it then automatically becoming the continuous) phase and obtain a water-in-water emulsion with corresponding very low interfacial tension between the two phases [47]. This ultra-low interfacial tension is caused by the fact that both sides of the interface contain mostly water and are not really very much distinguishable. This interface is inherently unstable. If one starts within a part of the phase diagram where one still has a one-phase system, one may destabilize that system by, for instance, changing the temperature and thereby inducing phase separation. This may occur starting from nucleation (dispersed phase) or from binodal decomposition (bicontinuous phase). If one jellifies one or both of the continuous phases during the spinodal decomposition phase, one ends up with bicontinuous gels [10].

The kinetics by which this separation takes place is determined by various parameters, and recent insights into the kinetics, depending on interfacial characteristics, among other things, allow control of the structure of such aqueous protein–polysaccharide mixtures. In turn, the interfacial characteristics are determined by the molecular properties of the constituents and the concentrations in both phases [47, 48]. Reversely, the distribution of molecular weights over the two phases turns out to follow a Boltzmann distribution as determined by the low interfacial tension between the two phases and the molecular weights of the compounds [45].

Several bicontinuous gels have recently been described in relation to their sensory attributes [49–54]. More specifically, the breakdown properties, as well as the excretion of liquid as a function of pressure, were reported, something significant when trying to mimic various types of foods during their mastication (think of artificial meat based on plant proteins, for example). Sensory attributes like crunchiness, crispiness, and juiciness were investigated in terms of the microstructure [49, 53].

Fibrils may be used as efficient flocculants in emulsions, leading to flocculated systems [55], but at lower concentrations they may also act like, for example, xanthan does in gelating salad dressings [56] leading to a weak gel.

Thus, knowledge on how to manipulate the nanoscale structure of mixed (bicontinuous) gels and emulsions helps to create desired textures of such gel-based foods.

8.7 Gastronomy and the Nanodomain: Molecular Gastronomy[1)]

8.7.1 Introduction

Let us start out with some definitions. The word "gastronomy" stems from two Greek words: *gastros*, meaning "stomach", and *nomos*, meaning "knowledge". Food and its exploration has actually been important for all times. For example, one may mention Parmentier (1737–1813), who introduced the potato into France; and Maillard, who investigated the reactions of glycerol and sugars with amino acids (1912). One famous person in gastronomy, who wrote a very interesting book, *Physiology of Taste*, is Brillat Savarin (1755–1826). In his book, he proposed to translate gastronomy literally as "the laws of the stomach". He defined gastronomy in broader terms as "the knowledge of all that is related to man as he nourishes himself". He also gave gastronomy an application purpose by stating: "Its purpose is to keep humankind alive with the best possible food ...". His book describes many different elements that are relevant to the sensory perception and physiological effects of food. As such, gastronomy is related to food science as we know it today.

A more modern definition of gastronomy often used today would be: "the art of enjoyably eating and drinking". This mainly refers to the pleasures of eating and drinking. Traditionally, though, the field of gastronomy encompasses more than pleasure of the food itself. It also involves the pleasures regarding the way the food is served (compare a fast-food restaurant to a three-star restaurant), as well as the environment in which the food is consumed (make the same comparison). Apart from that, if one talks about knowledge, many disciplines are in fact involved, as already noticed by Brillat Savarin. For instance, the understanding involves physics, chemistry, mathematics, economy, agriculture, psychology, physiology, cultural aspects, nutrition, and so on. Traditionally, according to Brillat Savarin and others, gastronomy is a very broad subject!

1) This section is adapted with permission from: E. van der Linden, Physical aspects of molecular gastronomy", Reader for the Wageningen University course on Molecular Gastronomy 1, 2009–2010.

Knowledge has been applied to scale up processes in order to provide food for many people. As a consequence, a food industry flourished and is developing still today. This same food industry has also invented and applied technologies. For example, for preparing novel food products, milk was spray-dried into a powder, in order to preserve the "milk" at high temperature in a simple way. By way of another example, we can mention freeze-dried instant coffee. Developments within an industrial context have thus resulted in novel foods and novel techniques becoming available. Interestingly enough, the home and restaurant kitchen have not experienced such a development to the same extent and with such broad acceptance as has been shown in industry. Some even note that the contemporary status is still not too far from what it was in the Middle Ages!

One of the turning points in stimulating the development of the home and restaurant kitchen occurred a few decades ago, when Nicholas Kurti, a low-temperature physicist, made a movie in 1969 together with the BBC, entitled "The physicist in the kitchen". To quote him: "It is a sad reflection that we know more about the temperature inside stars than inside a soufflé." Scientific issues in the kitchen were also picked up a little later by Hervé This, a French scientist, who sought collaboration with Kurti, when working on his PhD study. Alongside them, Harold McGee had also picked up the science of the kitchen in his famous book, *On Food and Cooking: The Science and Lore of the Kitchen*, which appeared for the first time in 1974, a new edition of which has appeared recently (2004) [5]. Other sources are currently the World Wide Web (about 255 000 hits), and blogs (see, e.g., http://www.khymos.org). Here, numerous examples are given for recipes that can only be successfully made following precise preparation methods, with accurate weighing procedures, accurate heating, and so on, just like what one is used to in a physical and chemical laboratory. The recipe and its procedures rely on nanoscale knowledge of the molecules and their interactions, in connection with transitions like gel transitions, heat sensitivity, mechanical stress sensitivity, irreversibility, concentration effects, and mixture effects – exactly the type of examples we treated in the previous sections.

As a result of their interest in the area of science in the kitchen, Hervé This and Nicholas Kurti [57] first coined the phrase "physical and molecular gastronomy", which can be summarized as "the science of enjoying food". More officially formulated a bit later [58] it was "the scientific exploration of culinary, and more generally, gastronomical transformations and phenomena, as described by cooks or by culinary books". Later still, the word "physical" was left out for the sake of brevity, as it was felt that "molecular" would also imply physical aspects. The word "molecular" implied that ultimately relations were sought between the molecular properties and the macroscopic properties of the food. Hence, molecular gastronomy (MG) focuses on home and restaurant aspects of culinary transformations and phenomena (formulated in relation to molecular properties) as well as home and restaurant aspects of eating phenomena, that is, aspects on gastronomy (also formulated in relation to molecular properties). In short, MG is the study of recipes and their details, and what they do to perception, all in terms of molecular properties. In this sense, within the field of molecular gastronomy, one tries to connect the nanoscale phenomena to gastronomy. The applications of molecular gastronomy provide a great illustration of how nanotechnology is ultimately applied to foods in a restaurant setting. One often assumes that gastronomy in the definition of molecular gastronomy actually refers only to "the aspects of pleasures while eating", but in principle the broader definition of Brillat Savarin still applies, thus obtaining in principle a very wide area of research.

In the definition of Kurti and This, several aspects and goals of molecular gastronomy are mentioned, which have been categorized and extensively described later on. Two relevant and illuminating papers on this categorization are by This [58, 59]. We summarize a few issues. The two most important aims of MG (free, after [60]) are:

a) to scientifically explore the recipes as they exist and try to describe the essence of the recipe;
b) to scientifically describe the necessary details that give the recipe its characteristic perception.

These two goals are both meant to be formulated in terms of molecular properties. The "essence of the recipe" is referred to as the "model", and the "details" are referred to as "precisions" (i.e., goal 2).

As one example [58], we mention the crackling skin of small roasted pigs. The story was that the skin would be more crackling right after preparation if one were to make a cut in the skin around the neck of the pig. Indeed, this turned out to be true, and can be explained by the fact that, in the case of making such a cut, the vapor within can disappear before it can condense and penetrate into the skin, making it soft.

As another example [58], it has been claimed that, in trying to get the egg yolk exactly in the middle of an egg after being cooked, one should use water that already has been boiling for a while before putting the eggs in. This may turn out to be true, but in general it is not a good proposition since an egg yolk has a lower density than the egg white, as is seen by its floating in the egg white when both are put together in a cylinder. One may argue that there is a certain binding between the yolk and the white while in the egg, therefore preventing floating in the egg, but a simple experiment of cooking eggs while holding them in place shows that the yolk always gets to the top of the egg. This is all due to gravity. The fact that boiling water may to a certain extent prevent floating of the yolk in the egg is that the boiling water moves the egg around to a large enough extent, thereby yielding no preferred direction any more for the yolk to move.

Apart from the two goals one has three applications in mind for molecular gastronomy:

a) introduction of new tools, methods and novel ingredients in the home or restaurant kitchen;
b) invention of new dishes based on investigations of classical recipes;
c) presentation to the general public.

Regarding the applications and technology, one has to keep in mind that part of making good food lies in craftsmanship, which is continuously connected to artistic issues of presentation of the food. A funny example of using creativity and an artistic approach is in making waffles that are partially open to one another, including a piece of ice cream, being entitled "mini-conversations", created by the three-star chef Pierre Gagnaire in 2003.

Regarding the application part of presenting to the public, it is clear that there are many activities worldwide already: workshops for scientists (Erice, others), primary and secondary school activities, cooking workshops for professionals and amateurs, and even sessions in food science symposia (Euro Food Chemistry XIV symposium in Paris, in 2007; and Delivery of Functionality symposium in Amherst, USA, in 2007) – see, for example, [61].

One issue we should mention is the relation to something referred to as "molecular cooking". Molecular gastronomy is not a synonym for molecular cooking. The term "molecular gastronomy", for example, also entails wine–cheese combinations and issues about how to avoid an astringent taste during the drinking of tea, while molecular cooking would refer only to cooking *per se*. Furthermore, the term "molecular cooking" is strange, since cooking automatically refers to a transformation on a molecular scale, in the end. Adding the adjective "molecular" to cooking is trivial and unnecessary.

The physical and chemical issues of food science and technology mainly refer to understanding and applying relations between the molecular properties of the food ingredients and the macroscopic properties and functionalities of the food, during making, storage, consumption. Molecular gastronomy, according to the definition of This and Kurti, then focuses on part of this larger area within food science and technology, which is the area described by chefs and in cookbooks. So, MG in that sense forms part of food science and technology, with focus on chefs and cookbooks, instead of, for example, industrial food preparation.

We note that recently issues regarding digestion and delivery of ingredients have also started to be included into food science and technology. This nutritional direction connects to the most general definition of gastronomy as given by Brillat Savarin: the knowledge of all that is related to man as he nourishes himself. Regarding the definition of This and Kurti, molecular gastronomy focuses on issues as described by chefs and in cookbooks. We refer to the introduction in the book of Hervé This [62] for more details on terms and usage as well as a historic overview of several aspects.

8.7.2
Recent Developments

One major resource for recent developments regarding molecular gastronomy is certainly the World Wide Web, where one is able to search for recipes, restaurants, books, activities, symposia, courses, and science-related activities. The number of chefs practicing components of molecular gastronomy, that is, using and applying the science in order to provide good and novel foods and dishes, is increasing. This activity is one of the application parts of molecular gastronomy.

One chef who deserves particular mention in applying science for novel recipes, dishes, and menus is Ferran Adrià, the chef of probably the world's currently most famous restaurant, *El Bulli*, in Roses, Catalonia, Spain. He has been the initiator, and is still one of the most successful chefs, and he collaborates with scientists and artists, among others. He maintains these collaborations in order to develop new dishes and concepts for the kitchen. He has inspired many to incorporate technological novelties and practices in the restaurant kitchen, and even to some extent into the home kitchen.

In order to get a flavor of what can come out of combining science, technology, and gastronomy, we will review shortly the history of his restaurant, the developments that it brought, and the philosophy. This short survey has been based mainly on an excellent new book, *A day at El Bulli* [63].

One of the cornerstones of his restaurant is creativity. The best definition yet according to Adrià has been given by the French chef, Jacques Maximin. He defined the term in 1987 in the following manner: "Creativity means you do not copy". When Adrià heard this during a conference in Nice, he decided to leave behind the cookbooks and seek his own identity. The restaurant closed in winter for six months, and together with his decision to seek his own identity, the journey for seeking creativity had begun.

When Ferran Adrià became chef of *El Bulli*, the kitchen was influenced greatly by the nouvelle cuisine. This was a movement as a reaction to the classical French cuisine, and it entailed a light cuisine and shorter cooking times. In addition, this nouveau cuisine was inspired by the traditional and regional dishes, ingredients, and techniques. Olive oil, for example, was first considered primitive, but this became popular in expensive restaurants.

8.7.2.1 Signatures of Creative Methods at *El Bulli*

A first signature of a creative method is the application of traditional cooking techniques to extraordinary and prestigious ingredients.

A second signature is in the influence of other regions. For example, after luxury Italian restaurants, pasta also appeared in French and Spanish restaurants. This aspect is broadened even further because of the global availability of many ingredients. In an extended version, one combines ingredients, classic recipes, and cooking techniques from all over the world on one plate. This is referred to as "fusion cooking". Whether such a recipe is really attractive depends on the craftsmanship of the chef and his/her ability to judge properly the quality.

A third signature is the establishment of new techniques and new concepts. This is the highest form of creativity found at *El Bulli*. A technique is a process, or a combination of processes, that makes a product edible or that induces another type of transition of the product. A concept is the basic idea behind a dish or recipe. One can work out a concept in many different ways, resulting in many different dishes. In fact, a concept is a way to represent an ingredient in a familiar way. For example, a concept can be carpaccio (thin slices of meat), a salad (greens with some dressing), or an omelet (egg loosely scrambled and heated, together with ingredients of choice).

At *El Bulli*, examples of new concepts were the creation of the frozen savory world (1994), ravioli of ingredients other than pasta (1994), and the liquid croquants (1994). Examples of new techniques are spherification (making very small and intensely tasting fruit and vegetable balls). Examples of combinations of concepts and techniques find themselves in the famous foams, warm gels, and the aires (very high fraction of air in foams). It was found that the taste of shellfish was not optimal during the longer traditional cooking times. Thus it was decided to cook them for very short times, maintaining the taste, and then serving them with a thin gel layer of their own cooking fluid. This finding resulted in a total new way of preparing shellfish.

The techniques applied are not important for the guest–the taste experience is the only thing that matters. However, knowledge of the techniques may add to the experience of the guest. At *El Bulli* there is a systematic way of working to reach the results with the high creativity feel for which that restaurant is known. For more details, see *A Day at El Bulli* [63].

In addition to these examples of dishes from *El Bulli*, we name a few other examples of other chefs. One is "Salada de L'Abbé Nolet", by Pierre Gagnaire, which is a vinaigrette salad that is completely dry, that is, without the vinaigrette in fluid form. The vinaigrette is jellified. Another is "Ravioli bras croisés", by Thierry Max, which involves ravioli with crossed arms, so to speak. It is based on smoked herring, Granny Smith apples, and pasta dough. Yet another example is called "Initiation karmique", which is based on the only available edible orchid (of about 30 000 species), which is covered by sugar crystals (by bathing the orchid in a concentrated sugar solution), and which holds a drop of milk as the symbol of the beginning of life. The milk is jellified by putting the drop of milk in a bath of alginate, due to which an elastic shell of calcium alginate develops around the milk droplet, since the calcium of the milk jellifies with the alginate solution. Another example of funny phenomena is the appearance of "smoke" out of somebody's nose as he eats something that has been cooled (in fact, it is the cooled air that runs out of the nose). This trick has been used, for example, by Heston Blumenthal in his restaurant *The Fat Duck*, Bray, Berkshire, UK, to get the people relaxed at the beginning of a meal to focus their minds on food and the dinner and forget about their daily worries as much as possible.

It is clear that chefs actually start to use more and more the nanoscale phenomena that are widely understood in the scientific world and within the world of industrially prepared food. Nonetheless, it is a "tasteful" and attractive way to talk about nanotechnology in relation to foods.

8.7.3
A Structured and Scientific Approach to Molecular Gastronomy: Back to Nano

We know that cooks and culinary books describe recipes, the resulting products, and their gastronomical value. The basic physics and chemistry behind the recipes and the phenomena occurring during preparation, storage, and consumption (relevant to the gastronomical value) forms the scientific part of molecular gastronomy.

The recipes contain ingredients and cooking techniques. In the evolution of molecular gastronomy, one can observe classical recipes, their improvements, and finally new recipes (containing either new combinations of ingredients, new ingredients, or new procedures). In all cases one has consider the basic physics and chemistry dealing with ingredients, recipes (be it classic, improved or novel), and the gastronomic value of the resulting products.

The first part of the recipe is the ingredient. Ingredients can be divided into roughly six classes. These classes are proteins, lipids, oils, water, air, and polysaccharides.

The second part of the recipe is the procedures and corresponding techniques. There are numerous different "cooking" techniques. They are, for example:

> barding, larding, dry and wet marinating, infusing, au bain marie, blanching, poaching, steaming, frying, roasting, grilling, stewing, stir-frying, deep-frying, drying, glazing (e.g., with honey), caramelizing, confiting, smoking, gravy making, salt crust cooking, artichoke cooking and fruit (pear) poaching.

In addition, one describes various cutting techniques, for example:

> brunoise, julienne, chiffonade, pocket creating, peeling "a vif", bouquet garni.

The procedures have the basic following physics behind them. They all have in common that they are a function of the principal parameters of temperature, density, time, and externally applied "fields" like a flow field (e.g., mixing). The cutting techniques deal with the way in which a material is fractured and to what extent a material is cohesive, versus the forces exerted on the material.

The gastronomic value of the product can be divided into the five senses: seeing, hearing, touching, tasting, and smelling. The physics of these deal, respectively, with color and transparency, sound- and fracture-related phenomena, thickness, elasticity, and so on, release and transport of taste molecules, and release and transport of volatiles. Digestive aspects refer to breakdown of structures and/or subsequent release and transport of nutrients.

The seeing also has strong chemical aspects, of course, in the sense that colors can change due to chemical reactions. The smell and taste likewise have strong chemical aspects to the chemical reactions that occur.

One of the tools to systemize knowledge on how ingredients and procedures lead to products with specific gastronomic properties is by considering the product already during its making, and also afterwards, in more detail. This can be done by considering a smaller scale, that is, the nanoscale. Most products exhibit a structure on this smaller scale. The knowledge relevant to recipes is how the properties of the ingredients determine the properties of the small structures, and how the properties of these structures in turn determine the product properties. The knowledge relevant to product properties versus gastronomic value is how the properties on the various scales of the product in the end relate to the response of one or more of the five senses. We should know how these aspects change as a function of the changing principal parameters temperature, flow and pressure, concentration and time. The link between molecular properties and gastronomy is facilitated by nanoscale descriptions. The systematic approach sketched above for describing how molecular properties and physical phenomena are relevant to gastronomy is summarized in Figure 8.2.

A directed approach on improvement of recipes, which often is desired in gastronomy, can take place when one has knowledge on how ingredients and procedures lead to specific products with specific gastronomic values. New recipes are found by newly combining known ingredients and/or known techniques. In addition, new recipes may be found by using novel ingredients (novel at least from the restaurant and home kitchen perspective) such as hydrocolloids. Also, in the case of new recipes, knowledge on how ingredients and procedures lead to specific products with desired gastronomic values is essential. Also there, the above structural breakdown of products and their respective properties is a helpful approach.

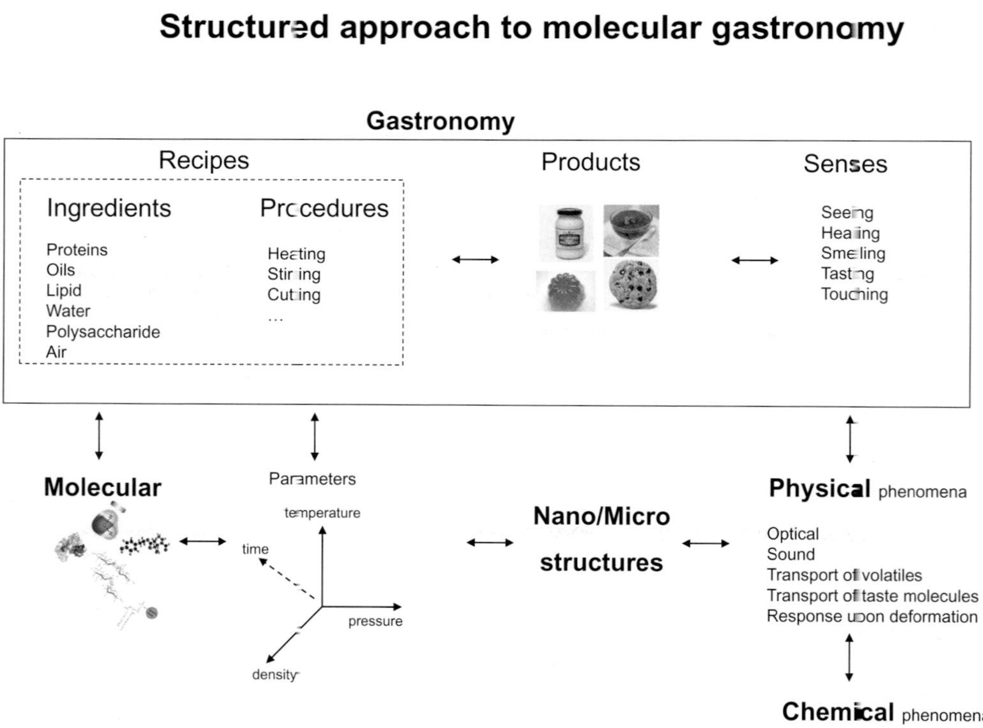

Figure 8.2 Relating molecular properties to physical product properties related to gastronomy.

8.8 Conclusions

Physics-based functionalities in sensory and nutritional aspects of foods are all coupled to the physics that is related to nanostructures within the food. Food technology is therefore facilitated by knowledge and technology in the nanodomain, that is, the area of nanotechnology. Reversely, the physics (and chemistry) of the nanodomain provides opportunities for co-tailoring the sensory and nutritional and health functionalities of foods. The physics (of nanostructures) connects food, in all its identities at the same time, with its molecules.

A "flavor" of how nanotechnology can act as the connector between molecular and macroscopic phenomena related to foods is given. This accounts for large-scale production of foods taking into account sustainability, affordability, and availability issues, as well as regarding the food as produced and enjoyed in a restaurant setting.

Examples are given of possible structural morphologies within foods, that is, fibrillar structures, plate-like structures, and spherical structures, in either water or oil phase, and bicontinuous structures. These examples are a small "sample platter" to "get a flavor" for what is possible in foods using nanotechnology with (edible) biomolecules, from both a structuring point of view and a functional point of view. We also treated developments within the restaurant setting that make use of knowledge and technologies from nanoscale science. The combination of examples also aimed to give an experience of the "nano-view" toward connecting the molecular world with foods, which is key to addressing the challenges ahead for food in the future. In fact, it is argued that nanotechnology provides tools for addressing the challenges ahead regarding food in the future.

References

1 Anderson, P.W. (1995) Physics: the opening to complexity. *Proc. Natl. Acad. Sci. USA*, **92** (15), 6653–6654.

2 Heertje, I., Roijers, E.C., and Hendrickx, H.A.C.M. (1998) Liquid crystalline phases in the structuring of food products. *FLWT–Food Sci. Technol. (Lebensm. Wissensch. Technol.)*, **31** (4), 387–396.

3 Veerman, C., Ruis, H.G.M., Sagis, L.M.C., and van der Linden, E. (2002) Effect of electrostatic interactions on the percolation concentration of fibrillar β-lactoglobulin gels. *Biomacromolecules*, **3** (4), 869–873.

4 Mezzenga, R., Schurtenberger, P., Burbidge, A., and Michel, M. (2005) Understanding foods as soft materials. *Nat. Mater.*, **4** (10), 729–740.

5 McGee, H. (2004) *On Food and Cooking: The Science and Lore of the Kitchen*, completely revised and updated edition, Scribner, New York.

6 Liu, A.J., and Nagel, S.R. (1998) Nonlinear dynamics–jamming is not just cool any more. *Nature*, **396** (6706), 21–22.

7 Joly, M., and Barbu, E. (1949) Etude par la biréfringence d'ecoulement de la dénaturation thermique de la serum albumine. *Bull. Soc. Chim. Biol.*, **31** (11-1), 1642–1655.

8 Joly, M. (1949) Etude par la biréfringence d'ecoulement de la gélification de la gélatine et de ses interactions avec

d'autres substances organiques. *Bull. Soc. Chim. Biol.*, **31** (1), 105–107.

9 Joly, M., and Barbu, E. (1950) L'action des électrolytes sur l'agrégation de la serum albumine–étude par la biréfringence d'ecoulement. *Bull. Soc. Chim. Biol.*, **32** (11-1), 849–850.

10 van der Linden, E., and Foegeding, E.A. (2009) Gelation: principles, models and applications to proteins, in *Modern Biopolymer Science: Bridging the Divide between Fundamental Treatise and Industrial Applications* (eds S. Kasapis, I.T. Norton, and J.B. Ubbink), Elsevier, Academic Press, London, pp. 29–91.

11 van der Linden, E., and Venema, P. (2007) Similarities in self-assembly of proteins and surfactants: an attempt to bridge the gap, in *Food Colloids: Self-Assembly and Material Science* (eds E. Dickinson and M.E. Leser), Royal Society of Chemistry, Cambridge, UK, pp. 57–67.

12 van der Linden, E., and Venema, P. (2007) Self-assembly and aggregation of proteins. *Curr. Opin. Colloid Interface Sci.*, **12** (4–5), 158–165.

13 Aymard, P., Durand, D., and Nicolai, T. (1996) A comparison of the structure of β-lactoglobulin aggregates formed at pH 7 and pH 2. *Int. J. Polym. Anal. Charact.*, **2** (2), 115–119.

14 Aymard, P., Nicolai, T., Durand, D., and Clark, A.H. (1999) Static and dynamic scattering of β-lactoglobulin aggregates formed after heat-induced denaturation at pH 2. *Macromolecules*, **32** (8), 2542–2552.

15 Weijers, M., Sagis, L.M.C., Veerman, C., Sperber, B., and van der Linden, E. (2002) Rheology and structure of ovalbumin gels at low pH and low ionic strength. *Food Hydrocoll.*, **16** (3), 269–276.

16 Mishra, R., Sorgjerd, K., Nystrom, S., Nordigarden, A., Yu, Y.-C., and Hammarstrom, P. (2007) Lysozyme amyloidogenesis is accelerated by specific nicking and fragmentation but decelerated by intact protein binding and conversion. *J. Mol. Biol.*, **366** (3), 1029–1044.

17 Akkermans, C., Venema, P., van der Goot, A.J., Gruppen, H., Bakx, E.J., Boom, R.M., and van der Linden, E. (2008) Peptides are building blocks of heat-induced fibrillar protein aggregates of β-lactoglobulin formed at pH 2. *Biomacromolecules*, **9** (5), 1474–1479.

18 Ipsen, R., Otte, J., and Qvist, K.B. (2001) Molecular self-assembly of partially hydrolysed alpha-lactalbumin resulting in strong gels with a novel microstructure. *J. Dairy Res.*, **68** (2), 277–286.

19 Doucet, D., and Foegeding, E.A. (2005) Gel formation of peptides produced by extensive enzymatic hydrolysis of β-lactoglobulin. *Biomacromolecules*, **6** (2), 1140–1148.

20 Akkermans, C., Venema, P., van der Goot, A., Boom, R., and van der Linden, E. (2008) Enzyme-induced formation of β-lactoglobulin fibrils by AspN endoproteinase. *Food Biophys.*, **3** (4), 390–394.

21 Kroes-Nijboer, A., Venema, P., Bouman, J., and van der Linden, E. (2009) The critical aggregation concentration of β-lactoglobulin-based fibril formation. *Food Biophys.*, **4** (2), 59–63.

22 Meersman, F., and Dobson, C.M. (2006) Probing the pressure–temperature stability of amyloid fibrils provides new insights into their molecular properties. *Biochim. Biophys. Acta – Proteins Proteomics*, **1764** (3), 452–460.

23 Gosal, W.S., Clark, A.H., Pudney, P.D.A., and Ross-Murphy, S.B. (2002) Novel amyloid fibrillar networks derived from a globular protein: β-lactoglobulin. *Langmuir*, **18** (19), 7174–7181.

24 Hamada, D., and Dobson, C.M. (2002) A kinetic study of β-lactoglobulin amyloid fibril formation promoted by urea. *Protein Sci.*, **11** (10), 2417–2426.

25 Veerman, C., Baptist, H.G.M., Sagis, L.M.C., and van der Linden, E. (2003) A new multistep Ca^{2+}-induced cold gelation process for β-lactoglobulin. *J. Agric. Food Chem.*, **51** (13), 3880–3885.

26 Veerman, C., Sagis, L.M.C., and van der Linden, E. (2003) Gels at extremely low weight fractions (0.07%) formed by irreversible self-assembly of proteins. *Macromol. Biosci.*, **3** (5), 243–247.

27 van der Linden, E., and Parker, A. (2005) Elasticity due to semiflexible protein

assemblies near the critical gel concentration and beyond. *Langmuir*, **21** (21), 9792–9794.

28 Odijk, T. (1983) The statistics and dynamics of confined or entangled stiff polymers. *Macromolecules*, **16** (8), 1340–1344.

29 Morse, D.C. (2001) Tube diameter in tightly entangled solutions of semiflexible polymers. *Phys. Rev. E*, **63** (3), 031502.

30 Bot, A., den Adel, R., and Roijers, E. (2008) Fibrils of γ-oryzanol + β-sitosterol in edible oil organogels. *J. Am. Oil Chem. Soc.*, **85** (12), 1127–1134.

31 Mensink, R.P., Zock, P.L., Kester, A.D.M., and Katan, M.B. (2003) Effects of dietary fatty acids and carbohydrates on the ratio of serum total to HDL cholesterol and on serum lipids and apolipoproteins: a meta-analysis of 60 controlled trials. *Am. J. Clin. Nutr.*, **77** (5), 1146–1155.

32 Bot, A., and Agterof, W.G.M. (2006) Structuring of edible oils by mixtures of gamma-oryzanol with beta-sitosterol or related phytosterols. *J. Am. Oil Chem. Soc.*, **83** (6), 513–521.

33 Pernetti, M., van Malssen, K.F., Floter, E., and Bot, A. (2007) Structuring of edible oils by alternatives to crystalline fat. *Curr. Opin. Colloid Interface Sci.*, **12** (4–5), 221–231.

34 Marangoni, A., Rietberg, M., Laredo, T., and Kim, D. (2009) Polymer gelation of edible oils, in Delivery of Functionality in Complex Food Systems, Physically-inspired approaches from nanoscale to microscale, Abstract, 3rd International Symposium, Wageningen, 18–21 October 2009, Wageningen Academic Publishers, Wageningen, p. 7.

35 Sagis, L.M.C., de Ruiter, R., Rossier Miranda, F.J., de Ruiter, J., Schroën, K., van Aelst, A.C., Kieft, H., Boom, R.M., and van der Linden, E. (2008) Polymer microcapsules with a fiber-reinforced nanocomposite shell. *Langmuir*, **24** (5), 1608–1612.

36 Akkermans, C., van der Goot, A.J., Venema, P., van der Linden, E., and Boom, R.M. (2008) Formation of fibrillar whey protein aggregates: influence of heat and shear treatment, and resulting rheology. *Food Hydrocoll.*, **22**, 1315–1325.

37 van der Linden, E., Hogervorst, W.T., and Lekkerkerker, H.N.W. (1996) Relation between the size of lamellar droplets in onion phases and their effective surface tension. *Langmuir*, **12** (13), 3127–3130.

38 van der Linden, E., and Dröge, J.H.M. (1993) Deformability of lamellar droplets. *Physica A Stat. Mech. Appl.*, **193** (3–4), 439–447.

39 van der Linden, E., and Buytenhek, C.J. (1997) Spontaneous formation of onion phases in a single surfactant system and their salt-induced transformation towards ordinary lamellar phases. *Physica A*, **245** (1–2), 1–10.

40 van der Linden, E. (1999) Mesoscopic physics and functional properties of foods, in *Supramolecular and Colloidal Structures in Biomaterials and Biosubstrates* (eds M. Lal, P.J. Lillford, V.M. Naik, and V. Prakash), Imperial College Press, Singapore, pp. 214–223.

41 Krebs, M.R.H., Devlin, G.L., and Donald, A.M. (2007) Protein particulates: another generic form of protein aggregation? *Biophys. J.*, **92**, 1336–1342.

42 Ruis, H.G.M., Venema, P., and van der Linden, E. (2007) Relation between pH-induced stickiness and gelation behaviour of sodium caseinate aggregates as determined by light scattering and rheology. *Food Hydrocoll.*, **21** (4), 545–554.

43 Ruis, H.G.M., van Gruijthuijsen, K., Venema, P., and van der Linden, E. (2007) Transitions in structure in oil-in-water emulsions as studied by diffusing wave spectroscopy. *Langmuir*, **23**, 1007–1013.

44 Schaink, H.M., van Malssen, K.F., Morgado-Alves, S., Kalnin, D.J.E., and van der Linden, E. (2007) Crystal network for edible oil organogels: possibilities and limitations of the fatty acid and fatty alcohol systems. *Food Res. Int.*, **40** (9), 1185–1193.

45 Edelman, M.W., Tromp, R.H., and van der Linden, E. (2003) Phase-separation-induced fractionation in molar mass in aqueous mixtures of gelatin and dextran. *Phys. Rev. E*, **67** (2), 021404.

46 Edelman, M.W., van der Linden, E., and Tromp, R.H. (2000) Phase separation in gelatine–dextran mixtures, in Proceedings

of the 5th International Hydrocolloids Conference.

47 Scholten, E., Visser, J.E., Sagis, L.M.C., and van der Linden, E. (2004) Ultralow interfacial tensions in an aqueous phase-separated gelatin/dextran and gelatin/gum arabic system: a comparison. *Langmuir*, **20** (6), 2292–2297.

48 Scholten, E., Tuinier, R., Tromp, R.H., and Lekkerkerker, H.N.W. (2002) Interfacial tension of a decomposed biopolymer mixture. *Langmuir*, **18** (6), 2234–2238.

49 van den Berg, L., Carolas, A.L., van Vliet, T., van der Linden, E., van Boekel, M.A.J.S., and van de Velde, F. (2008) Energy storage controls crumbly perception in whey proteins/polysaccharide mixed gels. *Food Hydrocoll.*, **22** (7), 1404–1417.

50 van den Berg, L., Klok, H.J., van Vliet, T., van der Linden, E., van Boekel, M.A.J.S., and van de Velde, F. (2008) Quantification of a 3D structural evolution of food composites under large deformations using microrheology. *Food Hydrocoll.*, **22** (8), 1574–1583.

51 van den Berg, L., Rosenberg, Y., van Boekel, M.A.J.S., Rosenberg, M., and van de Velde, F. (2009) Microstructural features of composite whey protein/polysaccharide gels characterized at different length scales. *Food Hydrocoll.*, **23**, 1288–1298.

52 van den Berg, L., van Vliet, T., van der Linden, E., van Boekel, M., and van de Velde, F. (2008) Physical properties giving the sensory perception of whey proteins/polysaccharide gels. *Food Biophys.*, **3** (2), 198–206.

53 van den Berg, L., van Vliet, T., van der Linden, E., van Boekel, M.A.J.S., and van de Velde, F. (2007) Serum release: the hidden quality in fracturing composites. *Food Hydrocoll.*, **21** (3), 420–432.

54 van den Berg, L., van Vliet, T., van der Linden, E., van Boekel, M.A.J.S., and van de Velde, F. (2007) Breakdown properties and sensory perception of whey proteins/polysaccharide mixed gels as a function of microstructure. *Food Hydrocoll.*, **21** (5–6), 961–976.

55 Blijdenstein, T.B.J., Veerman, C., and van der Linden, E. (2004) Depletion–flocculation in oil-in-water emulsions using fibrillar protein assemblies. *Langmuir*, **20** (12), 4881–4884.

56 Parker, A. (2009) Time dependence in jamming and unjamming, PhD Thesis, Wageningen University, Wageningen.

57 Kurti, N., and This-Benckhard, H. (1994) Chemistry and physics in the kitchen. *Sci. Am.*, **270** (4), 44–49.

58 This, H. (2005) Modelling dishes and exploring culinary "precisions": the two issues of molecular gastronomy. *Br. J. Nutr.*, **93** (Suppl. 1), S139–S146.

59 This, H. (2005) A Scientific Feast [Book Review] On food and cooking: the science and lore of the kitchen, 2nd edition, by Harold McGee. *Nature*, **433** (7028), 802.

60 This, H. (2002) Molecular gastronomy. *Angew. Chem. Int. Edn.*, **41** (1), 83–88.

61 van der Linden, E., McClements, D.J., and Ubbink, J. (2008) Molecular gastronomy: a food fad or an interface for science-based cooking? *Food Biophys.*, **3**, 246–254.

62 This, H. (2005) *Molecular Gastronomy: Exploring the Science of Flavor*, Columbia University Press, New York.

63 Adrià, F., Soler, J., and Adria, A. (2008) *A Day at El Bulli*, Phaidon Press, New York.

9
Products and Their Commercialization

Betty Bugusu, Ursula Vanesa Lay Ma, and John D. Floros

9.1
Introduction

Nanotechnology has been touted as the next Industrial Revolution, with potential to impact various sectors of the economy, including food and agriculture, medicine, energy, environment, and defense. The resulting nanosciences and nanotechnologies are expected to influence all aspects of science and technology, industry, environment, and human life in general. The economic and societal promise has led to substantial and sustained investments worldwide. Nanotechnology has shown the exceptional ability to attract great interest from governments, industries, and non-governmental entities all at once. Billions of dollars have been invested into research to advance the field in all sectors. The level of investment varies among different countries. Currently, the USA is considered as the leading investor, although competition for global leadership is intensifying as countries and industries around the world increase their investments. The level of investment also varies among different economic sectors, currently led by energy and defense.

Besides governmental efforts, private industry is believed to be conducting research in nanotechnology. However, the private industry information is not publicly available, in part because of the uncertainty in consumer and public acceptance and within the regulatory arena. Current regulation of nanomaterials in food worldwide is on a "case-by-case basis" [1]. Industry is looking to regulatory bodies to provide guidance on requirements for approval of nanomaterial usage in food, while the regulators need data to establish science-based safe use recommendations. There is a general lack of knowledge on the safety of nano-derived materials and potential effects on human health and the environment.

This chapter presents information on the current use of nanoscience and nanotechnology in foods, particularly as it pertains to the state of research, development, and commercialization of new food products, processes, packages, and related applications. Our scope is to discuss the growing public and private investment in nanotechnology in general and in particular in food and agriculture, to review recent innovative research with potential applications, to evaluate the state

Nanotechnology in the Agri-Food Sector: Implications for the Future, First Edition. Edited by Lynn J. Frewer, Willem Norde, Arnout Fischer, Frans Kampers.

of product commercialization and its challenges, and finally to briefly examine current and emerging markets and market strategies for new products and technologies. Public acceptance of nanotechnology in foods is important for its success, and will also be discussed.

9.2 Investment in Nanotechnology Research

In many countries around the world, governments are supporting programs for nanotechnology research and development. In 2005, the total world funding for nanotechnology research reached $10 billion [2]. Major national efforts in nanotechnology need to be directed toward improvement of efficiency in manufacturing and use of energy resources, reduction of industry and transportation environmental impact, enhancement of healthcare, production of better pharmaceuticals, improvement of agriculture and food production, and expansion of information technologies' capabilities [3]. One of the most beneficial applications is anticipated to be in the medical field [4, 5], which is perceived by the public with great interest [5, 6]. Jackson and others [7] considered nanotechnology not as a single market, but as a collection of technologies that can change the functionality of the product and solve high-value problems in every industry by applying nanoscale materials, or processes at the nanoscale level, in product innovation.

The US government funding is coordinated through the National Nanotechnology Initiative (NNI), established in 2001 as a unified Federal funding mechanism for nanotechnology research and development [8]. The NNI has enjoyed yearly budget increases since its inception, from approximately $464 million in 2001 to an estimated $1.5 billion for 2009, and projected $1.6 billion for 2010 [9]. It is believed that private industry is investing at least as much as government. The majority of US government funding is allocated to advancing research in areas of defense and energy technologies. Allocations for food and agriculture research are minimal, an estimated total funding of about $8 million for the year 2009. The US Department of Agriculture oversees funding for food and agriculture research projects.

Canada has heavily invested in nanotechnology research through the Networks of Centres of Excellence (NCE) programs, whose purpose is to mobilize Canada's research talent in the academic, public, and private sectors for economic development and improvement of quality of life. The Advanced Foods and Materials Network (AFMNet) is responsible for the food nanotechnology research and other food-related research topics. The purpose of the network is to develop knowledge and technology that result in foods and food processes that are commercially viable, socially acceptable, and value-added [10]. They work in partnership with industry, government, not-for-profit organizations, and national and international research institutions. AFMNet has been awarded a total budget of about Canadian $39 million for the years 2003 to 2010 [11].

The European Union (EU) has been supporting nanotechnology research for many years. Between 1994 and 1998, the EU invested approximately €30 million per year in nanotechnology projects, and between 1998 and 2002, this amount increased to €45 million per year. As of 2005, public funding in Europe for nanotechnology was estimated at €400 million per year. It is estimated that the total funding may be as much as €1.2 billion if regional and private funding is considered. In addition to the EU efforts, several individual countries have invested heavily in nanotechnology, with Germany leading the way [12]. In Germany in 2005, the annual government and industry funding was about €144 million and €44 million, respectively. In 1998, the German Federal Ministry of Education and Research established six competence centers to support nanotechnology research, to communicate with the public, to connect industry and universities, to stimulate technology transfer, and to enable the commercialization of nanotechnology by domestic manufacturers. As of 2005, Switzerland was considered as the European country with the highest *per capita* funds in nanotechnology, with more than SFr40 million per year for nanotechnology research. Switzerland had the TopNano21 program from 2000 to 2003 to support entrepreneurs. The National Competence Network "Nanoscale Science" is coordinated by the University of Basel and involves public and private partners [12].

Nanotechnology research and development in the Netherlands is organized around a national nanotechnology initiative known as NanoNed,[1] which is made up of nine partners consisting of the main nanotechnology institutes in the country. The NanoNed program is expected to run until 2010 with a total budget of €235 million. The Nano4Vitality program for food and health systems was also launched in 2007 with a total budget of €12million for four years. The goal of the program is to create demand-driven nanotechnologies for these markets [13]. The program has four major themes, food safety and quality, active packaging, process technology, and encapsulation and delivery systems. Each theme has an underlying business case with an aim to apply the results within three years. Most recently, the Netherlands government invested €12 million in the national facility for research and innovation in nanotechnology, NanoLab NL [14]. The goal of the facility is to bring together public and privately funded research infrastructure and be available to outside users.

France has likewise established various programs to support nanotechnology. For example, in 1999, the French Ministry of Education and Research funded the web service "Réseau de Recherche en Micro et Nano Technologies" to encourage collaboration between the public and private sectors. In 2000, the "Action Concertée Incitative" in nanostructures was installed by the Ministry for Research. In 2003, the National Center for Scientific Research created a funding program for nanotechnology and materials development [12]. In the UK, two interdisciplinary research collaborations in nanotechnology led by the Universities of Oxford and Cambridge were awarded in 2001 with more than £18 million for six years. These funds were available through government research councils and the Ministry of

1) See http://www.nanoned.nl/.

Defence. Similarly, other European countries, such as Denmark, Finland, Italy, and Ireland, have established nanotechnology programs [12].

Asian countries have also established comprehensive investment strategies in nanotechnology. In Japan, the Ministry of Education, Culture, Sports, Science, and Technology (MEXT) is responsible for research and development and for building government–industry–academia cooperative platforms that promote nanotechnology. The Ministry of Economics, Trade, and Industries (METI) and the Ministry of Health, Labor, and Welfare (MHLW) are responsible for developing standards and regulations [15]. Investment in food nanotechnology research was conducted in part through the Ministry of Agriculture, Forestry, and Fisheries (MAFF). The Ministry has carried out a five-year project "Development of Nanotechnology and Materials for Innovative Utilizations of Biological Functions" since 2002.

The Chinese Academy of Sciences and the Ministry of Education in China have co-founded the National Center for Nanoscience and Technology (NCNST), which consists of several divisions engaged in basic and applied research in nanoscience. The divisions include: Laboratory of Nanoprocessing and Nanodevices, Laboratory of Nanomaterials and Nanostructures, Laboratory of Nanomedicine and Nanobiotechnology, Laboratory of Nanostructure Characterization and Testing, and Coordination Laboratory.[2] In 2005, the Commission on Nanotechnology Standardization, affiliated with the NCNST, was established and was given the responsibility to develop national standards, including terminology, protocol, and safety requirements for nanomaterials and nanodevices. The Commission governs and guides the assessment and authorization of nanoproducts, enabling nanotechnology industries to improve product quality, and reduce health risks associated with new product development.[2]

There are several other government efforts around the world. In 2008–2009, the Australian government supported various nanotechnology-related agencies and programs with over Australian $100 million [16].

With all these investments in place, the market shared returns for application of nanotechnologies is estimated at $1.5 trillion by the year 2015 [17]. Although the largest portion of the investment is in non-food research such as chemicals, energy, defense, electronics, and health, the food and agriculture research and development area is experiencing increasing growth. Major food companies are beginning to embrace nanotechnology. For example, in 2000 Kraft established a Nanotek Consortium, which is a collaboration of 15 universities and national research laboratories [18].

9.3 Innovations in Food and Agriculture Nanotechnology

The interest in nanotechnology from a scientific standpoint is increasing in all areas. According to Youtie and colleagues [19], there is evidence of increased

2) See http://english.nanoctr.cas.cn/.

research activity as seen in the growing trends in the number of new publications, led by the USA and followed by China, Japan, and Germany. Similarly, new patents in the last decade have increased at the rate of about 20% per year [17], with the USA in the lead, followed by Japan, Germany, Republic of Korea, China, France, UK, and Taiwan [19].

Research shows great potential for application of nanosciences and nanotechnologies in food and agriculture. In the field of agriculture, there is potential for various applications such as pesticide reduction, release systems for pesticides and fertilizers, animal tracking and identification, and so on. Nanotechnology could be used in all areas of the food supply chain to improve food quality, safety and shelf-life, to improve food processing and packaging, and to improve nutrition.

The area of food packaging is most advanced, with applications spanning across four categories as described by Chaudhry and colleagues [20]: (i) packaging materials containing nanoparticles with improved properties, such as barrier properties, temperature and/or moisture stability, and mechanical properties, which in turn may also reduce the amount of plastics used in food packaging [21]; (ii) "active" food packages incorporating nanoparticles with antimicrobial or oxygen-scavenging properties; (iii) "intelligent" food packages incorporating nanosensors to monitor and report the condition of the food; and (iv) biodegradable polymer nanomaterial composites.

Among the first nanocomposite materials in the market for food packaging applications are polymer composites incorporating clay nanoparticles. Incorporation of these particles in the packaging materials results in improved gas barrier properties because the layered structure of the nanoclay creates a more tortuous path and retards the permeation of gases through the nanocomposites [20]. Similarly, when the clay layers are completely separated and dispersed in the polymer matrix, water molecules will follow a tortuous path to migrate through the polymer matrix, resulting in reduced water permeability. In addition, the dispersed clay layers improve the overall mechanical properties of the packaging material [22].

Other polymer nanocomposite materials have been developed by incorporating metal or metal oxide nanoparticles, including silver, gold, zinc oxide, silica, titanium dioxide, alumina, and iron oxides that provide ultraviolet (UV) light absorption, abrasion resistance, or antimicrobial properties. For example, nano-silver has been used in various applications (e.g., food packaging material, inner surface of domestic refrigerators, antibacterial kitchenware and tableware) because of its antimicrobial properties. Other metal oxide nanoparticles that have shown antimicrobial properties are nano-zinc oxide and magnesium oxide [23].

Carbon nanotubes have been used mainly for non-food applications [22], but can also be used in food packaging applications to improve the mechanical properties of packaging materials [24]. Carbon nanotubes have also shown strong antimicrobial properties possibly due to severe damage to the cell membrane [25]. Carbon nanotubes have also been used to develop nanosensors, for example, nanosized carbon tubes coated with deoxyribonucleic acid (DNA) strands for detection of odors and tastes [24] and carbon nanotube-based sensors for measurement of capsaicinoid levels in chili peppers [21]. Other types of nanosensors have been developed for detection of chemicals, pathogens, and toxins present in food [24]. Some examples include: array biosensors for detection of food-borne contaminants [21], microfluidic devices for detection of pathogens [26], nanoporous silicon-based biosensor for detection of *Salmonella* and *Escherichia coli*, DNA biochips for detection of pathogens, and DNA barcodes that fluoresce under UV light in a combination of colors for simultaneous detection of different pathogens [27].

Association colloids, such as micelles and liposomes, can be used for encapsulation and delivery of polar, non-polar, and amphiphilic ingredients [22, 28]. The encapsulated compound can be located either in the core or as part of the membrane of these structures. Micelles can range in size between 5 and 100 nm, and have the advantage of being thermodynamically favored, formed by self-assembly, and produce a typically transparent solution [22]. Molecules that are not soluble or are scarcely soluble in water can be encapsulated in micelles, making possible their use in aqueous systems [28]. However, because micelle formation depends on surfactant concentration, spontaneous dissociation can occur if the solution is diluted. Another disadvantage of these structures is that the large amount of surfactant necessary for their formation may raise issues related to flavor, cost, and regulations [22]. Liposomes can range between 20 nm and a few micrometers and can be used for encapsulation of water- and lipid-soluble compounds [28].

Nano-emulsions can also be used as delivery systems. Nano-emulsions with droplet sizes of less than 100–500 nm can be produced using high-pressure valve homogenizers or microfluidizers [22]. The small droplets do not scatter light in the visible region, resulting in a clear appearance of the emulsion. Another advantage of the small droplet size is that creaming – the formation of a concentrated oil-droplet layer on the top of the emulsion – is prevented [28].

Functional components can be encapsulated in biopolymeric nanoparticles. Self-association or aggregation of single biopolymers, or phase separation of mixed biopolymer systems, can be used to form these nanoparticles. The release of the encapsulated component can be triggered as a response to specific environmental conditions to induce complete dissolution or changes in porosity of the particles [22]. For example, amylose has been studied to encapsulate conjugated linoleic acid (CLA), which is known to have various physiological properties, including anti-adipogenic, anticarcinogenic, and antidiabetogenic properties [29, 30]. The use of amylose for molecular entrapment of certain guest molecules is based on the ability of amylose chains to form a single helix with a central cavity that is large enough to include certain lipids and small molecules. These amylose–CLA inclusion complexes protected CLA from oxidation and only released it upon enzymatic hydrolysis of amylose. It was suggested that this type of encapsulation may protect polyunsaturated fatty acids during processing and storage, and only release them in the intestine during digestion by enzymatic hydrolysis [29, 30].

Product development and ingredient functionality hold the greatest potential for growth and success [31]. The drivers for the food industry are currently considered to be food quality and stability, health and nutrition, sustainability and environmental issues, and food safety [31]. Nanotechnology has a great potential to enhance flavor and help to extend the shelf-life of products, to improve nutrient delivery, and to reduce the amount of plastic in food packages. The most promising technologies for commercialization in the near future are in the area of active and smart packaging, for example, nanosensors that monitor functional indicators such as pH, quality changes in food products, and contamination by microorganisms and other contaminants.

9.4 Nanotechnology Commercialization

Nanotechnology has increased innovation in many fields. Research in nanoscience and nanotechnology has increased in the past few years, but the transition from the laboratory to commercial products with enhanced and unique properties and functionality is difficult and takes a long time. The development of manufacturing technology that allows such transition is the key to the growth of the market of "nano-enabled products". Before these products can be commercialized, governments and companies need to meet regulatory requirements, and satisfy societal needs and concerns. Thus, scale-up to commercial quantities, together with evidence of environmental and human safety during manufacture, use, and disposal of nano-enabled products, are required for successful commercialization [32]. Various factors such as creativity of individual research, training of students in nanoscale science and engineering, connections between organizations, patent regulations, physical infrastructure, legal aspects, State and Federal policies, and the international context will determine the success of nanotechnology commercialization [33].

Emerging technologies can be classified in two categories: evolutionary and disruptive technologies [34]. Evolutionary technologies arise as a stream of continuous innovation originated from existing core technical competencies and customers' suggestions or requests, resulting in replacements or improved products to satisfy existing needs [34].

Disruptive technologies are usually derived from "new science". The commercialization of products manufactured with this type of technology is frequently called "radical or discontinuous innovations" [34], where breakthroughs and significant modifications among a wide array of technologies occur during the innovation process [2]. A change in the behavior or thinking of consumers is necessary for the commercialization of products made with this type of technology. To overcome consumer resistance, companies must show the advantages provided by such technology, for example, cost reduction or improved performance [34]. In many cases, nanotechnology innovations will likely be associated with disruption [32].

To reach the commercial marketplace, research results from universities and research centers need to be transferred into the commercial sector. Industry–university collaboration, patenting, creation of start-ups, and industry-sponsored research are some ways to transfer university research to commercialization for public benefit [35]. Any commercialization strategy will require a suitable infrastructure that supports research, product development, and manufacturing of an end-product [2]. The strengths of the USA and some European infrastructure are related to the development of close groups of companies and institutions in a particular field associated by common technologies and skills [2]. In order to transform ideas into inventions, these groups need [2] to:

- generate knowledge, for example in research centers or universities;
- transform knowledge into products or new services through companies;
- provide critical components or equipment through suppliers;
- deliver the product to customers through marketing and distribution firms.

The commercial success of nanotechnology will be achieved when compelling applications are developed [4]. In 2006, Kuzma and VerHage [36] reported that 55% of research projects in agri-food nanotechnology were in applied research, 28% in basic research, and 17% in development. The same authors estimated the time to commercialization of those projects: around 20% were estimated to be commercialized in 0–5 years, around 30% in 5–10 years, another 30% approximately in the next 10–15 years, and the remaining 20% in more than 15 years. Development of new products can either be created to satisfy a market need ("pulled by the market") or to create a new market need ("push the market") [37].

9.4.1 The Path to Commercialization

9.4.1.1 Ideas and Concepts

An idea or concept is the starting point for every new product or system, innovation or invention. Sometimes, ideas or concepts are driven by need, by inspiration, by financial motivation, or simply by chance. However, most ideas do not cross the gap to become a proven concept, and research is needed to make an idea conceptually viable [2].

9.4.1.2 Research and Product Development: Design, Modeling, and Simulation

Many nanotechnology applications are still at an early stage of research and development, and, before a viable product can be developed, more basic research is necessary [4]. Knowledge of material properties and behavior at a nanoscale, where large surface-to-volume ratio dominates, is necessary for the accuracy of simulation and modeling. Design, modeling, and simulation software packages are helpful tools that can decrease the time for product development, reduce cost, and improve performance of final products. However, even though packages for molecular modeling exist, very few useful packages are available for practical purposes of manufacturing at the nano-level. Thus new modeling software and design methodologies are required for atomic- and molecular-scale processes [2].

Universities and research centers are usually the places where most new technologies are developed. Fundamental intellectual property develops also at this stage [2]. Earlier in the research and development process, the formation of partnerships between venture capitalists, industry, academia, national laboratories, and funding agencies can increase synergy in nanotechnology development. In order to achieve success in the market, major projects should include early participation of social scientists, economists, and public proponents. This will improve the potential for successful applications and ensure that the contribution and benefits are equitably distributed [3, 35].

9.4.1.3 Standardization

Standardization plays an essential role in the commercialization of emerging technologies [38]. New manufacturing methodologies cannot progress without a consensus of standards [2, 39]. However, the nanotechnology community has not yet come to an agreement on the need for standards [38], and therefore the development of internationally acceptable standards will be a challenge [2, 39]. Commercialization of previous emerging technologies, such as information and communication technology, has been favored by the development of anticipatory standards. The creation of standards before a new technology is released can increase the confidence in a new technology, promoting a more rapid adoption [38].

9.4.1.4 Safety Assessment and Regulatory Issues

Nanotechnology is currently being used in many industries without many constraints from regulatory bodies [40]. For example, in the textile industry, fabrics with nanostructured textile coatings that make the fabric wrinkle-proof or stain-repellent are currently on the market [41]. However, in other products, such as food and medical applications, approval processes and regulations need to be adhered to, and should be addressed [40], which can be lengthy and laborious [21].

Research on potential applications of nanotechnology is increasing around the world. At the same time, the number of studies suggesting toxic effects of certain engineered nanomaterials on animals and cell cultures are growing [42]. One of the main concerns is that the current information on the effect of very small nanoparticles on toxicity is limited [43]. Therefore, funding available to understand the consequences of nanomaterials on human health and the environment should increase [3, 35, 44].

In addition to studies on the toxicity of nanomaterials, risk management strategies to prevent worker exposures and to avoid broader public health problems should be developed by research institutions, industries, and government agencies [42]. Legal or policy issues need to be addressed on a global scale [3]. Given the dependence of material properties on size, governments should review whether the current regulatory environment for nanomaterials is adequate [3, 35]. As nanotechnology applications reach the marketplace, the public should be confident that governments are taking the necessary steps to protect the environment and human health, while allowing the development of new products and technologies [3, 35].

Research on the safety of nanotechnology and the development of regulations for nanotechnology is focusing primarily on non-food applications [15]. However, as more nano-enabled food-related products reach the market, concerns about potential health risks related to nanotechnology applications in food are increasing [45]. Unfortunately, the lack of published scientific studies addressing the safety of nanomaterials in food applications remains [46].

In the mean time, the Institute of Food Science and Technology (IFST) [43], a professional organization with base in London, has declared in an information statement that size matters, and the nanoparticles used for food applications should be treated as new, potentially harmful, materials. Therefore, testing to determine whether they are safe or harmful is necessary. The concern arises partly because nanoparticles might be able to reach regions within cells or tissue that their macroscopic counterparts cannot reach, leading to an increase in toxicity. Hence, according to IFST, pre-market safety evaluation of these nanoparticles is necessary, even if they are made from a compound already approved for use in food applications.

In the USA, the Food and Drug Administration (FDA) is the entity responsible for overseeing the safety of foods, food additives, and dietary supplements. Currently, in regard to FDA regulation of nanotechnology products, the FDA states that the Agency "regulates on a product-by-product basis" through pre-market and/or post-market regulations. The FDA also states that the Agency "regulates products and no technologies" and that "particle size is not the issue" because FDA has traditionally regulated products containing particles at the nanoscale. However, new tests will be required as new toxicological risks arise from new materials [47]. In a report released in July 2007, an FDA Task Force concluded that FDA regulations are generally comprehensive for products requiring "pre-market" approval, such as food and color additives. However, FDA's oversight capacity is less thorough for products that do not require pre-market authorization, such as food ingredients generally recognized as safe and dietary supplements [48]. However, the Task Force recognizes that, as size changes within the nanoscale, the safety and effectiveness of products may vary, increasing the complexity of product review [49].

Currently, there are no regulations to specify on the label that food products and food packaging contain nanoparticles. Potential benefits of the application of nanotechnology in the food industry may be jeopardized if nano-foods are allowed to come to the market without a clear definition, proper regulations, a comprehensive understanding of the risks associated with nanoparticles, and an evaluation of food safety [15, 43]. Therefore, regulatory standards for nanotechnology-based food applications should be developed, and research on the safety of nanotechnology applications in the food sector should increase [15].

9.4.1.5 **Manufacturing – Scale-Up**

Before nano-enabled products can reach the market, the challenge of achieving a robust production and manufacture at a large scale must be overcome [4]. The cost and scale of the manufacturing ramp can dramatically affect the path to commercialization. Partnership can accelerate market entry [35]. For example, in nanomedicine, large pharmaceutical corporations can help start-up companies with the high costs of drug development and manufacture, and bring products to the market [50]. Currently, in the food sector, the formation of partnerships or licensing agreements between start-up companies and larger firms has been used as a strategy to accelerate the path to commercialization (see Section 9.5).

Nanomaterials can be produced with a "top-down" or "bottom-up" approach [39, 51]. Top-down manufacturing involves processes such as etching, milling, diamond cutting, electrical discharge, and lithography to produce materials at the nanometer scale [39, 51]. Bottom-up manufacturing involves manipulation at the molecular or atomic level, by chemical synthesis, self-assembly, and positional assembly [39]. Currently, most commercially important nanoparticles, such as titanium dioxide, zinc oxide, silicon dioxide, aluminum oxide, zirconia, and iron oxide, are produced with a bottom-up approach by chemical synthesis [39]. The formation of some nanostructures in foods, such as the organization of casein micelles, protein–polysaccharides coacervates, and liposomes, is by self-assembly [51].

9.4.1.6 Final Product Realization and Marketing

For a variety of products, such as electrical or mechanical systems, normal marketing practices can be used because buyers are mainly interested in product performance and cost, but not on the technology to produce it [2]. However, when it comes to consumer products such as pharmaceuticals, cosmetics, or food, consumers are more concerned with product ingredients and how the products were manufactured. Companies are cautious about advertising the technology used to manufacture their products to avoid any adverse publicity caused by problems that occurred with other products that used the same technology [2]. For example, there are many new sunscreens currently in the market containing nanoparticles, but only their advantages are advertised, often without making any reference to the technology used to produce them [2]. A common example associated with nanotechnology is the case of biotechnology and genetically modified organisms. The problems experienced by the negative publicity of genetically modified products discouraged the use of that term in many food and healthcare products. Some of the marketing strategies used to commercialize new technologies will be described in Section 9.5.1.

9.4.1.7 Intellectual Property

Patenting is a mechanism to protect individuals' and companies' intellectual property [2]. The invention of a novel process or method, or a useful piece of equipment, can be protected by utility patents. However, patents may also make the rest of the world aware of the idea. In the USA, a patent is protected for 20 years, so others cannot make or sell the patented invention. As of 2006, more than 15 000 patents containing the word "nano" had been issued by the US Patent Office [7]. Patents are of particular importance for start-up and small companies. Patents are a way of validating a company's foundational technology [52], which may attract investors, and protect against larger corporations [52, 53].

The use of trademarks is another method to protect intellectual property. Unless trade secrets are publicly disclosed, they can be maintained in secret indefinitely, unless they are revealed by using reverse engineering, which may be difficult to do with a nanotechnology product. As of 2005, the number of registered and pending trademarks containing the word "nano" was approximately 1800 [7]. Even though trade secrets may have many advantages, venture capitalists are less likely to invest in a start-up company that relies on trade secrets instead of patents. Investments from venture capitalists will be more likely to happen if a start-up company has been able to construct adequate defenses around its intellectual property [52].

9.4.2 Challenges to Commercialization

The use of nanotechnology in the food sector is in its early stages. Currently, some commercial products are already in the market, and the market for nano-food is expected to reach US $20.4 billion in 2010 [54]. However, the introduction of new food products associated with nanotechnology will face serious challenges. Many factors will determine the commercialization of nano-related food products. Similar to the commercialization path of other nano-related products such as in nanomedicine, early stages of commercialization will face challenges such as availability of people with knowledge in nanotechnology [2], large-scale production, high production costs, intellectual property licensing, public concern and hesitation, absence of early regulatory guidelines [55], and potential environmental, health, and safety risks.

One of the first challenges to commercialization of nanotechnology is related to human resources. Because the underlying physics and engineering that regulate the behavior of nano-products differ from their counterparts at the macro-level [2], the availability of personnel with an understanding of material properties and the nanoscale is necessary.

Another challenge is the transition from scientific innovation to a productive cost-efficient technology [53] and a robust large-scale production process [4]. Partnerships can alleviate the hurdle of high cost for small and start-up companies and accelerate the path of nano-products to the marketplace.

Because nanotechnology is in its early stages, most original patents are still in effect and manufacturing companies need to negotiate numerous licenses in order to produce a product [2]. In some cases, intellectual property has to be licensed from various sources to implement an invention [52], which creates a significant burden [2]. In other cases, companies may not be able to develop new products because of another company having a patent that dominates a technology [32].

Generation of knowledge and databases about environmental and health risks of nanomaterials is critical [56]. Currently, there is little information on the properties of nanoparticles and their potential toxic effects [43]. Previous concerns about transgenic organisms and the unpredicted environmental impact of materials such as asbestos and plastics support the request for an exhaustive analysis of the environmental impact of this technology [4]. Unfortunately, the lack of information on the potential risks of nanomaterials has resulted in requests by some organizations to temporarily prohibit the use of nanomaterials in foods and cosmetics [46].

9.4.2.1 Public Acceptance and Societal Implications

One of the major challenges that commercialization of nanotechnology will face is public fear, mostly because of misinformation [56]. The survival of a new technology requires the acceptance [4], confidence, and trust of the public [20]. Public perceptions and attitudes have shaped the direction and pace of scientific activity in a number of fields, including nuclear power and genetically modified organisms [57]. Similarly, development of nano-products could be inhibited by lack of societal acceptance or rejection of nanotechnologies [58].

The International Risk Governance Council summarized several studies on the public perception of nanotechnology, in which respondents were cautious, but generally in favor of the development of nanotechnology [46]. Distrust of industry and governments to act in the public interest in terms of risk management or regulation was also common among participants, while more credibility was attributed to scientific and consumer organizations. The positive perception toward nanotechnology is not stable, and if negative information is received and believed, this attitude may change [46]. As with genetic engineering, consumers could perceive that the technology is imposed without an adequate need, understanding of risks and benefits, and regulatory control [43]. Oversimplification of the risks and magnification of concerns through fear and uncertainty of a technology can delay its commercialization, similar to what happened with the public rejection of genetically modified foods [4].

Public mistrust and suspicion arise from the lack of credible information about nanotechnology products, their potential health and environmental implications, and the oversight risk management processes [58]. In the food industry, lack of communication about what companies do and know can result in growing concerns and distrust. However, the industry's credibility can be increased or regained by an active participation of companies in the nanotechnology debate [46].

A balanced approach between societal benefits and unexpected risks and benefits should be used when judging the societal implications of nanotechnology [33]. For example, drug synthesis and delivery, medical visualization, and tissue regeneration and replacement can be significantly advanced by using nanotechnology. However, the possible risk of unwanted nanoparticles entering cells or bioincompatibility of nanostructured tissues need to be investigated [33]. Public education on the potential benefits and risks should be addressed together, so the public can have an informed outlook of nanotechnology without a polarized perception, avoiding potential overreactions [57].

Persuasion of public opinion is more difficult once a certain point of view has been established. Risks are generally less acceptable if perceived to be: involuntary, emerged from an unfamiliar source, originated from man-made rather than natural origins, causing hidden or irreversible damage, poorly understood by science, and/or subject to contradictory statements from trustworthy sources [58].

Research on public opinion about nanotechnology may reveal what different groups in society want to know about nanotechnology and its implications in their daily lives, what their concerns are, and to whom they are looking for answers [59]. This information will help to achieve an effective communication. Societal implications have been considered as an integral part of research efforts in nanotechnology. In 2003, the US National Nanotechnology Initiative annual investment in research with educational and societal implications was estimated at about $30 million, and research with environmental implications at about $50 million [33].

In a study on the public acceptance of nanotechnology in foods and food packaging [60], it was reported that it is more acceptable to the consumer if the nanotechnology application is in the packaging rather than in the food itself. However, regardless of where nanotechnology was used (i.e., food or packaging), participants were hesitant to purchase the product. Even though assessment of food nanotechnology is affected by the perceived benefits, benefits alone cannot determine the willingness of consumers to purchase nano-foods [60], or their likelihood of using nanotechnology for health or environmental applications [57]. Participants seem to be uncertain or unwilling to use or to purchase nanotechnology-based consumer products even with low perceived risk levels [57, 60].

More research is needed to understand the public perception of nanotechnology. To communicate issues like nanotechnology successfully to consumers and the general public, scientists and social scientists should work together to determine what and how to communicate [59], and to understand and address public concerns [33].

9.5
Current and Emerging Markets

Nanotechnology is moving rapidly towards the marketplace, with more than $50 billion in nano-enabled product sales worldwide in 2006 [61]. Early in 2008, new consumer products involving nanotechnology were coming to the market at a rate of three or four per week [62]. There are several applications of nanotechnology in the market, although most are from the non-food sector. Some of these products include sunscreens, cosmetics, stain-resistant fabrics, composite materials for vehicles and sports equipment, medical devices and diagnostics, drug delivery systems, fire- and water-resistant coatings [39], dietary supplements, food products, and food packaging. Most major applications are however still a few years out and food contact materials represent the largest share of the current and predicted market shares in the sector [63]. Overall, the USA seems to be the market leader, having at least three times more nano-based products on the market than other countries. The most promising areas in the food sector include active and smart packaging, health foods, and functional foods.

Even though large food corporations and more than 200 companies worldwide are involved in nanotechnology research and development [54], and the nano-food market is expected to reach $20.4 billion in 2010 [54], very little scientific data on material characterization is provided by most companies. Therefore, an actual assessment of the current market of nanotechnology applications in the food industry is challenging. Worldwide, the sales of nanotechnology products to the beverage and packaging sector jumped from $150 million in 2002 to $860 million in 2004 [15]. The value of the nano-food market doubled between 2003 and 2005, from $2.6 billion to $5.3 billion [31]. China and other Asian markets with more than 50% of the world's population are expected to have the largest growth potential.

The Project on Emerging Nanotechnologies of the Woodrow Wilson International Center for Scholars (WWICS) and the Pew Charitable Organization maintain an inventory of products in the marketplace claiming to contain nanomaterials [64]. The number of nanotechnology products introduced each year is growing tremendously. The inventory increased 279% from March 2006 to August 2008. As of 2008, the inventory had a total of 803 products spread over several categories. However, only 10% of the products fall into the category of "food and beverage". In a report in March 2008 [65], Friends of the Earth identified only 101 food-related products produced using nanotechnology, including food packaging, kitchen and cooking equipment, foods and beverages, food additives, and food and health supplements. These reports are based only on nanotechnology-related products that have been identified as such by the manufacturers. Therefore, these reports cannot provide an accurate estimate of food nanotechnology applications currently in the market [46].

Based on products currently in the market, some of the biggest areas of nanotechnology applications in food are dietary supplements [5] and food packaging [20]. Development of nanotechnology applications for the food sector focuses on: optimization or modification of sensory properties (e.g., color, flavor, texture or consistency), control of flavor and nutrient release, enhancement of nutrient or nutriceutical absorption, extension of shelf-life, nanofiltration, development of functional foods, development of foods that can be modified by the consumers depending on nutritional needs or flavor preferences, improvement of traceability and safety, and development of improved packaging materials by incorporating nanoparticles or nanosensors [20].

Because nanotechnology is considered a high-risk business, buying developed products (e.g., ingredients, packaging material) from other companies will likely be the strategy of many food companies [66]. For example, Miller Brewing is currently using plastic beer bottles produced by nanocomposite barrier technology. However, the development of this technology was a joint effort of Nanocor Inc.,[3] one of the leading nanoclay suppliers for nanocomposite plastics, and Eastman Chemical Co.,[4] one of the world leaders in polyester and copolyester, and a center of many food and beverage container innovations. Other nanoclay composites in the marketplace are Durethan® (silicate nanoparticles in polyamide), produced by Bayer for use in multilayer bottles and films, beer bottles, and so on, and Imperm®,

3) See http://www.nanocor.com/.
4) See http://www.eastman.com/Markets/Food_Beverage/Packaging.htm.

produced by Nanocor Inc. Some of Nanocor's materials are approved by the FDA and the EU.

In other cases, start-up and smaller companies are the ones providing and licensing innovations to the food industry. For example, OilFresh, a company founded in 2005 in California, uses nanoceramic catalytic pellets to make an oil conditioning device that doubles the shelf-life of the oil used in deep fryers [66].

Some start-up companies are forming partnerships or signing agreements with established firms to bring their technology to the market. For example, EcoSolutions Intl., a start-up company that uses a natural nanocapsule technology to reduce the amount of plastic material used in shopping bags, signed an agreement in early 2009 with Hymopack Ltd, the largest plastic bag manufacturer in Canada, that supplies leading retailers such as Wal-Mart and McDonalds [67].

Another example of a start-up company is NutraLease, which was created by scientists from the Hebrew University of Jerusalem [66]. NutraLease uses micelles of about 30 nm in size to encapsulate various nutriceuticals such as coenzyme Q10, lutein, lycopene, phytosterol, and vitamins D and E in order to improve their solubility and bioavailability, for use in foods and beverages. Based on the company's website, other stakeholders of NutraLease are: Yissum, the Hebrew University business arm; Ashkelon Technological Industry, one of the largest life-science investment groups in Israel; Peerless Ltd, an Australian manufacturer of premium-quality edible oils, fats and margarines; and Adumim Food Ingredients, a specialty ingredient company responsible for scaling up the process and manufacturing the products. NutraLease[5] is also working in cooperation with Shemen Industries Ltd, the largest oil manufacturer and supplier in Israel to develop a canola oil containing free phytosterols, which are known to reduce cholesterol in the blood.

In the area of encapsulation and delivery technology, another commercially available product is NovaSOL®, developed by the German company Aquanova.[6] NovaSOL is an amphiphilic liquid product containing micelles of less than 30 nm that can carry a variety of ingredients and bioactive compounds, such as vitamins, omega-3 fatty acids, phyto extracts, preservatives, colors, flavonoids, carotenoids, and so on. Because of the small size of the micelles (smaller than the wavelength of light), NovaSOL can produce clear solutions (in fact, it uses "Crystal Clear Solutions" in its product information). In addition, Aquanova claims that bioactive compounds delivered with NovaSOL have higher bioavailability than with other commercial formulations, and therefore it offers potential opportunities for functional foods and drinks.

Nanocomposite materials are considered to be at the forefront for food packaging developments, with a predicted market of one billion pounds in 2010 [68]. These materials are already in use in other industries, such as the automobile industry, for their high thermal and mechanical properties [69]. Various packaging compa-

5) See http://www.nutralease.com/index.asp.

6) See http://www.aquanova.de/media/public/pdf_produkte%20unkosher/NovaSOL_OVERVIEW.pdf.

nies have developed these materials and are ready to venture into the market. The FresherLonger™ "miracle" storage container by Sharper Image (USA) retailers is an example of a food contact product in the WWICS 2009 inventory. The product is claimed to have silver nanoparticles infused into the polypropylene base material for inhibition of microbial growth such as molds and fungi [70]. FresherLonger plastic storage bags by Sharper Image are also claimed to help delay food spoilage by decreasing the growth of micro-organisms. Other food containers developed using nano-silver that claim to have antibacterial properties are "Food Container Nano-Silver" by A-DO Global Co. Ltd (Korea), "Nano Silver Baby Mug Cup" and "Nano Silver Baby Milk Bottle" from Baby Dream Co. Ltd (Korea), "Nano-Silver Storage Box Baoxianhe" from Quan Zhou Hu Zheng Nano Technology Co. Ltd (China), and "BlueMoonGoods Fresh Box Silver Nanoparticle Food Storage Containers" from BlueMoonGoods, LLC (USA) [64] .

9.5.1 Market Strategies for New Technology Products

As described in Section 9.4, emerging technologies can be either evolutionary or disruptive, and the market strategy of products derived from each technology will be different.

9.5.1.1 Market Strategies for Evolutionary Technologies

The continuous flow of innovations derived from evolutionary technologies can be either "pulled by the market" or "pushed by the technology". Innovations pulled by the market usually originate from consumer feedback requesting a new product or improvement of an existing product. Thus, the potential buyer and their needs are known. On the other hand, innovations pushed by the technology arise from research and development of an existing technology. In these cases, the needs of the customers are identified by the manufacturer, but may not be apparent to the buyers. The invention must then be advertised as a major improvement or a new product with major advantages such as improved quality or significantly lower cost [34], or more nutritious.

9.5.1.2 Market Strategies for Disruptive Technologies

Disruptive technologies may face substantial public resistance because they emerge from "new science". Hence, considerable time, effort, and money are necessary before the disruptive technology product can succeed.

As with evolutionary technologies, market strategies for disruptive technologies can be "market pulled" or "technology pushed". Market strategies for radical innovations are more complex and time consuming than continuous innovations, because there is no existing relationship with the potential consumers. However, potential buyers of disruptive technology products may be customers of other suppliers, who may be interested in forming an agreement with a start-up company pursuing an interesting disruptive technology [34]. An example is the agreement of EcoSolutions Intl. with Hymopack Ltd, a major supplier of plastic bags, described above. Sharing of technologies between new companies and established firms has become a common practice in the pharmaceutical industry, allowing the commercialization of disruptive technology with some financial support [34]. The same strategy appears to be applied by some start-up companies in the food industry.

On the other hand, if a relationship with potential buyers already exists, companies need to identify the needs of consumers for substitute or replacement products and demonstrate the significant improved benefits. In this case, a "buyer pulled" strategy can be used [34].

In the food industry, it appears that some food companies may not be developing nanotechnology within the company. Rather, they purchase ingredients or packaging materials made using nanotechnology, and use them in their products [66]. This way the innovation may be treated as a continuous innovation, where there is an existing relationship with potential buyers, and more simple marketing strategies for evolutionary technologies can be used.

9.6 Conclusions

The use of nanoscience and nanotechnology in food and agriculture is in its early stages. Worldwide, research in a wide range of potential applications can be found in food processing, encapsulation and delivery of ingredients, sensing of pathogens, chemicals or other substances, smart, active or otherwise improved packaging, traceability, and ingredient technology. However, the transition from the laboratory to the marketplace is a long and tortuous path. Collaboration between universities, research centers, industry, funding agencies, and venture capitalists can accelerate this transition. Partnership with venture capitalists or established firms is a market strategy that start-up companies are using to bring their technology to the marketplace. Most products currently in the market or at near-commercialization stage are in the areas of food packaging, and encapsulation and delivery systems. Before products can be manufactured and commercialized, many challenges must be overcome, including: availability of personnel with a deep understanding of nanotechnology, licensing of patents, scaling-up the manufacturing process, absence of regulatory guidelines, potential environmental and health risks, and public acceptance. Public perception can shape the direction of nanotechnology commercialization, particularly in food and agriculture. Open communication between industry, government, academia, and consumer groups, development of proper regulatory standards, and more research on environmental and health risks, are some of the factors that can increase public trust and acceptance of nanotechnology applications in food and agriculture.

References

1 Bugusu, B., Mejia, C., Magnuson, B., and Tafazoli, S. (2009) Global regulatory policies on food nanotechnology. *Food Technol.*, **63** (5), 24–28.

2 Tolfree, D., and Mehalso, R. (2008) The path to commercialization, in *Commercializing Micro-Nanotechnology Products* (eds D. Tolfree and M.J. Jackson), CRC Press, Boca Raton, FL, pp. 1–28.

3 Roco, M.C., and Bainbridge, W.S. (2005) Societal implications of nanoscience and nanotechnology: maximizing human benefit. *J. Nanopart. Res.*, **7**, 1–13.

4 Mazzola, L. (2003) Commercializing nanotechnology. *Nat. Biotechnol.*, **21** (10), 1137–1143.

5 Rejeski, D. (2006) *FDA-Regulated Products Containing Nanotechnology Materials*, Woodrow Wilson International Center for Scholars, Rockville, MD.

6 Cobb, M.D., and Macoubrie, J. (2004) Public perceptions about nanotechnology: risks, benefits and trust. *J. Nanopart. Res.*, **6** (4), 395–405.

7 Jackson, M.J., Kirchhoff, B.A., and Whitfield, M.D. (2008) Technology transfer of nanotechnology products from US universities, in *Commercializing Micro-Nanotechnology Products* (eds D. Tolfree and M.J. Jackson), CRC Press, Boca Raton, FL, pp. 71–79.

8 Roco, M. (2001) From vision to implementation of the US National Nanotechnology Initiative. *J. Nanopart. Res.*, **3** (1), 5–11.

9 National Nanotechnology Initiative (2010) Funding. Available at: http://www.nano.gov/html/about/funding.html (accessed 8 November 2010).

10 Yada, R.Y., and Sheremata, L. (2008) An overview of food-related nanoscience in the Advanced Foods and Materials Network (AFMNet) and in Canada. Available from: http://www.worldfoodscience.org/cms/?pid=1004074 (accessed 8 November 2010).

11 Networks of Centres of Excellence (2010) Advanced Foods and Materials Network – AFMNet. Available from: http://www.nce.gc.ca/nces-rces/afmnet_e.htm#1 (accessed 8 November 2010).

12 Saxl, O. (2005) Nanotechnology in Europe, in *Nanotechnology Global Strategies, Industry Trends and Applications* (ed. J. Schulte), John Wiley & Sons, Ltd, Chichester, UK, pp. 45–77.

13 Kampers, F.W.H. (2008) Food nanoscience in the Netherlands. Available from: http://www.worldfoodscience.org/cms/?pid=1004071 (accessed 8 November 2010).

14 NanoNed (2010) About NanoLab NL. Available from: http://www.nanoned.nl/nanolab-nl/about-nanolab-nl.html (accessed 8 November 2010).

15 Chau, C.-F., Wu, S.-H., and Yen, G.-C. (2007) The development of regulations for food nanotechnology. *Trends Food Sci. Technol.*, **18** (5), 269–280.

16 Australian Office of Nanotechnology (2010) National Nanotechnology Strategy Annual Report 2008–09. Available from: http://www.innovation.gov.au/Industry/Nanotechnology/NationalEnablingTechnologiesStrategy/Documents/National_Nanotechnology_Strategy_AR2008-09.pdf (accessed 8 November 2010).

17 Smalley, L.W. (2009) New patent class may spur nanotech growth. *Rochester Bus. J.*, **24** (43). Available from: http://www.harrisbeach.com/files/RBJ_Smalley_Nanotech_01_16_09.pdf (accessed 8 November 2010).

18 AZoNanotechnology (2004) Food industry tuning in to nanotechnology. Available from: http://www.azonano.com/details.asp?ArticleID=855 (accessed 8 November 2010).

19 Youtie, J., Shapira, P., and Porter, A.L. (2008) Nanotechnology publications and citations by leading countries and blocs. *J. Nanopart. Res.*, **10** (6), 981–986.

20 Chaudhry, Q., Scotter, M., Blackburn, J., Ross, B., Boxall, A., Castle, L., Aitken, R., and Watkins, R. (2008) Applications and implications of nanotechnologies for the food sector. *Food Addit. Contam. A*, **25** (3), 241–258.

21 Sozer, N., and Kokini, J.L. (2009) Nanotechnology and its applications in the food sector. *Trends Biotechnol.*, **27** (2), 82–89.

22 Weiss, J., Takhistov, P., and McClements, D.J. (2006) Functional materials in food nanotechnology. *J. Food Sci.*, **71** (9), R107–R116.

23 FOODproductiondaily.com (2005) Nanotech discovery promises safer food packaging. Available from: http://www.foodproductiondaily.com/Packaging/Nanotech-discovery-promises-safer-food-packaging (accessed 8 November 2010).

24 Brody, A.L., Bugusu, B., Han, J.H., Sand, C.K., and McHugh, T.H. (2008) Innovative food packaging solutions. *J. Food Sci.*, **73** (8), R107–R116.

25 Kang, S., Pinault, M., Pfefferle, L.D., and Elimelech, M. (2007) Single-walled carbon nanotubes exhibit strong antimicrobial activity. *Langmuir*, **23** (17), 8670–8673.

26 Baeumner, A. (2004) Nanosensors identify pathogens in food. *Food Technol.*, **58** (8), 51–55.

27 Steele, B. (2005) Researchers make synthetic DNA "barcodes" to tag pathogens, providing an inexpensive, off-the-shelf monitoring system. Cornell University News Service. Available from: http://www.news.cornell.edu/stories/June05/Luo.barcodes.ws.html (accessed 8 November 2010).

28 Chen, H., Weiss, J., and Shahidi, F. (2006) Nanotechnologies in nutraceuticals and functional foods. *Food Technol.*, **60** (3), 30–36.

29 Lalush, I., Bar, H., Zakaria, I., Eichler, S., and Shimoni, E. (2005) Utilization of amylose–lipid complexes as molecular nanocapsules for conjugated linoleic acid. *Biomacromolecules*, **6**, 121–130.

30 Shimoni, E. (2008) Starch as an encapsulation material to control digestion rate in the delivery of active food components, in *Delivery and Controlled Release of Bioactives in Foods and Nutraceuticals* (ed. N. Garti), CRC Press, Boca Raton, FL, pp. 279–293.

31 Groves, K. (2008) Potential benefits of micro and nanotechnology for the food industry: does size matter? *New Food*, **11** (4), 49–52.

32 Helmus, M.N. (2006) How to commercialize nanotechnology. *Nat. Biotechnol.*, **1** (3), 157–158

33 Roco, M.C. (2003) Broader societal issues of nanotechnology. *J. Nanopart. Res.*, **5**, 181–189.

34 Kirchhoff, B.A., and Walsh, S.T. (2008) Entrepreneurship's role in commercializing micro-nanotechnology products, in *Commercializing Micro-Nanotechnology Products* (eds D. Tolfree and M.J. Jackson), CRC Press, Boca Raton, FL, pp. 29–49.

35 Roco, M.C., and Bainbridge, W.S. (eds) (2008) *Nanotechnology: Societal Implications I – Maximizing Benefits for Humanity*, Springer, Dordrecht, The Netherlands.

36 Kuzma, J., and VerHage, P. (2006) Nanotechnology in agriculture and food production: anticipated applications. Project on Emerging Nanotechnologies, Woodrow Wilson International Center for Scholars, Washington, DC. Available from: http://www.nanotechproject.org/process/assets/files/2706/94_pen4_agfood.pdf (accessed 8 November 2010).

37 Osman, T.M. *et al.* (2006) The commercialization of nanomaterials: today and tomorrow. *JOM J. Min. Met. Mater.*, **58** (4), 21–24

38 Rashba, E. (2003) Anticipatory standards and the commercialization of nanotechnology. *J. Nanopart. Res.*, **5** (3), 401.

39 Tolfree, D. (2006) Commercialising nanotechnology. Concepts–products–markets. *Int. J. Nanomanuf.*, **1** (1), 117–133.

40 Flynn, T., and Wei, C. (2005) The pathway to commercialization for nanomedicine. *Nanomed. Nanotechnol. Biol. Med.*, **1** (1), 47–51.

41 Paull, R., Wolfe, J., Hebert, P., and Sinkula, M. (2003) Investing in nanotechnology. *Nat. Biotechnol.*, **21** (10), 1144–1147.

42 Powell, M.C., and Kanarek, M.S. (2006) Nanomaterial health effects – Part 1: Background and current knowledge. *Wisc. Med. J.*, **105** (2), 16–20.

43 Institute of Food Science and Technology (2006) Information Statement: Nanotechnology. Available from: http://www.ifst.org/science_technology_resources/for_food_professionals/information_statements/ (accessed 8 November 2010).

44 Walsh, S., Balbus, J.M., Denison, R., and Florini, K. (2008) Nanotechnology: getting it right the first time. *J. Clean. Prod.*, **16**, 1018–1020.

45 Powell, M.C., and Kanarek, M.S. (2006) Nanomaterial health effects – Part 2: Uncertainties and recommendations for the future. *Wisc. Med. J.*, **105** (3), 18–23.

46 Grobe, A., Renn, O., and Jarger, A. (2008) *Risk Governance of Nanotechnology Applications in Food and Cosmetics*, International Risk Governance Council, Geneva, Switzerland.

47 Food and Drug Administration (2009) FDA Regulation of Nanotechnology Products. Available from: http://www.fda.gov/ScienceResearch/SpecialTopics/Nanotechnology/NanotechnologyTaskForce/ucm115441.htm (accessed 8 November 2010).

48 Food and Drug Administration (2007) *Nanotechnology. A Report of the US Food and Drug Administration Task Force*, US Food and Drug Administration, Rockville, MD. Available from: http://www.fda.gov/ScienceResearch/SpecialTopics/Nanotechnology/NanotechnologyTaskForceReport2007/default.htm (accessed 8 November 2010).

49 Food and Drug Administration (2007) FDA Nanotechnology Task Force Report outlines scientific, regulatory challenges. Fact Sheet. Available from: http://www.fda.gov/ScienceResearch/SpecialTopics/Nanotechnology/NanotechnologyTaskForce/ucm110934.htm (accessed 8 November 2010).

50 Wagner, V., Dullaart, A., Bock, A.-K., and Zweck, Z.A. (2006) The emerging nanomedicine landscape. *Nat. Biotechnol.*, **24** (10), 1211–1217.

51 Sanguansri, P., and Augustin, M.A. (2006) Nanoscale materials development – a food industry perspective. *Trends Food Sci. Technol.*, **17** (10), 547–556.

52 Bawa, R., Bawa, S.R., Maebius, S.B., Flynn, T., and Wei, C. (2005) Protecting new ideas and inventions in nanomedicine with patents. *Nanomed. Nanotechnol. Biol. Med.*, **1** (2), 150–158.

53 Liota, T., and Tzitzios, V. (2006) Investing in nanotechnology. *Nanotechnol. Law Bus.*, **3**, 521–531.

54 Helmut Kaiser Consultancy (2004) Study: Nanotechnology in Food and Food Processing Industry Worldwide 2008–2010–2015. Available from: http://www.hkc22.com/nanofood.html (accessed 8 November 2010).

55 Bawa, R. (2005) Will the nanomedicine "patent land grab" thwart commercialization? *Nanomed. Nanotechnol. Biol. Med.*, **1** (4), 346–350.

56 Meyyappan, M. (2008) Nanotechnology: challenges and the way forward, in *The Yearbook of Nanotechnology in Society* (eds E. Fisher, C. Selin, and J.M. Wetmore), Springer, Dordrecht, The Netherlands, pp. 227–239.

57 Currall, S.C., King, E.B., Lane, N., Madera, J., and Turner, S. (2006) What drives public acceptance of nanotechnology? *Nat. Biotechnol.*, **1**, 153–155.

58 Tegart, G. (2006) Critical issues in the commercialization of nanotechnologies. *Innovat. Manag. Pol. Pract.*, **8** (4-5), 338–347.

59 Scheufele, D.A. (2007) Nano doesn't have a marketing problem ... yet. *Nano Today*, **2** (5), 48.

60 Siegrist, M., Cousin, M.-E., Kastenholz, H., and Wiek, A. (2007) Public acceptance of nanotechnology foods and food packaging: the influence of affect and trust. *Appetite*, **49** (2), 459–466.

61 LuxResearch (2007) Nanotechnology moves from discovery to commercialization: $50 billion in 2006 product sales, $12 billion in funding. Available from: http://www.luxresearchinc.com/press/2007-lux-research-nanotech-report-5.pdf (accessed 8 November 2010).

62 Woodrow Wilson International Center for Scholars (2008) New nanotech products hitting the market at the rate of 3–4 per week. Project on Emerging Nanotechnologies. Available from: http://

www.nanotechproject.org/news/archive/6697/ (accessed 8 November 2010).

63 Food Standards Agency (2006) Assessment of current and projected applications of nanotechnology for food contact materials in relation to consumer safety and regulatory implications. Project A03063. Available from: http://www.food.gov.uk/science/research/contaminantsresearch/contactmaterials/a03prog/a03projlist/a03063/ (accessed 8 November 2010).

64 Woodrow Wilson International Center for Scholars (2009) Consumer products: an inventory of nanotechnology-based consumer products currently on the market. Project on Emerging Nanotechnologies. Available from: http://www.nanotechproject.org/inventories/consumer/ (accessed 8 November 2010).

65 Miller, G., and Senjen, D. (2008) Out of the Laboratory and On to Our Plates: Nanotechnology in Food and Agriculture, Friends of the Earth, Australia, Europe and USA. Available from: http://www.foeeurope.org/activities/nanotechnology/Documents/Nano_food_report.pdf (accessed 8 November 2010).

66 Wolfe, J. (2005) Safer and guilt-free nano foods. Forbes/Wolfe Nanotech Report, Forbes.com. Available from: http://www.forbes.com/2005/08/09/nanotechnology-kraft-hershey-cz_jw_0810soapbox_inl_print.html (accessed 8 November 2010).

67 Pryweller, J. (2009) Can a new resin formulation from Japan deflate the need for too much material in bags? Packaging Strategies newsletter. Available from: http://www.flexpackmag.com/CDA/Articles/Newsworthy/BNP_GUID_9-5-2006_A_10000000000000568231 (accessed 8 November 2010).

68 AZoNanotechnology (2004) Nanotechnology and food packaging. Available from: http://www.azonano.com/Details.asp?ArticleID=857 (accessed 8 November 2010).Ray et al 2006 to go in here (renumbered) when author supplies details

69 Ray, S., Easteal, A., Quek, S.Y., and Chen, X.D. (2006) The potential use of polymer-clay nanocomposites in food packaging. *Int. J. Food Eng.*, **2**(4), 1–11.

70 Nano Science and Technology Institute (2006) Sharper Image introduces FresherLonger™ miracle food storage containers. Business Wire. Available from: http://www.nsti.org/press/PRshow.html?id=867 (accessed 8 November 2010).

Part Four
Nanotechnology and Society

10
Toxicology of Nanomaterials in Food

Bernadene A. Magnuson and Hans Bouwmeester

10.1
Introduction

Research on the potential applications of nanotechnology and engineered nanomaterials in the areas of food-borne pathogen detection, antimicrobial activity, food packaging, food processing, food ingredient development, and nutritional studies have demonstrated the potential for nanotechnology to provide significant benefits for the consumer. These benefits may include improved food safety through enhanced detection and control of the pathogens responsible for food poisoning, enhanced shelf-life and quality of food products, and superior health-promoting or nutritional properties of foods. However, as is the case with the development of any new food processing technology, food ingredient or food packaging material, there must also be adequate studies to demonstrate that these potential benefits of nanotechnology and engineered nanomaterials designed for use in foods are not accompanied by any undesirable adverse health effects. Thus evaluation of the potential hazards related to exposure to nanomaterials and nanotechnology-based products has emerged as an important area in toxicology and risk assessment.

10.2
What Makes Nanomaterials Special?

An engineered nanomaterial (ENM) is any material that is deliberately created such that it is composed of discrete functional and structural parts, either internally or at the surface, many of which will have one or more dimensions of the order of 100 nm or less [1, 2]. Compared to bulk-scale materials, nanomaterials have a very large relative surface area (i.e., surface area per unit mass) and high particle number per unit mass, and the ratio of surface area to total number of atoms or molecules increases exponentially with decreasing particle size. The physicochemical properties of ENMs make them different from micro- and macroscale material and from dissolved chemical of the same material. So, besides

Nanotechnology in the Agri-Food Sector: Implications for the Future, First Edition. Edited by Lynn J. Frewer, Willem Norde, Arnout Fischer, Frans Kampers.

offering a wide range of novel applications, this may also give rise to altered kinetics and toxicity profiles. This will be discussed in subsequent paragraphs.

The decreased size of nanomaterials results in increased specific surface area, until the properties of the surface molecules dominate. This very high surface area has several consequences and renders them, for example, more reactive, thus generating more effective catalysts in a variety of applications [3, 4]. However, when considering potential health implications, reactive groups on the surface of nanomaterials are likely to influence their biological and/or toxicological effects. As the surface of a nanoparticle provides the initial interaction between the particle and a biological system, and therefore is a crucial determinant of particle response, these unique properties need to be investigated and understood from a physicochemical and toxicological standpoint.

10.3
Characterization of Engineered Nanomaterials

An understanding of the physicochemical properties of ENMs that impact upon their interaction with biological systems can only be gained with sufficient measurement and reporting of these properties in studies that subsequently assess biological activity. Lack of adequate nanomaterial characterization limits the value and significance of a given study and renders it impossible to compare studies and to recognize parameters that might influence biological activity (i.e., desirable effects) or toxicity (i.e., undesirable effects) [5, 6]. Thus, there has been considerable effort by the scientific community to define a minimal set of characteristics of ENMs recommended for research studies. As the outcome of the European FP7 project NanoimpactNet, a set of minimal characteristics and metrics is recommended for every field of research investigating the health impact of nanomaterials [7]:

- size distribution (of primary particles);
- chemical composition;
- nanomaterial surface (i.e., surface area, surface charge, and surface chemistry);
- structure (agglomeration state);
- shape;
- persistence.

This list is very similar to the lists of parameters that have been recommended by other authors and scientific organizations [3, 8–13]. In addition, it is generally recommended to fully characterize the nanomaterials in order to understand the potential toxicity of the nanomaterials and relate toxicity to the physicochemical properties [3, 8, 14, 15].

It is also imperative to document these parameters in the experimental exposure media (cell culture media, oral dosing solution, etc.) to the greatest extent possible, as many physicochemical parameters differ depending on whether determined in experimental media or in the bulk dry (i.e., "as-received") state.

Analytical methods are available to determine most if not all characteristics of nanomaterials [16–18]. The issue, however, is that these methods are normally only able to determine one single characteristic, making the process of full characterization very labor intensive. In addition, these methods cannot be employed to determine the characteristics directly in the food matrix. This leads to the conclusion that currently not all ENM characteristics can be readily determined. Furthermore, different techniques that are available to measure the same nanomaterial characteristic can produce contrasting results (e.g., reported sizes of ENMs) – the variations typically emerge as a result of intrinsic biases and modeling assumptions of the techniques. Agreement on standard testing methods is lacking and the comparability between various methods to assess a specific metric is still being evaluated. The challenge is initially to prioritize some metrics based on biological dose–response relations and then to develop less labor-intensive analytical methods for characterizing ENMs in biological matrices. Additionally, harmonized sample preparation procedures need to be developed.

10.3.1 Unique Issues for Characterization of Engineered Nanomaterials for Food Applications

Current and foreseen nanotechnology applications in the agri-food production chain are focused on the development of nano-sized food ingredients and additives, delivery systems for bioactive compounds, and innovative food packaging [19]. Nanomaterials in food may appear in suspension (mostly solid in liquids) or emulsion (two liquid phases). Within the agri-food chain, metal or metal oxide nanomaterials (e.g., nano-Ag, nano-ZnO, nano-Cu, nano-TiO_2) are applied in, for example, food packaging materials. Each of these different types of nanomaterial requires a different characterization approach.

The focus for safety evaluation will be on persistent nanomaterials, that is, non-soluble or non-biodegradable particles, since potential risks are predominantly associated with these types of particle. But another category of nanotechnology application in the food sector is represented by nano-encapsulates. It is particularly challenging to detect these types of nanomaterial within the food matrix and to differentiate between naturally occurring micelles and liposomes (e.g., in milk) and the deliberately created nano-encapsulates. Some analytical methods are available for this [17].

Engineered nanomaterials in food may encompass many forms. It is likely that nanomaterials are used in foods in an agglomerated form, but it cannot be excluded that these agglomerates may break down, and that the consumer may ultimately be exposed to free nanomaterials. Owing to their specific physicochemical properties, it is to be expected that nanomaterials could interact with proteins, lipids, carbohydrates, nucleic acids, ions, minerals, and water in food, feed, and biological tissues. For nanomaterials present in food, their interactions with proteins are important [20, 21]. Therefore, it is important that the nanomaterials are characterized in the relevant food matrix [22, 23].

10.4 Safety Assessment of Oral-Exposure Engineered Nanomaterials for Food Application

There are several approaches to assessing the safety of ENMs. These include: (i) investigating the toxicokinetics of nanomaterials to determine if they are absorbed into the body, and how they are handled within the body after absorption; (ii) investigating the toxicodynamics of nanomaterials to determine how they interact with tissues, cells, and cellular components; and (iii) conducting classic oral toxicity studies. A brief review of studies conducted in each of these areas will be presented below. However, to put these studies into context, it is important for the reader first to be presented with several important considerations for toxicology studies on nanomaterials.

10.4.1 Experimental Design Considerations for Toxicology Studies

The basic tenet of the study of toxicology is from Paracelsus, who wrote: "All substances are poisons; there is none that is not a poison. The right dose differentiates a poison from a remedy. The dose makes the poison." In other words, at some level of exposure, all compounds will illicit an adverse effect. Thus, to demonstrate clearly the reported toxicological properties, evidence of a dose–response is required. This means that multiple doses of the materials must be given, to see that, with increasing dose, the magnitude or incidence of an adverse effect also increases. Ultimately this leads to the derivation of a concentration at which no significant effect is observed. Traditionally, this point is called the "no observed adverse effect level" (i.e., the dose at which no adverse effects are observed), though nowadays the more efficient "benchmark dose" is often derived [24]. These reference points are used further in the risk assessment and the establishment of toxicological safety values.

Unfortunately, many of the reported toxicology studies of nanomaterials either do not provide sufficient information on the doses used, do not use more than one dose, or conclude that, because adverse effects are observed at one dose, the nanomaterial is "toxic". This is unfortunate and provides limited useful data for risk assessment purposes. In order to put the results of toxicology experiments into perspective for human health implications, more doses need to be investigated, and a rationale for the doses chosen, or a comparison with likely human exposures, should be provided when possible.

Up to now it has not been possible to establish a single dose-describing parameter that best describes the possible toxicity. It is likely that mass alone is not a good metric [25]. As discussed earlier, the characterization of the exposure is crucial. As long as it is not known which metrics should be used to describe the dose (e.g., particle size distribution, number of particles, particle charge, total surface) [26, 27], the used doses should be expressed using different dose-describing parameters. A proper definition and dose metrics will help researchers to compare study results and will help regulators to formulate health-based limit values. It will also enable risk assessors to compare and combine exposure and hazard information and to conclude on the likelihood of health risks.

The importance of good experimental design for toxicological studies of nanomaterials cannot be over-emphasized. The following factors must be considered to assess the quality of the experimental design and data resulting from experiments.

- As discussed above, adequate nanomaterial characterization in general and specifically within the surrounding matrix is clearly needed to establish metrics other than mass alone that are relevant to the toxicity of nanomaterials.
- Inclusion of positive and negative controls, as in every scientific experiment, is obviously required. Importantly, in nanotoxicology the administration of nanomaterials to the testing system needs to be accompanied by larger-sized materials and conventional forms of the materials (i.e., ions). Without these experimental groups, the studies have very much less added value to the scientific literature and are not useful for risk assessment purposes.
- Nanomaterials are known to interfere with optical and other detection measurements and to adsorb essential growth factors and nutrients from the growth medium, leading to non-specific indirect growth inhibition and apparent cytotoxicity. Therefore, adequate controls need to be used to eliminate potential interference with colorimetric and fluorometric dyes as used in cell cytotoxicity assays, interference with assays for measurement of reactive oxygen species, and alteration of the nutritional properties of the growth medium. Several authors have discussed the limitations and high likelihood of false positives of these assays [28–32], indicating that improvements in sensitivity, reliability, and sophistication, and a clear correlation with *in vivo* activity is needed in order for *in vitro* assays to yield informative data.
- One of the most important questions for the safety assessment is the sensitivity and validity of currently used test assays [25]. The question of appropriate test methods for evaluating nanomaterials has been addressed by the Organisation for Economic Co-operation and Development (OECD) in a recently published document [33]. This provides a starting point from which researchers across the globe can design testing strategies that would standardize the testing of nanomaterials. Currently, there are 118 published OECD testing guidelines covering physicochemical characterization, effects on biotic systems, degradation and/or accumulation, health effects, and other endpoints. In general, the OECD guidelines were judged to be applicable for investigating the health effects of nanomaterials [33]. An important caveat was that additional consid-

eration needs to be given to the physicochemical characteristics of the material tested (17 physicochemical properties have been suggested as the necessary prerequisite for toxicological testing), including such characteristics in the actual dosing solution. Additional pathology following certain tests was also suggested.

- Assessment for endotoxin contamination, which is exceedingly common due to the ubiquitous nature of endotoxins, is a critical step in this cascade. As endotoxin contamination generates a cellular inflammatory response, it is necessary to establish whether any inflammatory response observed in biological systems exposed to nanomaterials is due to endotoxin contamination or the nanomaterial or both [28, 29, 34].
- High variability and the cost of manufacturing a sufficient amount of nanomaterials for animal studies with uniform characteristics represents a significant hurdle for toxicity testing of some nanomaterials. The stability of nanomaterials during storage and dosing formulation must also be considered.

10.4.2 Toxicokinetics

The ability of micro- and nanomaterials to cross over the intact healthy gastrointestinal tract in humans has been recognized for over 100 years, as citations of absorption date back to the early 1900s (see review by Florence [35]). Absorption of nanomaterials through the gastrointestinal tract has also been reported in the mouse, rat, sheep, pig, and cow (see review by Florence [36]). Thus the study of the pharmacokinetics and toxicokinetics of orally administered particles is not new, and, owing to improved methods for developing nanomaterials with very specific characteristics, there is greater need for understanding the specific parameters of nanomaterials that affect their pharmacokinetics and toxicokinetics.

The absorption, distribution, metabolism, and excretion (ADME) of orally administered nanomaterials is influenced by their characteristics, such as shape, size, hydrophobicity, surface charge, and functionalized groups [26, 36–39]. However, it is unclear to what extent the different physicochemical characteristics of nanomaterials contribute to their kinetics. In this section, current knowledge on the ADME characteristics of nanomaterials that may be relevant to oral exposures is discussed.

10.4.2.1 Absorption

The gastrointestinal (GI) tract represents a port of entry for nanomaterials, not only through the ingestion of food, dietary supplements, drugs, and water that may contain nanomaterials, but also by way of ingestion of the inhaled nanomaterials that are cleared by the respiratory tract [40].

Nanomaterials may gain entry into the body by crossing the intestinal wall through the M-cells, through normal enterocytes, and/or through paracellular spaces. Uptake in M-cells, which are specialized phagocytic enterocytes found in

Peyer's patches, occurs through adsorptive endocytosis involving clathrin-coated pits and vesicles, endocytosis, and phagocytosis [36]. Uptake of particles can also occur through normal enterocytes at the apical side of the intestinal epithelial cells (by endocytosis), transport through cells, and subsequent release at the basolateral side of the epithelial cells into the lymphatic system [36, 38, 40].

Another possible uptake route for nanomaterials is via the paracellular pathway or passage between the cells [41–43]. In this pathway, also known as persorption, nanomaterials rely on the gaps and the tight junctions between the endothelial cells to pass through the epithelial cell layer. Studies have shown that the permeability of the tight junctions between the endothelium to nanomaterials can be modulated by synthetic peptides such as E-cadherin-derived peptides, which can act on the aqueous-filled pores of the paracellular pathways and expand the tight junctions [44, 45].

Uptake of nanomaterials in the GI tract depends on a variety of factors, including the diffusion of particles through mucous, initial contact with the GI epithelium, cellular trafficking, and various uptake and translocation processes, which are governed at least in part by the characteristics of the nanomaterials. Specific characteristics of nanomaterials – including particle size, surface charge, attachment of ligands or coating with surfactants, shape and elasticity, and physical and chemical stability – have been shown to influence the transcellular uptake of particles in the GI tract [43, 46–50]. Protein adsorption to engineered nanomaterials may enhance membrane crossing and cellular penetration [51–53].

Studies have demonstrated that diffusion of nanomaterials across the mucus layer depends on the size of the particles: the smaller the particle diameter, the faster they may diffuse through GI secretion to reach the colonic enterocytes [37, 54–56]. For example, when polystyrene microspheres ranging from 50 nm to 3 μm were fed by gavage to female rats at a dose of 1.25 mg per kilogram body weight, the absorption rates were highest for 50 nm particles, lower for 100 nm particles, and particles larger than 300 nm were not detectable in the blood, indicating no absorption [37]. However, increased absorption with decreased size is not always observed. For example, no significant differences in absorption or accumulation was observed in guinea pigs administered customary (10 000–90 000 nm) or nanosized (200–300 nm) sitosterol in the diet for two weeks [57]. Concentrations were measured in plasma, blood cells, bile, liver, kidney, jejunal mucosa/serosa, cecum, colon, and feces.

Diffusion across the mucous layer also depends on the surface charge of the nanomaterial. Anionic or repulsive nanomaterials have been shown to reach the epithelial surface [46], while cationic particles became entrapped in the negatively charged mucus [55]. For example, polystyrene latex nanomaterials (14 nm) rapidly adhered to the mucosal layer of the rat intestine but did not enter the epithelial cells, and were observed to move further away from epithelial cell surfaces over time [55]. A comparative study of the uptake of nanomaterials in a human intestinal cell culture model (Caco-2 cells), in a mucus-secreting cell line (NTX-E12), and *in vivo* using intra-duodenal delivery in the rat, clearly illustrated that the mucous layer of the intestine has a profound effect on uptake of certain nanomaterials [58]. Thus, the GI tract can also act as a significant barrier to systemic exposure for many nanomaterials [5].

10.4.2.2 **Distribution**

Following absorption from the GI tract, nanomaterials can reach the systemic circulation, distribute to various tissues and organs, or potentially interact with various blood components, such as plasma proteins, red or white blood cells, coagulation factors, and platelets [34, 59–62]. Variables that can affect distribution and tissue localization of nanomaterials include flow in the lymph vessels, entrapment in lymph nodes, rate of transport between lymph and blood, blood flow, adhesion to capillary walls, extravasation and movement into tissues, and cellular components within tissues [36, 39].

A number of studies report a size-dependent distribution of nanomaterials to various tissues and organs, following their uptake from the GI tract [43, 63, 64]. For example, for gold nanomaterials, clear differences in biodistribution have been observed for 10 nm compared to 50 and 250 nm nanoparticles. While the 10 nm nanoparticles were present in the liver, spleen, kidney, testis, thymus, heart, lung, and brain following an intravenous administration to rats, the 50 and 250 nm nanoparticles were present only in the liver and spleen [63] . Others have found similar trends, where smaller gold nanomaterials (15 nm) showed greater biodistribution compared to larger gold nanomaterials (50, 100, and 200 nm) [64]. Hillyer and Albrecht [43] demonstrated that, following oral administration of metallic colloidal gold nanomaterials of different sizes (58, 28, 10, and 4 nm) to mice, the smallest particles (4 nm) were identified in the kidney, liver, spleen, lungs, and brain, while the largest particles (58 nm) were detected almost solely inside the GI tract.

Although direct comparisons between the above-mentioned studies may not be possible, as there were several differences in the experimental conditions, ports of entry, or study designs, it appears that there is a trend for greater biodistribution for smaller-sized nanomaterials as compared to larger-sized nanomaterials. The extent to which nanomaterials can cross the blood–brain barrier (BBB) is not well known. Although the permeability of the BBB is highly restricted to lipophilic molecules and actively transported or small soluble molecules, evidence exists that this distribution might be relevant for some nanomaterials, as low concentrations of gold were found in the brain after oral administration of gold nanomaterials [43]. In addition, widespread distribution was observed in females administered 60 nm silver nanomaterials in a 28-day subchronic studies [65].

Binding of proteins to the surface of nanomaterials has been shown to have a significant effect on the distribution and excretion of nanomaterials, and, therefore, to influence their potential toxic effects [66]. For example, binding of nanomaterials to serum proteins resulted in a reduction in cytotoxicity of silica [59] and quantum dots [60].

10.4.2.3 **Metabolism**

There is little known on the metabolism of nanomaterials. It is unlikely that inert nanomaterials, such as gold and silver particles, fullerenes, and carbon nanotubes, can be metabolized effectively upon absorption. However, there are some indications that functional groups added to inert nanomaterials may be susceptible to metabolism. For instance, the protein cap of a functionalized quantum dot could be cleaved by proteases [67].

10.4.2.4 Excretion

Similarly, there is limited information on the excretion of orally administered nanomaterials. Clearly, nanomaterials that are not absorbed are eliminated from the body in the feces. Renal clearance was reported for fullerenes and single-walled carbon nanotubes [68, 69]. Ogawara *et al.* [70] reported that, following intravenous administration in rats, 4% of the dose of polystyrene nanomaterials (50 nm) was excreted into bile; however, larger polystyrene microparticles (500 nm) were not transported to the bile. Similarly, liposomal-based nanomaterials have been reported to be primarily eliminated through the hepatobiliary system [71, 72].

In conclusion, assessing the toxicokinetic properties of nanomaterials as compared to larger macro- or bulk-state materials can be useful in predicting the likelihood that the nano-form of the materials will have altered biological effects. For example, an increase in absorption or a change in the distribution pattern may result in increased dose of the nanomaterial at the target site for toxicity and/or may change the target site. Currently, there is an insufficient number of well-conducted studies on oral exposure to various nanomaterials to develop accurate predictive models. Studies have demonstrated that the qualitatively different physicochemical characteristics of nanoparticles, such as their relatively large and active surface area, can result in altered absorption and body distribution compared with that of bulk materials, although this depends also on surface chemistry, charge, and the specific nanomaterial under investigation.

10.4.3 Toxicodynamics

Knowledge on the potential toxicity of nanomaterials is limited but rapidly growing. There is a body of review papers available [3, 73–75] that suggest that nanomaterials may have different toxicity profiles from their bulk equivalents. The most important question for risk assessment is the sensitivity and validity of currently existing test systems. It is generally thought that the standard battery will suffice, but special attention is needed for specific endpoints [76]. Stern and McNeil [5] point out that current data support the need for a material-specific risk approach, as a generalized risk paradigm for nanomaterials is not emerging from studies evaluating biological properties in which careful and adequate characterization of materials has been reported.

Most of the work that has been done so far addresses primarily the occupational hazards associated with the manufacture and handling of nanostructured materials. Some nanomaterials may initiate catalytic reactions and increase their fire and explosion potential and could potentially present a higher risk than similar quantities of a coarser material with the same chemical composition [77, 78]. Experimental studies in rodents and cell cultures have shown that the toxicity of nanomaterials may be greater than that of the same mass of larger particles of similar chemical composition, although it is often not clear if this is truly due to the interaction of the nanomaterials with the cell or due to the interference of the nanomaterial with the assay or measurement. In addition to particle surface area, other particle characteristics may influence the toxicity, including solubility, shape, and surface chemistry [3, 78, 79].

10.4.3.1 *In Vivo* Toxicity

There are only a limited number of published *oral* toxicity studies using ENMs, mostly using insoluble metals and metal oxides. Acute, subacute, and subchronic toxicity following oral exposure have been investigated in rodents for several different nanoparticles (e.g., silver, copper, selenium, zinc and zinc oxide, and titanium dioxide nanoparticles). There is a great demand for studies using chronic oral exposure to nanomaterials combined with a broad screen for potential effects [80]. The results of the available oral toxicity studies indicate that, depending on the particle size, coating, and chemical composition of the nanoparticles, acute toxicity at high doses may occur [81–86]. In a subchronic 28-day study of 60 nm silver nanomaterial given by oral gavage at dose levels of 0, 30, 300, or 1000 mg per kilogram body weight per day, modest effects on body and organ weights, and on some blood parameters were observed in the mid- and high-dose groups. Histological studies revealed dose-dependent hyperplasia of the ventral vein in the liver [65]. The *in vivo* micronucleus test revealed no effects upon exposure. In a follow-up study, the same group [87] reported data on gender-specific silver accumulation in the kidneys of rats; however, interpretation of the findings of the study is difficult, as the treatment groups received different doses of carboxymethylcellulose and the dose of nano-silver was not reported.

It is not only the ENM itself that might trigger biological effects. Since ENMs can absorb or bind different compounds on their surfaces [88], including proteins [21], it has been speculated that a so-called "Trojan horse" effect is possible, where ENMs can act as carriers of potentially harmful chemicals and foreign substances into the organism [1]. The use of nano-encapsulates to increase the bioavailability of bioactive compounds raises similar concerns. These carrier systems might introduce unintended macromolecules, for example, undigested or unmetabolized compounds across the GI tract, leading to unknown distribution and accumulation and ultimate toxicological effects. However, clear demonstration of such an effect has yet to be reported for nanomaterials designed for food-related applications.

10.4.3.2 *In Vitro* Toxicity

Numerous *in vitro* studies using various nanomaterials are available in the scientific literature. It is beyond the scope of the current chapter to discuss all these studies. Recent reviews suggest that nanomaterials *in vitro* can trigger the release of reactive oxygen species and cause oxidative stress and subsequent inflammation by means of interaction with the reticulo-endothelial system [74, 75, 89–91]. While these results are useful for hazard identification of nanomaterials, caution has to be exercised when extrapolating results or mechanisms for the hazard characterization and subsequent human risk assessment [74]. Especially for the *in vitro* studies, a solid description and understanding of the interactions of nanomaterials with the cell culture medium is required. In addition, colorimetric techniques are frequently used as read-out systems. This might be problematic because of the interaction of the nanomaterials with the dyes used in these assays [92, 93]. Thus, while *in vitro* studies might be useful in a tiered screening approach, development of validated assays and assessment of sublethal changes, for example by means of profiling studies, are recommended [76, 94].

10.4.3.3 Study Reliability

Only a very limited number of repeated-dose oral-exposure studies are available. The quality of many studies, however, is disputable, severely limiting the use of this information for risk assessment purposes [1, 95]. For example, in most studies, only a single-sized, poorly characterized nanoparticle is used, or nanomaterials are administered at unrealistically high doses, or a narrow range of effects are generally studied [74]. Evaluation of the quality of a study and thus the reliability of the data reported is critical for risk assessment of a nanomaterial. A two-step method to objectively evaluate the reliability of safety studies of nanomaterials has recently been developed [95]. The first step utilizes a publicly available tool to rank the reliability of the study based on adequacy of design and documentation of methods, materials, and results, providing a "study score". The second step determines the completeness of physicochemical characterization of the nanomaterial(s) assessed within the study, providing a "nanomaterial score". This approach is encouraged to promote the notion that, for studies conducted with nanomaterials, the combination of a reliable study and sufficient nanomaterial characterization is of significantly greater value than either of these alone.

In addition, when evaluating the plethora of *in vitro* studies with nanomaterials, caution has to be exercised when extrapolating their results or mechanisms for hazard characterization to subsequent human risk assessment [74]. The *in vitro* studies might be suitable in searching for mechanistic explanations of toxic effects, or as screening methods in combination with profiling studies in a tiered hazard assessment approach [76, 94].

10.5 Conclusions

It is the added functionality of nanomaterials – due to a combination of their small size, physiochemical properties, chemical composition, and surface structure – that makes these materials different not only from natural small-sized particles, but also from their conventional counterparts [8, 75, 89, 96]. Because of this, unexpected toxicological effects might occur. The introduction of nanomaterial-based consumer products into the marketplace in various industrial sectors increases the urgency for a better understanding of the potential negative impacts that nanomaterials may have on biological systems. The main concerns stem from the lack of knowledge about the potential effects and impacts of nano-sized materials on human health and the environment [3, 80]. In addition to scientific risk assessment-related concerns, consumer concerns regarding a new technology such as nanotechnology application in food products are mainly related to safety issues [97].

On the other hand, potential beneficial effects of nanotechnologies are generally well described. Nanotechnologies used to improve certain properties of food products can range from the use of so-called soft nanomaterials like micelles and vesicles to encapsulate nutrients and deliver them to specific locations in the gastrointestinal tract, to the use of nano-formulated substances to improve the flow behavior of powdered foodstuffs. It is generally agreed among toxicologists that the supramolecular structures that are designed to break down within the gastrointestinal tract constitute relatively low risks, assuming that the molecules used to make these structures are safe. Also, nanomaterials that easily dissolve in water or are biodegradable will most likely not be very hazardous.

Most of the concerns of applications of nanotechnologies in food are focused on insoluble, free, and persistent nanomaterials that potentially can pass certain barriers and enter the body, and subsequently enter certain tissues or even individual cells. Because of their persistent nature, they can stay there for prolonged periods and induce harmful effects. A special, food-related case of concern is represented by nano-formulations designed to increase the bioavailability of the bulk equivalent. This might impact on the toxic profile of these compounds, and needs to be assessed. Importantly, future studies on the safety of nanomaterials must address the considerations discussed earlier in this chapter, including adequate characterization of the nanomaterial, dose metrics, method validation, and study design, to facilitate interpretation of the data and comparison of results from study to study. Only when sufficient studies of high quality are available will we achieve a greater understanding of the biological effects of nanomaterials.

Techniques in biotechnology, X-omics, and next-generation sequencing might offer valuable instruments to generate an understanding of the mechanism of biological action of nanomaterials, offering a battery of responses from biological systems (e.g., a fingerprint of the ENM in a biological matrix). In addition, combining physiochemical properties integrated with dose–response information from biokinetic and biodynamic studies should be combined in cross reading approaches, like quantitative structure–activity relationships (QSARs), to allow the prediction of the toxicity of a substance using a computer model. These *in silico* approaches are still under development for conventional chemicals and are driven by the European REACH (Registration, Evaluation, Authorization, and Restriction of Chemicals) initiative.

Globally, the scientific and industrial communities need to come together to resolve the key issues of safety of the use of nanomaterials in food. At this stage of lack of knowledge of nanotoxicology, it is unavoidable that risk assessors need as much information as possible about nanoparticles and their appearance and behavior in biological matrices and organisms. This is a prerequisite to fully exploit the benefits of nanomaterials without exposing the public to harm.

References

1 EFSA (2009) The potential risks arising from nanoscience and nanotechnologies on food and feed safety (Scientific Opinion). *EFSA J.*, **7**, 958 . Available at: http://www.efsa.europa.eu/en/efsajournal/doc/958.pdf (accessed 9 November 2010).

2 SCENIHR (2007) Opinion on the Scientific Aspects of the Existing and Proposed Definitions Relating to Products of Nanoscience and Nanotechnologies, Scientific Committee on Emerging and Newly Identified Health Risks, European Commission, Health and Consumer Protection Directorate-General, Brussels, Belgium. Available at: http://ec.europa.eu/health/ph_risk/committees/04_scenihr/docs/scenihr_o_012.pdf (accessed 9 November 2010).

3 Oberdörster, G., Maynard, A., Donaldson, K., Castranova, V., Fitzpatrick, J., Ausman, K., *et al.* (ILSI Research Foundation/Risk Science Institute Nanomaterial Toxicity Screening Working Group) (2005) Principles for characterizing the potential human health effects from exposure to nanomaterials: elements of a screening strategy. *Part. Fibre Toxicol.*, **2** (8), 1–35.

4 Fadeel, B., Kagan, V., Kagan, V., Krug, H., Shvedova, A., Svartengren, M., *et al.* (2007) There's plenty of room at the forum: potential risks and safety assessment of engineered nanomaterials. *Nanotoxicology*, **1** (2), 73–84.

5 Stern, S.T., and McNeil, S.E. (2008) Nanotechnology safety concerns revisited. *Toxicol. Sci.*, **101** (1), 4–21.

6 Warheit, D.B., Sayes, C.M., Reed, K.L., and Swain, K.A. (2008) Health effects related to nanoparticle exposures: environmental, health and safety considerations for assessing hazards and risks. *Pharmacol. Ther.*, **120** (1), 35–42.

7 Bouwmeester, H., Lynch, I., Marvin, H., Dawson, K., Berges, M., Braguer, D., *et al.* (2010) Minimal analytical characterization of engineered nanomaterials needed for hazard assessment in biological matrices. *Nanotoxicology*. doi: 10.3109/17435391003775266.

8 Powers, K.W., Brown, S.C., Krishna, V.B., Wasdo, S.C., Moudgil, B.M., and Roberts, S.M. (2006) Research strategies for safety evaluation of nanomaterials. Part VI. Characterization of nanoscale particles for toxicological evaluation. *Toxicol. Sci.*, **90** (2), 296–303.

9 Characterization Matters, The Minimum Information for Nanomaterial Characterization (MINChar) Initiative (2008) Recommended minimum physical and chemical parameters for characterizing nanomaterials on toxicology studies. Developed at Workshop on Ensuring Appropriate Material Characterization in Nanotoxicity Studies, October 28–29, Washington, DC. Available at: http://characterizationmatters.files.wordpress.com/2008/11/minchar-parameters-list.pdf (accessed 9 November 2010).

10 Gonzalez, L., Lison, D., and Kirsch-Volders, M. (2008) Genotoxicity of engineered nanomaterials: a critical review. *Nanotoxicology*, **2** (4), 252–273.

11 Warheit, D.B. (2008) How meaningful are the results of nanotoxicity studies in the absence of adequate material characterization. *Toxicol. Sci.*, **101** (2), 183–185.

12 Card, J.W., and Magnuson, B.A. (2009) Proposed minimum characterization parameters for studies on food and food-related nanomaterials. *J. Food. Sci.*, **74** (8), vi–vii.

13 NCI/NCL (2009) Assay Cascade Protocols. National Cancer Institute

(NCI), Nanotechnology Characterization Laboratory (NCL), Frederick, MD. Available at: http://ncl.cancer.gov/working_assay-cascade.asp (accessed 9 November 2010).

14 Royal Society and Royal Academy of Engineering (2004) Nanoscience and Nanotechnologies: Opportunities and Uncertainties. The Royal Society, London, UK. Available at: http://www.nanotec.org.uk/finalReport.htm (accessed 9 November 2010).

15 Thomas, K., and Sayre, P. (2005) Research strategies for safety evaluation of nanomaterials. Part I: Evaluating the human health implications of exposure to nanoscale materials. *Toxicol. Sci.*, **87** (2), 316–321.

16 Hassellov, M., Readman, J.W., Ranville, J.F., and Tiede, K. (2008) Nanoparticle analysis and characterization methodologies in environmental risk assessment of engineered nanoparticles. *Ecotoxicology*, **17** (5), 344–361.

17 Luykx, D.M., Peters, R.J., van Ruth, S.M., and Bouwmeester, H. (2008) A review of analytical methods for the identification and characterization of nano delivery systems in food. *J. Agric. Food Chem.*, **56** (18), 8231–8247.

18 Tiede, K., Boxall, A.B.A., Tiede, D., Tear, S.P., David, H., and Lewis, J. (2009) A robust size-characterisation methodology for studying nanoparticle behaviour in real environmental samples, using hydrodynamic chromatography coupled to ICP-MS. *J. Anal. At. Spectrom.*, **24** (7), 964–972.

19 Chaudhry, Q., Scotter, M., Blackburn, J., Ross, B., Boxall, A., Castle, L., *et al.* (2008) Applications and implications of nanotechnologies for the food sector. *Food Addit. Contam. A Chem. Anal. Control. Expo. Risk Assess.*, **25** (3), 241–258.

20 Linse, S., Cabaleiro-Lago, C., Xue, W.F., Lynch, I., Lindman, S., Thulin, E., *et al.* (2007) Nucleation of protein fibrillation by nanoparticles. *Proc. Natl. Acad. Sci. USA*, **104** (21), 8691–8696.

21 Lynch, I., and Dawson, K.A. (2008) Protein–nanoparticle interactions. *Nano Today*, **3** (1/2), 40–47.

22 Murdock, R.C., Braydich-Stolle, L., Schrand, A.M., Schlager, J.J., and Hussain, S.M. (2008) Characterization of nanomaterial dispersion in solution prior to *in vitro* exposure using dynamic light scattering technique. *Toxicol. Sci.*, **101** (2), 239–253.

23 Boverhof, D.R., and David, R.M. (2010) Nanomaterial characterization: considerations and needs for hazard assessment and safety evaluation. *Anal. Bioanal. Chem.*, **396** (3), 953–961.

24 EFSA (2009) Use of the benchmark dose approach in risk assessment (Scientific Opinion). *EFSA J.*, **7**, 1150. Available at: http://www.efsa.europa.eu/en/efsajournal/doc/1150.pdf (accessed 9 November 2010).

25 SCENIHR (2007) Opinion on the Appropriateness of the Risk Assessment Methodology in Accordance with the Technical Guidance Documents for New and Existing Substances for Assessing the Risks of Nanomaterials, Scientific Committee on Emerging and Newly Identified Health Risks, European Commission, Health and Consumer Protection Directorate-General, Brussels, Belgium. Available at: http://ec.europa.eu/health/ph_risk/committees/04_scenihr/docs/scenihr_o_010.pdf (accessed 9 November 2010).

26 Hagens, W.I., Oomen, A.G., De Jong, W.H., Cassee, F.R., and Sips, A.J. (2007) What do we (need to) know about the kinetic properties of nanoparticles in the body? *Regul. Toxicol. Pharmacol.*, **49** (3), 217–229.

27 Oberdörster, G., Oberdörster, E., and Oberdörster, J. (2007) Concepts of nanoparticle dose metric and response metric. *Environ. Health Perspect.*, **115** (6), A290.

28 Dobrovolskaia, M.A., Aggarwal, P., Hall, J.B., and McNeil, S.E. (2008) Preclinical studies to understand nanoparticle interaction with the immune system and its potential effects on nanoparticle biodistribution. *Mol. Pharm.*, **5** (4), 487–495.

29 Jones, C.F., and Grainger, D.W. (2009) *In vitro* assessments of nanomaterial toxicity. *Adv. Drug Deliv. Rev.*, **61** (6), 438–456.

30 Landsiedel, R., Kapp, M.D., Schulz, M., Wiench, K., and Oesch, F. (2009) Genotoxicity investigations on nanomaterials: methods, preparation and characterization of test material, potential artifacts and limitations – many questions, some answers. *Mutat. Res.*, **681** (2/3), 241–258.

31 Park, M.V., Lankveld, D.P., van Loveren, H., and De Jong, W.H. (2009) The status of *in vitro* toxicity studies in the risk assessment of nanomaterials. *Nanomedicine (London)*, **4** (6), 669–685.

32 Dobrovolskaia, M.A., Clogston, J.D., Neun, B.W., Hall, J.B., Patri, A.K., and McNeil, S.E. (2008) Method for analysis of nanoparticle hemolytic properties *in vitro*. *Nano Lett.*, **8** (8), 2180–2187.

33 OECD (2009) Preliminary Review of OECD Test Guidelines for their Applicability to Manufactured Nanomaterials, ENV/JM/MONO(2009)21. Working Party on Chemicals, Pesticides and Biotechnology, Organisation for Economic Co-operation and Development (OECD), Paris, France. Available at: http://www.oecd.org/officialdocuments/displaydocument/?doclanguage=en&cote=env/jm/mono(2009)21 (accessed 9 November 2010).

34 Hall, J.B., Dobrovolskaia, M.A., Patri, A.K., and McNeil, S.E. (2007) Characterization of nanoparticles for therapeutics. *Nanomedicine (London)*, **2** (6), 789–803.

35 Florence, A.T. (1997) The oral absorption of micro-and nanoparticulates: neither exceptional nor unusual. *Pharm. Res.*, **14** (3), 259–266.

36 Florence, A.T. (2005) Nanoparticle uptake by the oral route: fulfilling its potential? *Drug Discov. Today Technol.*, **2** (1), 75–81.

37 Jani, P., Halbert, G.W., Langridge, J., and Florence, A.T. (1990) Nanoparticle uptake by the rat gastrointestinal mucosa: quantitation and particle size dependency. *J. Pharm. Pharmacol.*, **42** (12), 821–826.

38 des Rieux, A., Fievez, V., Garinot, M., Schneider, Y.J., and Préat, V. (2006) Nanoparticles as potential oral delivery systems of proteins and vaccines: a mechanistic approach. *J. Control. Release*, **116** (1), 1–27.

39 Li, S.D., and Huang, L. (2008) Pharmacokinetics and biodistribution of nanoparticles. *Mol. Pharm.*, **5** (4), 496–504.

40 Hoet, P., Bruske-Hohlfeld, I., and Salata, O. (2004) Nanoparticles – known and unknown health risks. *J. Nanobiotechnol.*, **2** (1), 12.

41 Volkheimer, G., Schulz, F.H., Aurich, I., Strauch, S., Beuthin, K., and Wendlandt, H. (1968) Persorption of particles. *Digestion*, **1** (2), 78–80.

42 Aprahamian, M., Michel, C., Humbert, W., Devissaguet, J.P., and Damge, C. (1987) Transmucosal passage of polyalkylcyanoacrylate nanocapsules as a new drug carrier in the small intestine. *Biol. Cell*, **61** (1/2), 69–76.

43 Hillyer, J.F., and Albrecht, R.M. (2001) Gastrointestinal persorption and tissue distribution of differently sized colloidal gold nanoparticles. *J. Pharm. Sci.*, **90** (12), 1927–1936.

44 Sinaga, E., Jois, S.D., Avery, M., Makagiansar, I.T., Tambunan, U.S., Audus, K.L., *et al.* (2002) Increasing paracellular porosity by E-cadherin peptides: discovery of bulge and groove regions in the EC1-domain of E-cadherin. *Pharm. Res.*, **19** (8), 1170–1179.

45 Salamat-Miller, N., and Johnston, T.P. (2005) Current strategies used to enhance the paracellular transport of therapeutic polypeptides across the intestinal epithelium. *Int. J. Pharm.*, **294** (1/2), 201–216.

46 Jani, P., Halbert, G.W., Langridge, J., and Florence, A.T. (1989) The uptake and translocation of latex nanospheres and microspheres after oral administration to rats. *J. Pharm. Pharmacol.*, **41** (12), 809–812.

47 Hillery, A.M., Jani, P.U., and Florence, A.T. (1994) Comparative, quantitative study of lymphoid and non-lymphoid uptake of 60 nm polystyrene particles. *J. Drug Target.*, **2** (2), 151–156.

48 Hussain, N., Jani, P.U., and Florence, A.T. (1997) Enhanced oral uptake of tomato lectin-conjugated nanoparticles in the rat. *Pharm. Res.*, **14** (5), 613–618.

49 Hussain, N., and Florence, A.T. (1998) Utilizing bacterial mechanisms of epithelial cell entry: invasin-induced oral uptake of latex nanoparticles. *Pharm. Res.*, **15** (1), 153–156.

50 Araujo, L., Lobenberg, R., and Kreuter, J. (1999) Influence of the surfactant concentration on the body distribution of nanoparticles. *J. Drug Target.*, **6** (5), 373–385.

51 John, T.A., Vogel, S.M., Tiruppathi, C., Malik, A.B., and Minshall, R.D. (2003) Quantitative analysis of albumin uptake and transport in the rat microvessel endothelial monolayer. *Am. J. Physiol. Lung Cell Mol. Physiol.*, **284** (1), L187–L196.

52 Pante, N., and Kann, M. (2002) Nuclear pore complex is able to transport macromolecules with diameters of about 39 nm. *Mol. Biol. Cell*, **13** (2), 425–434.

53 John, T.A., Vogel, S.M., Minshall, R.D., Ridge, K., Tiruppathi, C., and Malik, A.B. (2001) Evidence for the role of alveolar epithelial gp60 in active transalveolar albumin transport in the rat lung. *J. Physiol.*, **533** (Part 2), 547–559.

54 Desai, M.P., Labhasetwar, V., Walter, E., Levy, R.J., and Amidon, G.L. (1997) The mechanism of uptake of biodegradable microparticles in Caco-2 cells is size dependent. *Pharm. Res.*, **14** (11), 1568–1573.

55 Szentkuti, L. (1997) Light microscopical observations on luminally administered dyes, dextrans, nanospheres and microspheres in the pre-epithelial mucous gel layer of the rat distal colon. *J. Control. Release*, **46** (2), 233–242.

56 Desai, M.P., Labhasetwar, V., Amidon, G.L., and Levy, R.J. (1996) Gastrointestinal uptake of biodegradable microparticles: effect of particle size. *Pharm. Res.*, **13** (12), 1838–1845.

57 Keller, S., Helbig, D., Härtl, A., and Jahreis, G. (2007) Nanoscale and customary non-esterified sitosterols are equally enriched in different body compartments of the guinea pig. *Mol. Nutr. Food Res.*, **51** (12), 1503–1509.

58 Behrens, I., Pena, A.I., Alonso, M.J., and Kissel, T. (2002) Comparative uptake studies of bioadhesive and non-bioadhesive nanoparticles in human intestinal cell lines and rats: the effect of mucus on particle adsorption and transport. *Pharm. Res.*, **19** (8), 1185–1193.

59 Barrett, E.G., Johnston, C., Oberdörster, G., and Finkelstein, J.N. (1999) Silica binds serum proteins resulting in a shift of the dose-response for silica-induced chemokine expression in an alveolar type II cell line. *Toxicol. Appl. Pharmacol.*, **161** (2), 111–122.

60 Lovric, J., Bazzi, H.S., Cuie, Y., Fortin, G.R., Winnik, F.M., and Maysinger, D. (2005) Differences in subcellular distribution and toxicity of green and red emitting CdTe quantum dots. *J. Mol. Med.*, **83** (5), 377–385.

61 Rothen-Rutishauser, B.M., Schurch, S., Haenni, B., Kapp, N., and Gehr, P. (2006) Interaction of fine particles and nanoparticles with red blood cells visualized with advanced microscopic techniques. *Environ. Sci. Technol.*, **40** (14), 4353–4359.

62 Aggarwal, P., Hall, J.B., McLeland, C.B., Dobrovolskaia, M.A., and McNeil, S.E. (2009) Nanoparticle interaction with plasma proteins as it relates to particle biodistribution, biocompatibility and therapeutic efficacy. *Adv. Drug Deliv. Rev.*, **61** (6), 428–437.

63 De Jong, W.H., Hagens, W.I., Krystek, P., Burger, M.C., Sips, A.J., and Geertsma, R.E. (2008) Particle size-dependent organ distribution of gold nanoparticles after intravenous administration. *Biomaterials*, **29** (12), 1912–1919.

64 Sonavane, G., Tomoda, K., Sano, A., Ohshima, H., Terada, H., and Makino, K. (2008) *In vitro* permeation of gold nanoparticles through rat skin and rat intestine: effect of particle size. *Colloids Surf. B Biointerfaces*, **65** (1), 1–10.

65 Kim, Y.S., Kim, J.S., Cho, H.S., Rha, D.S., Kim, J.M., Park J.D., *et al.* (2008) Twenty-eight-day oral toxicity, genotoxicity, and gender-related tissue distribution of silver nanoparticles in Sprague–Dawley rats. *Inhal. Toxicol.*, **20** (6), 575–583.

66 Dutta, B.J., Fifield, L.S., Jacobs, J.M., *et al.* (2007) Adsorbed proteins influence the biological activity and molecular targeting of nanomaterials. *Toxicol. Sci.*, **100** (1), 303–315.

67 Hardman, R. (2006) A toxicologic review of quantum dots: toxicity depends on physico-chemical and environmental factors. *Environ. Health Perspect.*, **114** (2), 165–172.

68 Rajagopalan, P., Wudl, F., Schinazi, R.F., and Boudinot, F.D. (1996) Pharmacokinetics of a water-soluble fullerene in rats. *Antimicrob. Agents Chemother.*, **40** (10), 2262–2265.

69 Singh, R., Pantarotto, D., Lacerda, L., Pastorin, G., Klumpp, C., Prato, M., *et al.* (2006) Tissue biodistribution and blood clearance rates of intravenously administered carbon nanotube radiotracers. *Proc. Natl. Acad. Sci. USA*, **103** (9), 3357–3362.

70 Ogawara, K.Y.M., Furumoto, K., Takakura, Y., Hashida, M., Higaki, K., and Kimura, T. (1999) Uptake by hepatocytes and biliary excretion of intravenously administered polystyrene microspheres in rats. *J. Drug Target.*, **7** (3), 213–221.

71 Ishida, T., Harashima, H., and Kiwada, H. (2002) Liposome clearance. *Biosci. Rep.*, **22** (2), 197–1224.

72 Longmire, M., Choyke, P.L., and Kobayashi, H. (2008) Dendrimer-based contrast agents for molecular imaging (abstract only). *Curr. Top. Med. Chem.*, **8** (14), 1180–1186.

73 Donaldson, K., Stone, V., Clouter, A., Renwick, L., and MacNee, W. (2001) Ultrafine particles. *Occup. Environ. Med.*, **58** (3), 211–216.

74 Oberdörster, G., Stone, V., and Donaldson, K. (2007) Toxicology of nanoparticles: a historical perspective. *Nanotoxicology*, **1** (1), 2–25.

75 Nel, A., Xia, T., Mädler, L., and Li, N. (2006) Toxic potential of materials at the nanolevel. *Science*, **311** (5761), 622–627.

76 Balbus, J.M., Maynard, A.D., Colvin, V.L., Castranova, V., Daston, G.P., Denison, R.A., *et al.* (2007) Meeting report: hazard assessment for nanoparticles – report from an interdisciplinary workshop. *Environ. Health Perspect.*, **115** (11), 1654–1659.

77 Oberdörster, G., Ferin, J., and Lehnert, B.E. (1994) Correlation between particle size, *in vivo* particle persistence, and lung injury. *Environ. Health Perspect.*, **102** (Suppl. 5), 173–179.

78 Duffin, R., Tran, C.L., Clouter, A., Brown, D.M., MacNee, W., Stone, V., *et al.* (2002) The importance of surface area and specific reactivity in the acute pulmonary inflammatory response to particles. *Ann. Occup. Hyg.*, **46** (Suppl. 1), 242–245.

79 Donaldson, K., Aitken, R., Tran, L., Stone, V., Duffin, R., Forrest, G., *et al.* (2006) Carbon nanotubes: a review of their properties in relation to pulmonary toxicology and workplace safety. *Toxicol. Sci.*, **92** (1), 5–22.

80 Bouwmeester, H., Dekkers, S., Noordam, M.Y., Hagens, W.I., Bulder, A.S., de Heer, C., *et al.* (2009) Review of health safety aspects of nanotechnologies in food production. *Regul. Toxicol. Pharmacol.*, **53** (1), 52–62.

81 Jia, X., Li, N., and Chen, J. (2005) A subchronic toxicity study of elemental nano-Se in Sprague–Dawley rats. *Life Sci.*, **76** (17), 1989–2003.

82 Zhang, J., Wang, H., Yan, X., and Zhang, L. (2005) Comparison of short-term toxicity between nano-Se and selenite in mice. *Life Sci.*, **76** (10), 1099–1109.

83 Chen, Z., Meng, H., Xing, G., Chen, C., Zhao, Y., Jia, G., *et al.* (2006) Acute toxicological effects of copper nanoparticles *in vivo*. *Toxicol. Lett.*, **163** (2), 109–120.

84 Wang, B., Feng, W.Y., Wang, M., Wang, T.C., Gu, Y.Q., Zhu, M.T., *et al.* (2008) Acute toxicological impact of nano- and submicro-scaled zinc oxide powder on healthy adult mice. *J. Nanopart. Res.*, **10** (2), 263–276.

85 Wang, B., Feng, W.-Y., Wang, T.-C., Jia, G., Wang, M., Shi, J.-W., *et al.* (2006) Acute toxicity of nano- and micro-scale zinc powder in healthy adult mice. *Toxicol. Lett.*, **161** (2), 115–123.

86 Wang, J., Zhou, G., Chen, C., Yu, H., Wang, T., Ma, Y., *et al.* (2007) Acute toxicity and biodistribution of different sized titanium dioxide particles in mice after oral administration. *Toxicol. Lett.*, **168** (2), 176–185.

87 Kim, W.Y., Kim, J., Park, J.D., Ryu, H.Y., and Yu, I.J. (2009) Histological study of gender differences in accumulation of

silver nanoparticles in kidneys of Fischer 344 rats. *J. Toxicol. Environ. Health A*, **72** (21/22), 1279–1284.

88 Šimon, P., and Joner, E. (2008) Conceivable interactions of biopersistent nanoparticles with food matrix and living systems following from their physicochemical properties. *J. Food Nutr. Res.*, **47** (2), 51–59.

89 Nel, A.E., Madler, L., Velegol, D., Xia, T., Hoek, E.M., Somasundaran, P., *et al.* (2009) Understanding biophysicochemical interactions at the nano–bio interface. *Nat. Mater.*, **8** (7), 543–557.

90 Donaldson, K., and Seaton, A. (2007) The Janus faces of nanoparticles. *J. Nanosci. Nanotechnol.*, **7** (12), 4607–4611.

91 Donaldson, K., Faus, S., Borm, P.J.A., and Stone, V. (2007) Approaches to the toxicological testing of particles, in *Particle Toxicology* (eds K. Donaldson and P.J.A. Borm), CRC Press/Taylor and Francis, London, pp. 299–316.

92 Worle-Knirsch, J.M., Pulskamp, K., and Krug, H.F. (2006) Oops they did it again! Carbon nanotubes hoax scientists in viability assays. *Nano Lett.*, **6** (6), 1261–1268.

93 Casey, A., Herzog, E., Davoren, M., Lyng, F.M., Byrne, H.J., and Chambers, G. (2007) Spectroscopic analysis confirms the interactions between single walled carbon nanotubes and various dyes commonly used to assess cytotoxicity. *Carbon*, **45** (7), 1425–1432.

94 Lewinski, N., Colvin, V., and Drezek, R. (2008) Cytotoxicity of nanoparticles. *Small*, **4** (1), 26–49.

95 Card, J.W., and Magnuson, B.A. (2010) A method to assess the quality of studies that examine the toxicity of engineered nanomaterials. *Int. J. Toxicol.*, **29** (4), 402–410.

96 Rogers, N.J., Franklin, N.M., Apte, S.C., and Batley, G.E. (2007) The importance of physical and chemical characterization in nanoparticle toxicity studies. *Integr. Environ. Assess. Manage.*, **3** (2), 303–304.

97 Siegrist, M., Stampfli, N., Kastenholz, H., and Keller, C. (2008) Perceived risks and perceived benefits of different nanotechnology foods and nanotechnology food packaging. *Appetite*, **51** (2), 283–290.

11
Nanomaterials in Food and Food Contact Materials – Potential Implications for Consumer Safety and Regulatory Controls

Qasim Chaudhry, Laurence Castle, and Richard Watkins

11.1
Background

Recent advances in nanosciences and nanotechnologies have led to great interest in the study and potential manipulation of the properties of materials and substances at the nanoscale. Like other sectors, the new technological developments are promising to revolutionize the food sector – from production to processing, packaging, distribution, storage and consumption. The main focus of research and development in the food processing area relates to the development of processed nanostructures in food, and nano-sized food additives. The use of engineered nanomaterials to improve properties of plastic polymers has opened up another major application area for the development of innovative food packaging materials. Similar developments in the agricultural sector, although mainly at research and development stage at present, could offer many more potentially large-scale applications of nanotechnologies for food production. Despite the promise of enormous benefits, such developments have also raised a number of safety, ethical, policy, and regulatory questions. In particular, the likelihood of consumer exposure to potentially harmful engineered nanomaterials through consumption of nano-enabled foods and drinks has led to calls for a moratorium, or an outright ban, on the use of nanotechnologies until they are proven to be safe to consumers and the environment [1–3].

As for conventional substances, any risk to consumers from the use of engineered nanomaterials will be dependent on the toxicological properties of the materials, as well as the likelihood, extent, and frequency of any exposure. This will inevitably depend on the properties of the engineered nanomaterials used, and the nature of each application. In some applications, for example in food packaging, engineered nanomaterials may be incorporated in a fixed, bound or embedded form, and thus may not pose a significant risk of exposure to the consumer. Other applications may, on the other hand, contain free nanoparticles and therefore pose a relatively greater risk to the consumer.

Nanotechnology in the Agri-Food Sector: Implications for the Future, First Edition. Edited by Lynn J. Frewer, Willem Norde, Arnout Fischer, Frans Kampers.

This chapter discusses the potential implications of nanotechnology applications for consumer safety in the light of the different existing and anticipated applications in the food and related sectors. The chapter also discusses the relevance of existing regulatory frameworks in relation to controlling the potential risks of nanotechnology applications to the consumer.

11.2 Nanomaterials Likely to be Used in Food and Related Applications

Based on the available information, the engineered nanomaterials likely to be used in nano-enabled food products fall into three main categories: inorganic, surface functionalized, and organic engineered nanomaterials [4]. In addition to the deliberately manufactured engineered nanomaterials, there is also a possibility that some micronized materials may also contain a nanoscale fraction due to a natural variation in the size range of manufactured materials [5]. Some engineered nanomaterials may also end up in food products as a result of environmental contamination, migration from packaging, contact with active surfaces, or from the use of nano-sized agrochemicals (e.g., pesticides or veterinary medicines).

11.2.1 Inorganic Nanomaterials

A number of inorganic engineered nanomaterials are known to be used in food and health food products and food packaging applications. These include engineered nanomaterials of transition metals (such as silver, titanium dioxide and iron), alkaline earth metals (such as calcium and magnesium), and non-metals (such as selenium and silicates) [4, 6]. Food packaging is currently the major area of application of inorganic metal and metal oxide engineered nanomaterials. Example applications include plastic polymers with nanoclay as a gas barrier, nano-silver and nano-zinc oxide for antimicrobial action, nano-titanium dioxide for ultraviolet (UV) protection, nano-titanium nitride as a processing aid, and nano-silica for surface coating.

Nano-silver is increasingly used as an antimicrobial, antiodorant, and (proclaimed) health supplement. Although the current use of nano-silver relates mainly to health food and packaging applications, its use as an additive in antibacterial wheat flour is the subject of a recent patent application [7].

Amorphous silica has been used for many years in food applications, such as in clearing of beers and wines, and as a free-flowing agent in powdered soups. The conventional bulk form of silica is a permitted food additive (SiO_2 INS 551). Porous silica is used in nano-filtration to remove undesired components in food and beverages – such as undesirable tastes in some plant extracts. Amorphous nano-silica is also known to be used in food contact surfaces and food packaging applications.

The conventional bulk form of titanium dioxide is already approved as an additive for food use (TiO_2 INS 171). Nano-titanium dioxide is currently used in a number of consumer products (e.g., paints, coatings, cosmetics, water treatment) but its use may extend to foodstuffs. For example, a US Patent (US 5741505) describes the potential application of nanoscale inorganic coatings directly on food surfaces to provide a barrier to moisture and oxygen to improve shelf-life, and/or the flavor impact of foods. The materials described for the nano-coatings, which are intended to be applied in a continuous process as a thin amorphous film of 50 nm or less, include titanium dioxide, along with silicon dioxide and magnesium oxide. The main intended applications described in the patent include confectionary products. However, to our knowledge, this technology has not so far been used in any commercial application.

Nano-iron is available commercially as a health supplement. Zero-valent nano-iron is also used in the treatment of contaminated water, where it is claimed to decontaminate water by breaking down organic pollutants and killing microbial pathogens. Nano-selenium is being marketed as an additive to a green tea product in China, with a number of (proclaimed) health benefits resulting from the enhanced uptake of selenium. Nano-calcium salts are subject to patent applications for intended uses in chewing gums (WO/2004/028262, and US Patent 20060034975 – Coated Chewing Gum – Sustech GmbH & Co. KG, Darmstadt, Germany). Nano-calcium and nano-magnesium salts are also available commercially as health supplements.

11.2.2
Surface-Functionalized Nanomaterials

Surface-functionalized nanomaterials are the second-generation engineered nanomaterials that can add certain functionality to the matrix, such as antimicrobial activity or a preservative action, for example, through absorption of oxygen. For food packaging materials, functionalized engineered nanomaterials are used to bind with the polymer matrix to offer mechanical strength or a barrier against movement of gases, volatile components (such as flavors) or moisture. One such example is the use of functionalized nanoclays to develop food packaging materials with enhanced gas barrier properties. The nanoclay mineral is mainly montmorillonite (also termed bentonite), which is a natural clay obtained from volcanic ash/rocks. Nanoclay has a natural nanoscale layer structure and is organically modified to bind to polymer matrices. Compared to unmodified engineered nanomaterials, the surface-functionalized engineered nanomaterials are more likely to react with different food components, and therefore become bound to food matrices. They are thus less likely to be available in free particulate forms in food, and also unlikely to migrate from packaging materials.

11.2.3
Organic Nanomaterials

A number of organic nano-sized materials, many of them naturally occurring, have been developed for use in food and feed products. These include vitamins, antioxidants, colorants, flavoring agents, and preservatives, which may be encapsulated in nano-delivery systems. The main proclaimed benefits of using a nano-sized organic additive over conventional forms are better dispersion of insoluble substances in foodstuffs without the need for additional fat, increased uptake and absorption, and improved bioavailability in the body. There are a wide range of available food additives (e.g., benzoic acid, citric acid, ascorbic acid) and supplements (e.g., vitamins A and E, isoflavones, β-carotene, lutein, omega-3 fatty acids, coenzyme-Q10). A synthetic nano-sized water-dispersible form of lycopene, a naturally occurring carotenoid in tomatoes, is also available commercially with a reported particle size in the range of 100 nm. Lycopene has been notified as of GRAS (generally regarded as safe) status to the Food and Drug Administration (FDA) in the USA (GRAS Notice GRN000119/2002), and a recent European Food Safety Authority (EFSA) opinion has also considered its use in food and beverages as safe [8]. However, the evaluations by both EFSA and the Joint FAO/WHO Expert Committee on Food Additives (JECFA) do not include the use of nano-sized forms. It is not known if the nano-form of the material is currently used in any food or beverage product.

11.3
Potential Consumer Safety Implications

It is known that materials manufactured at a nanometer scale can behave differently from their conventional equivalents, in terms of physicochemical properties, behavior, and interactions with other substances. For example, properties of engineered nanomaterials in the lower range of the nanometer scale are likely to be influenced by quantum effects compared to larger-sized equivalents. It is also possible that such changes in properties can lead to a significant deviation in the anticipated effects and impacts of engineered nanomaterials on biological systems. Studies have already suggested a change in the toxicity profile for some engineered nanomaterials compared to conventional equivalents [9, 10]. For example, exposure to some engineered nanomaterials has been shown to cause induction of oxyradical generation in both *in vitro* and *in vivo* studies, which may lead to oxidative stress and inflammatory reactions [11–13].

Another important aspect in relation to the potential harmful effects of engineered nanomaterials is their ability (especially of free nanoparticles) to penetrate biological membranes that act as barriers to the entry of particulate substances into cells and tissues [14, 15]. This adds a new dimension to the toxicology of particulate materials, as certain insoluble and potentially reactive or biopersistent nanoparticles may reach new targets in the body, where the entry of larger equivalents would be restricted [16–22].

Owing to their enormous surface free energies, engineered nanoparticles can adsorb or bind different compounds and moieties on their surfaces [23]. This, combined with their ability to cross cellular barriers, poses a potential risk of such particles acting as a carrier of harmful substances into the circulatory system, and to other unintended organs in the body. Depending on the surface chemistry, it is also known that systemically introduced engineered nanomaterials can interact with various biological entities, such as plasma proteins, platelets, and cells [23, 24]. In biological environments, nanoparticles may become coated with different biomolecules, especially proteins [25]. Such coatings may also direct them to specific parts of the body. For example, coating of nanoparticles with apolipoprotein E has been associated with their transport to the brain [26]. This suggests that engineered nanomaterials can undergo various interactions and transformations in both food and biological systems, which can influence their absorption, distribution, metabolism, and excretion (ADME) properties [27], and may lead to a deviation in their biological effects.

While there is a growing literature on the inhalation toxicity of engineered nanomaterials, only a limited number of studies have been carried out so far on the translocation and distribution of nanoparticles to various organs and tissues, and their effects following oral administration. The diffusion rate of particulate materials through the gastrointestinal (GI) mucus is reported to be dependent on a number of factors, such as size, charge [28], and surface coating [29]. The translocation from the gastrointestinal tract has been found to be greater for nano-sized particles than for larger ones [20, 30]. Engineered nanomaterials in the smaller nanometer range have also been found to cross the mucus layer faster than the larger ones [17, 18]. Within the gastrointestinal tract, the rate of uptake of nanoparticles has been reported to be between 2 and 200 times greater in the Peyer's patches compared to that in the enterocytes [20], despite the fact that Peyer's patches only represent around 1% of the total intestinal surface area.

Jani *et al.* [22] demonstrated that titanium dioxide nanoparticles (rutile, 500 nm) translocated to systemic organs, such as the liver and spleen, following oral gavage (forced feeding) for 10 days to female Sprague–Dawley rats. The nanoparticles were also detected in the lungs and peritoneal tissues, but not in the heart and kidney. Oral administration of colloidal gold nanoparticles (58, 28, 10, and 4 nm) to mice has been shown to result in an increasing distribution of smaller nanoparticles to different organs [17]. Following repeated oral administration of nanosilver (60 nm) at different dose levels, accumulation of the nanoparticles has been observed in the GI tract of Sprague–Dawley rats, followed by the kidney and liver, lungs, testes, brain and blood [31].

The insoluble, biopersistent engineered nanomaterials are typically taken up by the M-cells of Peyer's patches and passed to underlying macrophages, where they accumulate and appear as pigmentation in cells at the base of human intestinal lymphoid aggregates [32]. The few studies carried out so far have not found a clear association between dietary particulates (micro- or ultra-fine) with the initiation or exacerbation of gut diseases, such as Crohn's disease or irritable bowel syndrome [33, 34]. In an *in vitro* study on human epithelial cell cultures, Chen and von Mikecz [35] have shown that fluorescently labeled silica nanoparticles smaller than 70 nm could enter cell nuclei. The study also found protein accumulation in the nuclei and indication for impairment of deoxyribonucleic acid (DNA) replication and transcription. However, the relevance of the findings to potential *in vivo* effects of orally ingested nano-silica is not certain. Some engineered nanomaterials, such as nano-silver, are also known to have strong antimicrobial activity. The ingestion of such engineered nanomaterials via food may have a deleterious effect on the gut natural microflora. However, there is no published research at present on the potential effects of nano-silver or other antimicrobial engineered nanomaterials on the gut microflora.

A healthy digestive system allows the absorption of nutrients from the gastrointestinal tract after digestion of the food components. The gut wall is designed to let the digested dietary nutrients through, but to prevent translocation of larger-sized materials or foreign substances into the circulatory system. Our food is composed of complex natural polymers, such as proteins, carbohydrates, and fats. Many of these food components either exist, or are metabolized, at a nanoscale. In that sense, it seems that our bodies are used to dealing with nanostructures all the time. However, most of these materials are those that are either digested in the gastrointestinal tract, or are excreted from the body, and hence are not biopersistent. Many engineered nanomaterials falling in such "soft" categories would be expected to be dealt with in the body in a similar way as the conventional equivalents, and therefore should not pose any different risk to the consumer. The exception may be those engineered nanomaterials that are composed of a harmful substance, but such materials are highly unlikely to be used for food applications.

The consumer safety concerns relate mainly to the potential exposure to those engineered nanomaterials that are insoluble, indigestible, and biopersistent. If such "hard" nanomaterials are used for food applications, consumption of the food products will provide a direct route of entry of the engineered nanomaterials into the body. Although the potential use of "hard" engineered nanomaterials has raised a number of consumer safety concerns, there are major knowledge gaps in this area at present to allow an adequate assessment of any risk. For example, the behavior, interactions, fate, and effects of most engineered nanomaterials inside and outside the gastrointestinal tract are currently not known. It is possible that many "hard" engineered nanomaterials, when added to food products, will not remain in a free particulate form due to agglomeration, aggregation, binding with other food components, transformations due to reaction with stomach acid or digestive enzymes, and so on, and hence will not be available for translocation from the gastrointestinal tract. The lack of validated methodologies for detection and characterization of engineered nanomaterials in food matrices is also currently a major barrier in regard to further developments in this area.

11.4
Current and Projected Applications for Food

A number of recent reports and reviews have highlighted the current and projected uses of engineered nanomaterials for food and food packaging applications [4, 5, 36–40]. A review by Chaudhry *et al.* [4] has identified the following main categories of known and projected applications of nanotechnologies for the food and related areas:

- processed nanostructures in foodstuffs;
- nano-sized food additives;
- incorporation of engineered nanomaterials into coatings, packaging materials, and (bio)nanosensors;
- nano-enabled pesticides, veterinary medicines and other agrochemicals.

11.4.1
Processed Nanostructures in Foodstuffs

The advent of new probe microscopy tools in the early 1980s, such as atomic force microscopy, enabled the study and better understanding of the structures of food components close to the molecular level. This further enabled the development of new food textures through rational design of natural nanostructures, rather than by empirical guesswork [41]. This knowledge is driving the development of innovative food products based on nanostructures (also termed nanotextures), in the form of stable nano-micelles and liposomes. The methods commonly used for this purpose involve development of nano-emulsions, surfactant micelles, emulsion bilayers, double or multiple emulsions or reverse micelles [42].

The nanostructured food products are expected to offer novel or improved food tastes, textures, and mouth sensations. A typical example of this technology can be envisaged in the form of a low-fat food, which because of nanostructuring has a creamy texture and taste that is similar to its full-fat equivalent. At present there is no known example of a commercially available nanostructured food, although a number of products are understood to be in the research and development pipeline – some of which may be near market. An example of a product currently under research and development is a mayonnaise that is composed of an emulsion that contains nano-sized droplets of water inside. The mayonnaise would offer taste and texture attributes similar to the full-fat equivalent, but with a significant reduction in the fat content [43]. It can be envisaged from the nature of these applications that developments in this area will be aimed at those food products that are traditionally high in fat content, such as spreads, mayonnaises, creams, ice creams, sauces, dressings, and so on. Depending on their safety, scale of market penetration, and acceptance by the consumers, the nanostructured food products could provide a useful means to the consumer to reduce their dietary intake of fat, while still enjoying tasty foodstuffs.

This area of application is expected to involve mainly the use of "soft" nanomaterials that are likely to be digested in the gastrointestinal tract. As discussed in Section 11.1, this application area should not raise any special consumer safety concerns, and as such need not be branded "nanotechnology". The safety evaluations for such applications should, however, consider whether nanoscale processing of some food ingredients could lead to a drastic change in the digestibility, uptake, and bioavailability of the resulting nanostructures in the body.

11.4.2
Nano-sized Food Additives

The development of nano-sized food additives and supplements represents an emerging area of nanotechnology applications, which could potentially exploit a much wider range of food and health food products. The applications involve the use of nano-sized or nano-encapsulated food additives, such as colors, preservatives, flavoring agents, and supplements. A number of nano-sized additives are already available in some countries. Examples include minerals, antimicrobials, vitamins, and antioxidants. Virtually all such additives and supplements claim improved absorption and bioavailability in the body compared with their larger-sized equivalents [4]. The technology employed for this purpose involves the development of nano-sized substances, or nano-encapsulating them in the form of micelles, liposomes or biopolymer-based carrier systems. These methods have also been used to develop delivery systems for additives and supplements for use in food and beverage products.

Nano-encapsulation offers benefits similar to micro-encapsulation, that is, in terms of preserving the ingredients and additives during processing and storage, masking unpleasant tastes and flavors, controlling the release of additives, improving the dispersion of water-insoluble food ingredients and additives, and improving the uptake of encapsulated nutrients and supplements. The concept of nano-delivery systems seems to have originated from medical research into targeted delivery of drugs and therapeutics. While the use of nano-carrier technology in food and related applications can offer a number of benefits, such as increased absorption and uptake, and improved bioavailability of nutrients and supplements, it also has the potential to alter the ADME characteristics of the substances in the body.

For example, using this approach, a water-soluble food additive can be rendered fat dispersible, or vice versa. Such transformations may not have an adverse health implication, provided that the nano-carrier breaks down and releases its contents in the gastrointestinal tract. In such a case, the risk of the encapsulated substance will not be any different from that of its conventional equivalent. However, if a nano-carrier is capable of delivering a substance to the circulatory system and other parts of the body, the altered ADME characteristics of some additives may pose an increased risk to the consumer's health. The safety considerations for this technology should also need to ensure that a nano-carrier does not act as a "Trojan horse", in terms of facilitating the translocation of potentially harmful or foreign materials from the GI tract to other unintended parts of the body.

As discussed in Section 10.1, the main risk to consumers from nano-additives in food is, however, expected to arise from the use of insoluble and biopersistent "hard" engineered nanomaterials.

11.4.3
Applications for Food Packaging

Nanotechnology applications for food contact materials (FCMs) and especially food packaging materials constitute the largest market share of the current and short-term predicted applications for the food sector [4, 44]. While most nanotechnology applications in the food and agriculture sectors are currently at research and development or near-market stages, the applications for food packaging seem to have become a commercial reality in some countries. The likely benefits of the technology include nano-enabled packaging materials that are lightweight but strong, and/or that can prolong shelf-life of the packaged foodstuffs. Considering the fixed or embedded nature of engineered nanomaterials in plastic polymers, this area of application is not expected to pose any significant risk to the consumer due to lack of migration into the packaged foodstuffs. A variety of nano-enabled packaging materials are currently available worldwide. The main applications in this area fall into the following broad categories [4, 45]:

- engineered nanomaterial–polymer composites (including biodegradable composites) with improved packaging properties in terms of flexibility, gas barrier properties, and temperature and moisture stability;
- active food contact materials incorporating engineered nanomaterials with antimicrobial or oxygen scavenging properties;
- intelligent or smart packaging concepts, incorporating nanosensors that can monitor and report food quality during transportation and storage.

Nanoclays have been incorporated into a variety of polymer composites for improved gas barrier properties. These include polyamides, polyolefins, polystyrene, ethylene–vinyl acetate copolymer, epoxy resins, polyurethane, polyimides, and polyethylene terephthalate. Known applications of the nanoclay–polymer composites include multilayer film packaging, bottles for beer and carbonated drinks, and thermoformed containers.

Metal and metal oxide engineered nanomaterial–polymer composites have been developed for a range of purposes, such as antimicrobial surfaces, abrasion resistance, ultraviolet absorption, or mechanical strength. The main engineered nanomaterials used in this area include nano-silver and nano-zinc oxide for antimicrobial action, nano-titanium dioxide for ultraviolet protection, nano-titanium nitride as a processing aid, and nano-silica for surface coating. For example, a number of "active" food contact materials incorporating nano-silver are available commercially that are claimed to preserve the food materials by inhibiting the growth of microorganisms on the food contact material surface. These include plastic food storage containers and bags. Nano-silver is also reported to have been incorporated into the plastic linings of many domestic refrigerators to prevent microbial growth, to maintain a clean environment, and to aid cleaning. In this regard, the discovery of antimicrobial properties of nano-zinc oxide and nano-magnesium oxide provides more affordable materials for applications in food packaging [46]. A plastic wrap containing nano-zinc oxide is currently available in Taiwan, which is claimed to sterilize under indoor lighting conditions. Any significant extension in the shelf-life of packaged food products should contribute toward reducing the waste of foodstuffs.

A range of coatings containing engineered nanomaterials is available for antimicrobial, scratch-resistant, anti-reflective, or corrosion-resistant surfaces. Examples of these include silver nano-coating on kitchenware, cutting boards, teapots, and other kitchen objects. Antibacterial nano-coatings on food preparation surfaces, such as meat cutting machinery in abattoirs, and food preparation and processing surfaces and conveyer belts, could also help to maintain hygiene during food processing. This may have special benefits for complex or hard-to-reach parts that are difficult to clean in place.

Nanotechnology has also enabled the development of nanosensors that can be applied as labels or coatings to add an intelligent function to food packaging in terms of ensuring the integrity of the package through detection of leaks (e.g., for foodstuffs packed under vacuum or inert atmosphere), indications of time–temperature variations (e.g., freeze–thaw–refreezing), or microbial safety (deterioration of foodstuffs). Food safety also requires confirmation of the authenticity of products. This is where application of nano-barcodes incorporated into printing inks or coatings has shown the potential for use in tracing the authenticity of the packaged product [47].

Any consumer safety concerns from nano-enabled food packaging and labels will only arise if engineered nanomaterials migrate into the packaged foodstuffs. Currently, there are only a few published studies on the migration of engineered nanomaterials from packaging materials. Avella *et al.* [48] determined the migration of Fe, Mg, and Si from a biodegradable starch–nanoclay nanocomposite film into packaged vegetables (lettuce and spinach). The results showed an insignificant increase in the levels of Fe and Mg in the packaged vegetables, while a consistent increase in Si (the main component of nanoclay) was noted.

A recent study by Šimon *et al.* [49] modeled the potential migration of engineered nanomaterials from different food contact materials on the basis of physicochemical parameters. The modeling predicted that any detectable migration of engineered nanoparticles from packaging polymers to packaged foodstuffs will take place only: (i) in the case of very small nanoparticles with a radius in the lower nanometer range; (ii) for polymers with a low dynamic viscosity such as polyolefins; and (iii) with no nanoparticle–polymer binding.

Another recent (unpublished) study by Bradley *et al.* (FERA, York) determined the migration of nanoclay components from commercial beer bottles that had a nanoclay composite embedded between polyethylene terephthalate (PET) layers. The study also determined the migration of nano-silver from commercial food containers made of polypropylene–nano-silver composite. The study found no detectable migration of nanoclay from PET bottles, and noted only a very low migration of silver (less than the limit of quantification) from food containers made of polypropylene–nano-silver composite. In either case, the presence of the engineered nanomaterials did not affect migration of other non-nano-components from the packaging materials. While these few studies provide some reassurance in the safety of nano-enabled food packaging materials, more tests will be needed to establish migration patterns for other engineered nanomaterial–polymer composites.

11.4.4 Applications in Food Production

The likely benefits of substituting active ingredients or carriers with nano-sized equivalents has opened up new avenues for research into potential applications of engineered nanomaterials to develop novel formulations of pesticides, veterinary medicines, and other agrochemicals, such as fertilizers and plant growth regulators [50, 51]. The anticipated benefits include a potential reduction in the use of certain agrochemicals, better dispersions, and control of dosage and applications of the nano-formulations in the field.

Theoretically, any nano-sized mineral, vitamin, or other additive or supplement developed for a food application can equally be used for animal feed. There are a few examples of available products where a nano-sized additive has been specifically developed (or is under development) for animal feed. For example, certain nano-grade vitamin mixes are available commercially for use in poultry and livestock feed. Examples of research and development in this area include a feed additive comprising a natural biopolymer from yeast cell walls that can bind mycotoxins to protect animals against mycotoxicosis, and an aflatoxin-binding nano-additive for animal feed, which is derived from modified nanoclay [52]. A polystyrene nanoparticle, with polyethylene glycol (PEG) linker and mannose targeting biomolecule, has also been developed that adheres to *Escherichia coli*. Administration of the nanoparticle through feed is likely to be helpful in removing food-borne pathogens in the gastrointestinal tract of the animals [53].

Research is also being carried out into the development of various nano-sized agrochemicals, such as fertilizers, pesticides, and veterinary medicines. The use of nano-sized active ingredients has been suggested to offer improved delivery of the agrochemicals in the field, better efficacy of pesticides, and better control over dosing of veterinary products. For example, nano-encapsulated and solid lipid nanoparticles have been explored for the delivery of agrochemicals [54], such as slow- or controlled-release fertilizers and pesticides. One example is a combined fertilizer and pesticide formulation encapsulated in nanoclay for the slow release of growth stimulants and bio-control agents [2]. Fertilizer compositions, claimed to contain nano-sized micronutrients, and micronized (volcanic) rock dust, are available commercially for remineralization of soil.

Despite a great deal of industrial interest in the use of nanotechnologies in the food production area, examples of the available products at present are very few and far between. Most of the developments seem to be currently at the research and development stage. However, such applications have the potential for large-scale use in the future, which is also likely to increase the potential exposure to agrochemicals used in food production.

11.5 Implications for Regulatory Frameworks

In many countries, regulatory frameworks exist for pre-market evaluation for food products. Despite some regulatory uncertainties [55, 56], the new developments in nanotechnology are taking place in a regulatory vacuum, as the potential risks will be controlled under the existing frameworks [57]. These relate to a plethora of regulatory frameworks on general food safety, food additives, novel foods, specific health claims, chemical safety, food contact materials, water quality, and other specific regulations on the use of certain chemicals in food production and protection, such as biocides, pesticides, veterinary medicines, and so on. Environmental regulations are also likely to capture the use of engineered nanomaterials in food packaging, and agri-food production applications.

Examples of general food laws include the Federal Food, Drug, and Cosmetic Act (the FDC Act) in the USA, which is administered by the Food and Drug Administration,[1] the European Commission's Food Law Regulation 178/2002, which sets down the general principles and requirements of food law within the European Union and provides for the establishment of the European Food Safety Authority (EFSA), and the Australia New Zealand Food Standards Code (the Food Standards Code).[2]

Most countries also have legislation relating to food contact materials, setting out (approved) materials and additives that can be used for food packaging, and acceptable levels of migration of substances from packaging into foodstuffs. The legislation takes different forms in different countries, but the principles – to help ensure consumer protection and avoid contamination of foodstuffs – are universal. The relevant regulations require that food contact materials should be made and used in such ways that they do not transfer constituents to food in quantities that could:

a) endanger human health;
b) bring about an unacceptable change in composition; or
c) bring about deterioration in organoleptic characteristics thereof.

There are also certain cross-cutting horizontal regulations that are relevant to nanotechnology applications for food and food packaging. An example of this in Europe is the Directive 2001/95/EC of 3 December 2001 on General Product Safety (in force since 14 January 2004, replacing Directive 92/59/EC). This legislation embodies the main principle that only safe products can be placed on the market. Briefly, a safe product is one that, under normal and reasonably foreseeable conditions of use, does not present any risk (or only the minimum acceptable risk), taking into account the characteristics, effects, presentation of the products, and the categories of persons at risks. Owing to its broad and horizontal scope, the Directive applies to risks that are not covered by other specific European Union

1) For the purposes of this chapter, other relevant legislative instruments include the Dietary Supplement Health and Education Act of 1994, the Food Additive Amendment Act of 1958, and the Code of Federal Regulations.

2) Available at: http://www.foodstandards.gov.au/foodstandards/foodstandardscode/.

provisions on products. Thus it applies to products containing engineered nanomaterials, with the onus of ensuring the safety of such products resting with the person who places them on the market. Another notable regulation relevant to nanotechnology applications for food packaging is the EU's chemicals regulation (EC 1907/2006, in effect from 1 June 2007) REACH (Registration, Evaluation, Authorization, and Restriction of Chemicals), which requires registration of all substances that are produced and/or marketed in the European Union above $1\,t\,yr^{-1}$ – as such, in preparations, or in articles.

A number of studies have assessed the relevance and adequacy of existing regulatory frameworks in relation to the increasing applications of nanotechnologies for the food sector [55, 58–61]. Findings of these reviews suggest that the current regulatory frameworks for food and food contact materials in different jurisdictions, such as the European Union, the USA, and Australia, are broad enough to "catch" nano-enabled food and food contact materials. A few uncertainties in regulatory frameworks, however, appear to arise from the current lack of understanding in relation to, for example, a clear definition that encompasses the distinctive properties of nano-ingredients and additives, a clearly defined responsibility and liability for relevant products and applications, appropriate permissible limits that relate to the (potential) effects of nano-substances in food, and an exclusive premarket approval system for nano-enabled food products. Nevertheless, a case-by-case assessment of the safety of intended applications by the manufacturers (as recommended in the recent EFSA opinion [5]) should ensure that only safe applications of the new technology are placed on the market.

In this regard, there are also some recent developments in the regulatory area. These include recasting of key European regulatory instruments, such as Regulation 258/97 (the Novel Foods Regulation), which requires safety assessment of any food product that does not have a significant history of use in the European Union, or that is produced using a new production process, or that gives rise to significant changes in the nutritional value, metabolism or level of undesirable substances of the foods or food ingredients. The legislation is currently being reviewed in Europe, and is expected to include a specific reference to foods modified by new production processes "such as nanotechnology and nanoscience, which may have an impact on food".

The use of food additives in the European Union is currently controlled by the Food Additives Framework Directive) and the subordinate legislation. Subject to adoption by the European Community, the Food Additives Framework Directive will be replaced by a common authorization system in 2010, which will provide for a common basis of controls on food additives (EC Regulation No. 1333/2008), food enzymes (EC Regulation No. 1332/2008), and food flavorings (EC Regulation No. 1334/2008). The adoption of the common authorization procedure will also bring together all of the existing food additive regulations, and will introduce comitology[3)] for the approval of the three categories of substances. The most relevant aspect in relation to the use of nano-sized food additives in the new

3) Comitology in the European Union refers to the committee system that oversees the delegated acts implemented by the European Commission.

Regulation is the re-evaluation of safety assessment, which will ensure that food additives, once permitted, are kept under continuous observation and re-evaluation. Therefore, under the new Regulation, producers or users of food additives that are "significantly different from those included in the risk assessment of the Authority or different from those covered by the specifications laid down" will be obliged to inform the Commission of any new information that may affect their safety assessment. Also, under the new Regulation, the EFSA will be invested with the power to re-evaluate a food additive on the basis of "new scientific information".

The commercial exploitation of nanotechnology is almost concurrent to that of the start of online marketing of consumer products via the Internet. Virtually all of the currently available nanotechnology-derived consumer products in the areas of food and health food can be bought via the Internet anywhere in the world. The global boundaries of online marketing have also raised questions over the applicability and effectiveness of national food laws to control risks from products that may be produced abroad but are bought by a consumer through the Internet for personal consumption. This results in the regulation of nanotechnology products to be applied at the global scale, together with establishing liabilities, which poses a challenge in that food laws in many countries may not conform to each other. As research clears some of the main scientific uncertainties in the coming years, issues like these will need resolving at the international level through the development of frameworks that relate to global trade agreements.

11.6 Conclusions

The overview of nanotechnology applications presented in this chapter shows a variety of benefits for the whole of the food chain – from new or improved tastes, textures, and mouth sensations, through potential reduction in the dietary intake of fat and various food additives, to enhanced absorption of nutrients, preservation of quality and freshness, better traceability, and security of food products. It is also clear that currently there are major knowledge gaps in our understanding of the properties, behavior, and effects of the engineered nanomaterials that may be used in food applications. While these knowledge gaps make it difficult to assess the risk of such applications to a consumer, a careful consideration of the materials and applications can provide a basis for a conceptual risk assessment.

For example, the use of "soft" nanomaterials may not require as detailed evaluations as the "hard" nanomaterials. As more research uncovers the basic rules that drive the properties, behavior, and effects of engineered nanomaterials, even some "hard" nanomaterials may not prove to be as harmful as feared. This does not, however, mean that an unexpected hazard or risk of some engineered nanomaterials will not come to surface in the future, but this applies equally to other (conventional) materials, processes, and products. The existence of stringent regulatory controls provides some reassurance that only safe products and applications of nanotechnologies will be permitted on the market. However, the industry needs to adopt a pragmatic approach – especially where intended applications relate to the use of "hard" engineered nanomaterials and carry a likelihood of consumer exposure – and perform a case-by-case safety evaluation of the intended products (as recommended in the recent EFSA opinion [5]) before placing them on the market.

References

1 ETC Group (2004) *Down on the Farm: The Impact of Nano-Scale Technologies on Food Agriculture*, ETC Group, Ottawa.

2 Miller, G., and Senjen, D. (2008) Out of the Laboratory and On to Our Plates: Nanotechnology in Food and Agriculture, Friends of the Earth, Australia, Europe and USA. Available from: http://www.foeeurope.org/activities/nanotechnology/Documents/Nano_food_report.pdf (accessed 8 November 2010).

3 Soil Association (2008) Press Release: Soil Association first organisation in the world to ban nanoparticles – potentially toxic beauty products that get right under your skin, 17 January. Soil Association, Bristol.

4 Chaudhry, Q., Scotter, M., Blackburn, J., Ross, B., Boxall, A., Castle, L., Aitken, R., and Watkins, R. (2008) Applications and implications of nanotechnologies for the food sector. *Food Addit. Contam.*, **25** (3), 241–258.

5 EFSA (2009) The potential risks arising from nanoscience and nanotechnologies on food and feed safety (Scientific Opinion). *EFSA J.*, **7**, 958. Available at: http://www.efsa.europa.eu/en/efsajournal/doc/958.pdf (accessed 9 November 2010).

6 Woodrow Wilson International Center for Scholars (2009) Consumer products: an inventory of nanotechnology-based consumer products currently on the market. Project on Emerging Nanotechnologies. Available from: http://www.nanotechproject.org/inventories/consumer/ (accessed 8 November 2010).

7 Park, K.H. (2005) Preparation method for antibacterial wheat flour by using silver nanoparticles, Korean Intellectual Property Office (KIPO), 1020050101529A, 24.10.2005.

8 EFSA (2008) Safety of synthetic lycopene (Scientific Opinion). *EFSA J.*, **6**, 676. Available at: http://www.efsa.europa.eu/en/efsajournal/doc/676.pdf (accessed 9 November 2010).

9 Donaldson, K., Stone, V., Clouter, A., Renwick, L., and MacNee, W. (2001) Ultrafine particles. *J. Occup. Environ. Med.*, **58**, 211–216.

10 Nel, A., Xia, T., Madler, L., and Li, N. (2006) Toxic potential of materials at the nanolevel. *Science*, **311**, 622–627.

11 Li, N., Sioutas, C., Cho, A., Schmitz, D., Misra, C., Sempf, J., Wang, M., Oberley, T., Froines, J., and Nel, A. (2003) Ultrafine particulate pollutants induce oxidative stress and mitochondrial damage. *Environ. Health Perspect.*, **111**, 455–460.

12 Donaldson, K., Stone, V., Tran, C.L., Kreyling, W., and Borm, P.J. (2004) Nanotoxicology. *Occup. Environ. Med.*, **619**, 727–728.

13 Oberdörster, E. (2004) Manufactured nanomaterials (fullerenes C_{60}) induce oxidative stress in the brain of juvenile largemouth bass. *Environ. Health Perspect.*, **112**, 1058–1062.

14 Geiser, M., *et al.* (2005) Ultrafine particles cross cellular membranes by nonphagocytic mechanisms in lungs and in cultured cells. *Environ. Health Perspect.*, **113** (11), 1555–1560.

15 Oberdörster, G., Sharp, Z., Atudorei, V., Elder, A., Gelein, R., Kreyling, W., and Cox, C. (2004) Translocation of inhaled ultrafine particles to the brain. *Inhal. Toxicol.*, **16**, 437–445.

16 Carr, K.E., Hazzard, R.A., Reid, S., and Hodges, G.M. (1996) The effect of size on uptake of orally administered latex microparticles in the small intestine and transport to mesenteric

lymph nodes. *Pharm. Res.*, **13**, 1205–1209.

17 Hillyer, J.F., and Albrecht, R.M. (2001) Gastrointestinal persorption and tissue distribution of differently sized colloidal gold nanoparticles. *J. Pharm. Sci.*, **90**, 1927–1936.

18 Hoet, P., Bruske-Hohlfeld, I., and Salata, O. (2004) Nanoparticles – known and unknown health risks. *J. Nanobiotechnol.*, **2**, 1–15.

19 Florence, A.T. (2005) Nanoparticle uptake by the oral route: fulfilling its potential? *Drug Discov. Today Technol.*, **2**, 75–81.

20 des Rieux, A., Fievez, V., Garinot, M., Schneider, Y.J., and Preat, V. (2006) Nanoparticles as potential oral delivery systems of proteins and vaccines: a mechanistic approach. *J. Control. Release*, **116**, 1–27.

21 De Jong, W.H., Hagens, W.I., Krystek, P., Burger, M.C., Sips, A.J., and Geertsma, R.E. (2008) Particle size-dependent organ distribution of gold nanoparticles after intravenous administration. *Biomaterials*, **29**, 1912–1919.

22 Jani, P.U., McCarthy, D.E., and Florence, A.T. (1994) Titanium dioxide (rutile) particle uptake from rat GI tract and translocation to systemic organs after oral administration. *Int. J. Pharm.*, **105**, 157–168.

23 Šimon, P., and Joner, E. (2008) Conceivable interactions of biopersistent nanoparticles with food matrix and living systems following from their physicochemical properties. *J. Food Nutr. Res.*, **47**, 51–59.

24 Nemmar, A., Hoet, P.H.M., Vanquickenborne, B., *et al.* (2002) Passage of inhaled particles into the blood circulation in humans. *Circulation*, **105**, 411–414.

25 Lynch, I., and Dawson, K.A. (2008) Protein–nanoparticle interactions. *Nano Today*, **3**, 40–47.

26 Michaelis, K., Hoffmann, M.M., Dreis, S., Herbert, E., Alyautdin, R.N., Michaelis, M., Kreuter, J., and Langer, K. (2006) Covalent linkage of apolipoprotein E to albumin nanoparticles strongly enhances drug transport into the brain. *J. Pharmacol. Exp. Ther.*, **317**, 1246–1253.

27 Dobrovolskaia, M. (2007) Immunological properties of engineered nanomaterials. *Nat. Nanotechnol.*, **2**, 469–478.

28 Szentkuti, L. (1997) Light microscopical observations on luminally administered dyes, dextrans, nanospheres and microspheres in the pre-epithelial mucus gel layer of the rat distal colon. *J. Control. Release*, **46**, 233.

29 Lai, S.K., O'Hanlon, D.E., Harrold, S., Man, S.T., Wang, Y.-Y., Cone, R., and Hanes, J. (2007) Rapid transport of large polymeric nanoparticles in fresh undiluted human mucus. *Proc. Natl. Acad. Sci., USA*, **104**, 1482–1487.

30 Desai, M.P., Labhasetwar, V., Amidon, G.L., and Levy, R.J. (1996) Gastrointestinal uptake of biodegradeable microparticles: effect of particle size. *Pharm. Res.*, **13**, 1838.

31 Kim, Y.S., Kim, J.S., Cho, H.S., Rha, D.S., Kim, J.M., Park, J.D., Choi, B.S., Lim, R., Chang, H.K., Chung, Y.H., Kwon, H., Jeong, J., Han, B.S., and Yu, J. (2008) Twenty-eight-day oral toxicity, genotoxicity, and gender-related tissue distribution of silver nanoparticles in Sprague–Dawley rats. *Inhal. Toxicol.*, **20**, 575–583.

32 Powell, J.J., Harvey, R.S., Ashwood, P., Wolstencroft, R., Gershwin, M.E., and Thompson, R.P. (2000) Immune potentiation of ultrafine dietary particles in normal subjects and patients with inflammatory bowel disease. *J. Autoimmun.*, **14** (1), 99–105.

33 Lomer, M.C., Grainger, S.L., Ede, R., Catteral, A.P., Greenfield, S.M., Cowan, R.E., Vicary, F.R., Jenkins, A.P., Fidler, H., Harvey, R.S., Ellis, R., McNair, A., Ainley, C.C., Thompson, R.P., and Powell, J.J. (2005) Lack of efficacy of a reduced microparticle diet in a multi-centred trial of patients with active Crohn's disease. *Eur. J. Gastroenterol. Hepatol.*, **17**, 377–384.

34 Lomer, M.C., Thompson, R.P., and Powell, J.J. (2002) Fine and ultrafine particles of the diet: influence on the mucosal immune response association with Crohn's disease. *Proc. Nutr. Soc.*, **611**, 123–130.

35 Chen, M.A., and von Mikecz, A. (2005) Formation of nucleoplasmic protein

aggregates impairs nuclear function in response to SiO_2 nanoparticles. *Exp. Cell Res.*, **305**, 51.

36 Kuzma, J., and VerHage, P. (2006) Nanotechnology in agriculture and food production: anticipated applications. Project on Emerging Nanotechnologies, Woodrow Wilson International Center for Scholars, Washington, DC. Available from: http://www.nanotechproject.org/process/assets/files/2706/94_pen4_agfood.pdf (accessed 8 November 2010).

37 Bouwmeester, H., Dekkers, S., Noordam, M., Hagens, W., Bulder, A., de Heer, C., ten Voorde, S., Wijnhoven, S., and Sips, A. (2007) Health impact of nanotechnologies in food production, Report 2007.014, RIKILT–Institute of Food Safety, and National Institute of Public Health and the Environment. Available at: http://www.rikilt.wur.nl/NR/rdonlyres/BDEEDD31-F58C-47EB-A0AA-23CB9956CE18/54352/R2007014.pdf (accessed 10 November 2010).

38 Food Safety Authority of Ireland (2008) The relevance for food safety of applications of nanotechnology in the food and feed industries, Food Safety Authority of Ireland, Dublin.

39 Groves, K. (2008) Potential benefits of micro and nano technology for the food industry: does size matter? *New Food Mag.*, **4**, 49–52.

40 Morris, V.J. (2008) Nanotechnology in the food industry. *New Food Mag.*, **4**, 53–55.

41 Morris, V.J. (2010) Natural food nanostructures, in *Nanotechnologies in Food, Nanotechnology Applications for Food Ingredients, Additives and Supplements* (eds Q. Chaudhry, L. Castle, and R. Watkins), Royal Society of Chemistry, London, pp. 50–68.

42 Weiss, J., Takhistov, P., and McClements, D.J. (2006) Functional materials in food nanotechnology. *J. Food Sci.*, **71** (9), R107–R116.

43 Clegg, S., Knight, A., Beeren, C., and Wilde, P. (2009) Fat reduction whilst maintaining the sensory characteristics of fat using multiple emulsions, in *International Symposium on Food Rheology and Structure (ISFRS)*, ETH, Zürich.

44 Cientifica (2006) Nanotechnologies in the Food Industry, Cientifica, London.

45 AZoNanotechnology (2004) Food packaging using nanotechnology methods: an overview of "smart packaging" and "active packaging". Available from: http://www.azonano.com/details.asp?ArticleID=1317 (accessed 10 November 2010).

46 Zhang, L., Jiang, Y., Ding, Y., Povey, M., and York, D. (2007) Investigation into the antibacterial behaviour of suspensions of ZnO nanoparticles (ZnO nanofluids). *J. Nanopart. Res.*, **9** (3), 479–489.

47 Han, M., Gao, X., Su, J.Z., and Nie, S. (2001) Quantum-dot-tagged microbeads for multiplexed optical coding of biomolecules. *Nat. Biotechnol.*, **19**, 631–635.

48 Avella, M., De Vlieger, J.J., Errico, M.E., Fischer, S., Vacca, P., and Volpe, M.G. (2005) Biodegradable starch/clay nanocomposite films for food packaging applications. *Food Chem.*, **93**, 467–474.

49 Šimon, P., Chaudhry, Q., and Bakoš, D. (2008) Migration of engineered nanoparticles from polymer packaging to food–a physiochemical view. *J. Food Nutr. Res.*, **47** (3), 105–113.

50 Liu, F., Wen, L.X., Li, Z.Z., Yu, W., Sun, H.Y., and Chen, J.F. (2006) Porous hollow silica nanoparticles as controlled delivery system for water-soluble pesticide. *Mater. Res. Bull.*, **41** (12), 2268–2275.

51 Wang, L., Li, X., Zhang, G., Dong, J., and Eastoe, J. (2007) Oil-in-water nanoemulsions for pesticide formulations. *J. Colloid Interface Sci.*, **314** (1), 230–235.

52 Shi, Y.H., Xu, Z.R., Feng, J.L., Hu, C.H., and Xia, M.S. (2005) *In vitro* adsorption of aflatoxin adsorbing nano-additive for aflatoxin B1, B2, G1, G2. *Sci. Agric. Sin.*, **38** (5), 1069–1072.

53 Kuzma, J., Romanchek, J., and Kokotovich, A. (2008) Upstream oversight assessment for agrifood nanotechnology. *Risk Anal.*, **28**, 1081–1098.

54 Frederiksen, H.K., Kristensen, H.G., and Pedersen, M. (2003) Solid lipid microparticle formulations of the pyrethroid gamma-cyhalothrin – incompatibility of the lipid and the pyrethroid and biological properties of

the formulations. *J. Control. Release*, **86** (2/3), 243–252.

55 Chaudhry, Q., George, C., and Watkins, R. (2007) Nanotechnology regulation: developments in the United Kingdom, in *New Global Frontiers in Regulation: The Age of Nanotechnology* (eds G. Hodge, D. Bowman, and K. Ludlow), Edward Elgar, Cheltenham, UK, pp. 212–238.

56 Hodge, G., Bowman, D., and Ludlow, K. (eds) (2007) *New Global Frontiers in Regulation: The Age of Nanotechnology*, Edward Elgar, Cheltenham, UK.

57 Gergely, A. (2007) Regulation of nanotechnology–within REACH? *Nano Now*, February, 44–46.

58 Food Standards Agency (2006) A review of potential implications of nanotechnologies for regulations and risk assessment in relation to food. Draft report of FSA regulatory review. Available at: http://www.food.gov.uk/multimedia/pdfs/nanotech.pdf (accessed 10 November 2010).

59 Taylor, M.R. (2006) Regulating the products of nanotechnology. Does FDA have the tools it needs? Project on Emerging Nanotechnologies, Woodrow Wilson International Center for Scholars, Washington, DC. Available at: http://nanotechproject.org/file_download/110 (accessed 10 Novemebr 2010).

60 Taylor, M.R. (2008) Assuring the safety of nanomaterials in food packaging: the regulatory process and key issues, Project on Emerging Nanotechnologies, Woodrow Wilson International Centre for Scholars, Washington, DC, p. 5. Available at: http://www.nanotechproject.org/publications/archive/nano_food_packaging/ (accessed 10 November 2010).

61 Ludlow, K.A. (2009) The readiness of Australian food regulation for the use of nanotechnology in food and food packaging. *University of Tasmania Law Review*, **26**(2), 177–203.

12
Environmental Considerations of and Societal Reactions to Nanotechnology in the Food Sector

Michael Siegrist, Bernd Nowack, and Hans Kastenholz

12.1
Introduction

Food products and food packaging containing engineered nanoparticles are already commercially available [1]. Industry and governments invest considerable amounts of money in employing this technology for new applications in such areas as food packaging, food processing, food safety, and agricultural production [2]. It is therefore expected that nanotechnology will be even more important in the near future [3].

Nanotechnology has great potential to generate new products in various domains. Because nanotechnology may affect so many aspects of human life, risk assessment errors may result in irreversible damage [4]. Although many studies examining possible applications of nanotechnology or nanoparticles have emphasized that the new technology may have adverse effects on health and environment, no conclusions could be reached due to lack of data. The life cycle assessment of nanotechnology applications is currently in its very infancy [5]. It is difficult, therefore, to assess the environmental and human health impacts of nano-based products and services. Overall, we currently know much more about the possible benefits of this technology than about the possible risks. This makes regulation very difficult, since there are few hard facts on which such regulations could be built [6–7].

Nanotechnology allows the creation of materials with new, desired properties. The very same properties that lead to potentially great benefits may also result in unwanted risks, however [8]. The novel properties of nanomaterials and the potentially broad introduction of nanomaterial-based products have raised many concerns over their consequences for human and environmental health [9]. Results of risk assessment studies suggest that some nanomaterials may have damage potential if they are exposed to humans or the environment [10]. It has been concluded that there is a lack of knowledge regarding the toxic effects of free nanoparticles, and that not enough is known about dosage and exposure for traditional risk analysis models [4]. The situation is similar to that for other new technologies.

Nanotechnology in the Agri-Food Sector: Implications for the Future, First Edition. Edited by Lynn J. Frewer, Willem Norde, Arnout Fischer, Frans Kampers.

Lack of data and understanding make it very difficult to reliably assess the potential and the risks of nanotechnology.

There are various pathways through which humans may be confronted with nanoparticles in foods. Figure 12.1 shows the ways humans could theoretically be exposed to food containing nanomaterials:

- consumption of foods or health supplements for which nanoparticles were used in the manufacturing process;
- consumption of nanoparticles that have migrated into foods from food packaging coated with nanomaterials;
- consumption of foods exposed to nanomaterials during farming practice.

In addition, there are indirect pathways by which nanoparticles can end up in food:

- nanoparticles in discarded packaging eventually get into the environment – ingested nanoparticles can be excreted again, removed from wastewater, and added to soil by sludge, or nanoparticles are not removed during wastewater treatment;
- nanoparticles can also enter the environment from non-food products and applications, either through wastewater, solid waste or direct input;
- once in the environment, nanoparticles can be taken up by foods and thus return back into the food cycle.

12.2
Life Cycle of Nanotechnology Food Products

The release of nanoparticles can occur throughout the whole life cycle of products [11], and it is thus important to take into account the possible exposures to

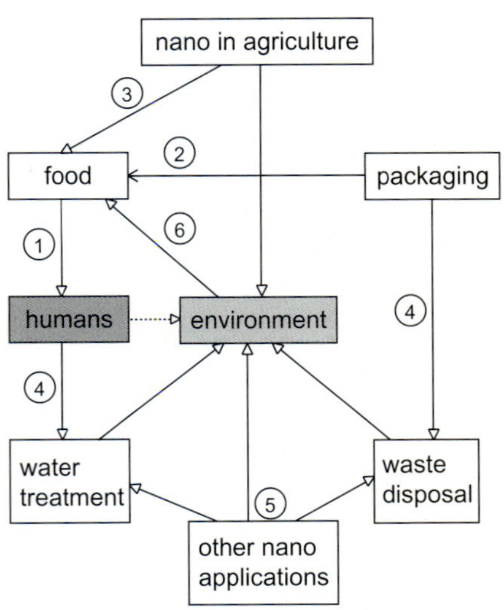

Figure 12.1 Nanoparticle flows between food, humans, and the environment.

nanoparticles in different phases of the life cycle of a product [12]. Critical points are during (i) production and shipping, where release into the air is most likely, (ii) production of the final product, (iii) use, and finally (iv) disposal or recycling. The amount of nanoparticles released by the different processes depends on several factors: the nanoparticle stock in the product, the product's lifetime, the way nanoparticles are incorporated into the material, and the actual use/usage of the product [11]. Products with a long lifetime, a loose incorporation of the nanoparticle and/or an intense use (e.g., through frequent cleaning) will most likely not contain any nanoparticles at the time of disposal. On the other hand, factors such as short lifetime, low usage, and strong fixation of nanoparticles increase the likelihood that particles will not be released until disposal. In the following subsection, a few examples from the food, food packaging, and agriculture sectors are presented. Benefits and possible risks for humans and for the environment are discussed.

12.2.1 Food

Acids present in soft drinks may have erosion effects on teeth. Through the use of nanotechnology, new functional soft drinks may be created, which have less erosion effects on teeth than conventional soft drinks [13]. Encapsulation and delivery systems are another example of the use of nanotechnology as a tool for new products in the food industry [14]. Vitamins or other supplements are packed in nanoparticles and infused into the foodstuffs. Encapsulation increases stability, and it allows for controlled release of the ingredients to specific places in the human body. Personalized beverages and foods could also be of some interest to consumers and producers [1]. After purchase of the food product, it is customized according to the preferences of the consumer. Ultrasonic frequency, heat or other triggers are used to release varying flavors, colors or nutrients from nano-emulsions in drinks or food products.

In all these applications, the life cycle of the products is clear because they are intended for complete human consumption. The entry point of the nanoparticle into the body or the environment after passage through the gut is therefore well defined and easily quantifiable.

12.2.2 Packaging

Nanotechnology promises new food packaging with great benefits. Nowadays, most packaging materials are produced from fossil fuels, and disposal therefore is often problematic. Nanocomposites may be used to create edible and biodegradable films that help to reduce packaging waste [15]. Antimicrobials and antioxidants incorporated in films should increase the shelf-life and quality of coated food [14]. In the future, nanosensors embedded in the packaging will inform the consumer as to whether the food is still good or already bad [15].

In these applications, release of nanoparticles can occur during the storage phase, and transfer to the food is possible. The nanoparticles that stay within the packing are then disposed of or recycled (e.g., with compostable plastics).

12.2.3 Agriculture

Almost no applications of nanomaterials in agriculture are on the market yet, but a lot of research activity is in progress. In Switzerland, for example, only one plant protection product containing nanoparticles was found [16]: a nano-silver-containing spray for indoor and outdoor use on plant leaves, with low sales volume [17]. There is some evidence that research is ongoing for new formulations of plant protection products with quantitatively high application potential. Without mentioning specific products, a Nanoforum Report states that many companies have products with nanoparticulate ingredients within a size range of 100–250 nm and that other companies suspend herbicidal or pesticidal nanoparticles of 200–400 nm size in oil or water [18]. The current research focus is on encapsulation, use of organo-clays, and improved storage.

Nanotechnology in fertilizers can be used in this field for slow-release mechanisms such as entrapping, encapsulating or dispersing the active agents in a matrix of biodegradable or inert material. To date, no indications of real applications have been found in the scientific literature, on the Web, and in feedback from experts or associations [16]. The main research focus for the application of nanoparticles in fertilizers is placed on slow and controlled release of fertilizers. In that group, various polymers and clays are mentioned repeatedly. For clays, the benefit of nanoparticulate size is not clear, as in most cases the interlayer distance matters, and in fact several patents were found using clay particles or platelets of larger size. Plants take up nutrients mainly by roots or via leaves, and slow and controlled release is advantageous for nutrient supply on both routes. Foliar fertilizers are often used to satisfy short-term nutrient supply, where a nanoparticulate nutrient may be better for efficient nutrient uptake.

12.2.4 Non-Food Sector

All nanoparticles that are used in any imaginable application may end up in the environment at some stage of their life cycle, and thus may be fed back into the food cycle. However, current information on the release of nanoparticles from products into the environment is scarce. The release of nano-titanium dioxide (TiO_2) from coatings on wood, polymer, and tile were the highest from coated tile, and ultraviolet light increased the release of particles [19].

Release of nanoparticles into the environment can also occur at the end of the life of nano-products, when they are dumped into landfills or burned in waste incineration plants. Although the particle filters of water incineration plants are very effective, low concentrations of nanoparticles can leave the stack and be distributed by air. However, the largest input of nanoparticles into the environment is most likely by products that are used up during use, for example, sunscreens. All nanoparticles contained in sunscreen will be present in water, either directly washed off the skin into open waters or removed during showering. This pathway is very important not only for nano-TiO_2 but also for nano-silver [20].

There are socks on the market with nano-silver as an antimicrobial agent. During the washing process, some of these nano-silver particles may end up in the sewage water. A model by Benn and Westerhoff [21] suggests that a typical wastewater treatment facility could treat a high concentration of silver stemming from socks or other textiles. The authors concluded, however, that increased consumption of textiles with nano-silver may restrict the use of bio-solids as a fertilizer for agricultural lands. However, silver is also released in ionic form from nanoparticles, and this was considered to be the major process of silver release from plastics and textiles [22]. Silver from nanoparticles was found to contribute only 0.5–15% to the total silver flow into the environment.

12.3 Occurrence of Engineered Nanoparticles in the Environment

There is a lack of information not only about the economic impact of nanotechnology, but also about the type and quantity of industrially used, manufactured nanoparticles [23]. As a result, it is nearly impossible to quantify the level of exposure for consumers and the environment. Results of an initial survey in Swiss industry suggests that the largest quantities of nanoparticles are used in the production of cosmetics, food, paints, and powders [24]. More information is needed about the number of people exposed to engineered nanoparticles and what amount of material these people are exposed to. It is clear that, with an increased use of engineered nanoparticles, the potential for unintended environmental consequences will also increase [25].

Because analytical measurements of engineered nanoparticles in the environment are lacking, the expected concentrations have to be modeled with the help of extrapolations and analogies. A recent study modeled the silver emissions from nano-silver containing biocidal products and compared the expected concentrations in the environment with a reference emission [22]. In this study, nano-silver served only as a silver ion (Ag^+) source, and no particulate silver emissions were considered; therefore, no concentrations of nano-silver in the environment were modeled.

Another study used a life cycle perspective to model the quantities of engineered nanoparticles released into the environment [20]. Three types of nanoparticle were studied: nano-silver (nano-Ag), nano-TiO_2 and carbon nanotubes. The quantification was based on a substance flow analysis from products to air, soil, and water in Switzerland. The following parameters were used as model inputs: estimated worldwide production volume, allocation of the production volume to product categories, particle release from products, and flow coefficients within the environmental compartments. To estimate a possible risk, the predicted environmental concentrations were then compared to the predicted no-effect concentrations derived from the literature.

The expected concentrations of the three nanoparticles in the different environmental compartments vary widely, caused by the different life cycles of the nanoparticle-containing products. The predicted environmental concentration values for nano-TiO_2 in water are 0.7–16 μg l^{-1}. The results of this study make it possible for the first time to carry out a quantitative risk assessment of nanoparticles in the environment and suggest further detailed studies of nano-TiO_2. The modeling suggests that currently nano-silver poses little or no risk to soil organisms. The risk quotient (predicted no-effect concentrations divided by predicted environmental concentrations) for water is less than one-thousandth. Also, in the high-exposure scenario, the modeling suggests that currently little or no risk is to be expected from nano-Ag in the soil compartment and the water in general. The modeling suggests that nano-TiO_2 may pose a risk to aquatic organisms, with a risk quotient between about 0.73 and 16 or more. By contrast, the risk quotient for air is smaller than 0.001.

A similar study has been done for the UK [26], although with a different approach. Based on assumed market penetrations of nano-products and the known usage of these products, concentrations in water, air, and soil were modeled. For the 10% market penetration model, which probably overestimates current exposure levels, concentrations of silver, aluminum, and fullerene concentrations were predicted to be in the range of nanograms per liter, whereas nano-TiO_2, zinc oxide, and hydroxyapatite are predicted to be in the micrograms per liter range.

12.3.1
Environmental Behavior of Nanoparticles

The main processes that are acting on nanoparticles in the environment and that are determining their environmental fate are aggregation–disaggregation and adsorption–desorption [27]. Nanoparticles interact among themselves and with other natural nanoparticles or larger particles. The formation of aggregates in natural systems can be understood by considering physical processes, that is, Brownian diffusion, fluid motion, and gravity. Aggregation is dependent on particle size and results in efficient removal of small particles in environmental systems [28]. To quantify the stability of nanoparticles in the environment, we have to predict the stability of their suspension and their tendency to aggregate or interact with other particles [29]. The nature of the nanoparticle is modified by adsorption processes [30], and especially the surface charge plays a dominant role [31, 32].

The movement of nanoparticle in porous media is impeded by two processes: straining or physical filtration, where the particle is larger than the pore and is trapped; and true filtration, where the particle is removed from solution by interception, diffusion, and sedimentation. However, particles removed from solution by such processes can readily become resuspended upon changes in the chemical or physical conditions (e.g., changes in pH, ionic strength, and flow rate [33, 34].

Several studies have investigated the transport of a wide range of engineered nanoparticles through porous media [35–37]. Particles smaller than 100 nm are predicted to have very high efficiencies of transport to collector surfaces due to Brownian diffusion. If all particle–collector contacts were to result in particle attachment to the collector, these small particles would be retained to a large extent by the porous medium. However, nano-sized silica particles were not appreciably removed, and also anatase nanoparticles were only removed between 55% and 70%, depending on the flow velocity [37]. The most efficient removal was observed for an iron oxide nanoparticle [36]. These studies show that the collector efficiency for nanoparticles can be very different and that especially the surface-modified nanoparticles displayed high mobilities. Also the environmental conditions are important, most important being the pH.

Owing to their high surface area, nanoparticles have a high sorption capacity not only for metal ions and anions [38, 39] but also for organic compounds [40–42]. Contaminant sequestration is accomplished mainly by surface complexation, but aggregation of particles may encapsulate sorbed surface species. This strong interaction of metal ions and oxide nanoparticles is very important for the behavior and cycling of metals in the environment [43]. The interaction of nanoparticles with toxic compounds can both amplify as well as alleviate the toxicity of the compounds. Nanoparticles can have an advantageous influence on toxicants in the environment by reducing the free toxicant concentration by adsorption onto their surfaces and hence reducing the toxicity of the pollutant.

12.3.2 Toxicology of Nanoparticles

The most important routes for nanoparticles entering the human body are through the gastrointestinal tract, the skin or the lungs [44]. It is obvious that, for food products, the most likely route is through the gastrointestinal tract. The distribution of the nanoparticles in the body is strongly determined by the nanoparticle's surface characteristics [45]. Engineered nanoparticles differ in respect to material, size, surface, and shape. It is not possible, therefore, to make general claims about the health risks of nanoparticles. As a consequence, it has been suggested that engineered nanoparticles need to be assessed on a case-by-case basis [46].

Concern has been raised over the safety of nanoparticles because they have properties that are clearly associated with pathogenicity in particles [47]. Several recent papers have highlighted this area of toxicology, the gaps in research, and possible testing strategies for nanoparticles [10, 12, 48]. The consistent body of evidence shows that nano-sized particles are taken up by a wide variety of mammalian cell types, and are able to cross the cell membrane and become internalized [49–51]. The uptake of nanoparticles is size dependent [52, 53]. In general, cells can survive low concentrations of nanoparticles ($<10\,mg\,l^{-1}$); however, at high doses, cytotoxic effects emerge in a dose- and time-dependent manner for many nanoparticles [48]. While the causes of the increase in cell death observed at higher concentrations and longer exposure times are material specific, the generation of reactive oxygen species is a common finding. The small particle size, a large surface area, and the ability to generate reactive oxygen species play major roles in the toxicity of nanoparticles [9]. Inflammation and fibrosis are effects observed on an organism level, whereas oxidative stress, antioxidant activity, and cytotoxicity are effects observed on a cellular level [10].

The potential effects of nanoparticles in the gastrointestinal tract are largely unknown [54]. A healthy digestive system only allows absorption of nutrients from the gut after digestion of foods. The gut wall is designed to ensure the passage of nutrients and to prevent the passage of larger or foreign material. Transport of particles across the epithelium can occur by the paracellular route (between cells) and the transcellular route [55]. The paracellular route is limited because of the very small surface area of the intercellular space and the tightness of the junctions between cells (pore diameter just 0.3–1 nm). Transcellular uptake of nanoparticles occurs by transcytosis, a process by which nanoparticles are taken up by cells. This transport depends on several factors [55]: (i) the physicochemical properties of the particles, (ii) the physiology of the gastrointestinal tract, and (iii) the animal model used to study the uptake. In general, the nanoparticle uptake increases as the particle diameter decreases.

12.4
How Should Society Deal with Uncertainty?

The use of nanotechnology may result in applications with numerous benefits. However, as outlined above, the very same properties that make nanotechnology or engineered nanoparticles so promising are also the properties that could be responsible for unwanted effects in humans and in the environment. Owing to the lack of available data related to toxicity, exposure, and life cycle of nanotechnology applications, regulatory decisions are in a state of ambiguity or high level of uncertainty. Too much regulation may result in forgoing the benefits of nanotechnology, and too relaxed regulation may result in damages [56]. Some have expressed fear that governmental agencies may not regulate engineered nanoparticles quickly enough, and that therefore the development and implementation of voluntary standards of care are important [57]. Others have called for a moratorium on the use of nanomaterials, especially on the further commercial release of food products, food packaging, food contact materials, and agrochemicals, until nanotechnology-specific safety laws are established and the public is involved in decision-making.

Studies have shown that perceptions on regulatory policy issues in the field of nanoparticulate materials differ among the involved stakeholders [46]. Industry, scientists, governmental bodies, and environmental advocacy groups find regulatory interventions useful, but they are of different opinions as to whether regulations should be evidence-oriented or precaution-oriented, voluntary or top-down controlled. Whereas regulatory bodies and industry do not see the need to regulate this area until more scientific evidence indicates that nanomaterials may be harmful, non-governmental organizations are asking for more proactive risk management strategies. However, companies are legally obliged to guarantee that their products are safe and that they do not cause any harm to human health and the environment.

In recent years, different integrative risk management frameworks for nanomaterials have been developed to overcome the apparent weaknesses of previous approaches [58–60]. Each of the frameworks shares common elements, including integration of hazard assessment, exposure assessment, risk management, and risk communication. However, it should be noted that the risk and safety research and management approaches of nanoparticulate materials are still mainly focusing on non-food nanomaterials and aspects. Appropriate risk governance strategies for nanoscaled materials in food products and food packaging are still in their infancy [7].

12.4.1
Public Perception of Nanotechnology

Several surveys have examined public perception of nanotechnology. Even results of recent studies suggest that public awareness of nanotechnology is low, and that knowledge about nanotechnology is limited at best [61–63]. These studies examined attitudes toward nanotechnology in the abstract, as opposed to attitudes toward realistic products. Based on these studies, it is difficult to predict how the public will react toward real products. It seems likely that perceived benefits largely determine willingness to buy nanotechnology applications.

Owing to the fact that most people do not have much knowledge about nanotechnology, and that they do not have clear ideas about the promises of this technology, study participants should be given some information about nanotechnology, and the applications should be briefly described to enable participants to create attitudes toward nanotechnology applications. Results of a Swiss study examining a broad set of nanotechnology applications, ranging from water sterilization to ammunition, suggest that lay people perceive the various nanotechnology applications differently [64]. More specifically, results showed that lay people perceived applications such as food packaging or water sterilization as more dreaded risks than applications that are not related to food products. On the surface, there seem to be some parallels to gene technology. Consumers are less likely to accept genetically modified (GM) food products compared with medical applications [65]. The research in the domain of nanotechnology further emphasizes that the public is especially concerned when new food technologies are introduced.

In two studies, we examined lay people's perceptions of different nanotechnology foods and nanotechnology food packaging applications [66–67]. Results suggest that lay people perceive nanotechnology packaging as being more beneficial and less risky than nanotechnology foods. Thus, consumers may be less likely to accept nanotechnology foods than innovations related to packaging.

In the study by Siegrist *et al.* [67], 19 different applications were examined. Lay people perceived individually modifiable foods as the most risky applications. Customization of the product, in which nanoparticles release varying flavors, colors or nutrients when warmed in the microwave, depending on the wavelength chosen, was not an accepted innovation for most participants. The second highest risk ratings were received by health-promoting feed and forage. In such an application, livestock feed and forage is infused with proteins encapsulated in nanoparticles. Based on lay people's risk ratings, nanoparticles used for removing toxins in the soil had the seventh highest risk rating of the 19 applications. Overall, results suggest that lay people based their risk assessments, not on the possible environmental impact of the applications, but rather on whether or not the nanoparticles are consumed. Participants may not have taken into consideration the possible migration of nanoparticles from the food packaging to the food. As a result, additional and new information may have changed lay people's perception of nanotechnology food and food packaging applications.

In several countries, public participation and focus group studies have been conducted, in which participants received information about nanotechnology in order to form attitudes toward this new technology [68, 69]. In Switzerland, focus groups were organized to facilitate public discussion and to help decision-makers in assessing nanotechnology [68]. Participants read a brochure about nanotechnology prior to the meetings, and they therefore had some basic knowledge about this enabling technology. Results of this study showed that Swiss citizens had a neutral attitude toward nanotechnology – they were neither enthusiastic about the technology, nor were they rejecting it.

Lay people differ in their acceptance of nanotechnology, and trust seems to be a factor that influences how lay people assess nanotechnology applications [64, 66, 67]. Participants having trust in the industry and in regulatory agencies assessed the nanotechnology application more positively than participants not having trust. The importance of naturalness seems also to be a factor that can affect the perceived risk and the perceived benefit of nanotechnology foods and nanotechnology food packaging [67]. It is likely that general perception of technological progress and attitudes toward technology shape attitudes toward nanotechnology. People often use such convictions in assessing new technologies, about which they have little knowledge [70].

Lay people may have difficulties in understanding the size scale and symbolism of nanotechnology [71]. It should be noted, however, that there may be no need for lay people to understand the principles of nanotechnology in order to accept or to reap the benefits of nanotechnology. The importance of lay people's scientific knowledge must not be overstated. Most people could not explain how a car works. Nevertheless, they drive a car and are willing to accept this technology.

The problems associated with the introduction of genetically modified products in some countries raise the question whether nanotechnology food products may be faced with the same difficulties. It has been argued that GM and nanotechnology are quite different food technologies, and therefore no premature generalizations should be made [72]. Genetic modification that involves the insertion of genes from another species produced a large drop in perceived naturalness [73]. The idea of tampering with nature [74] seems to be an important reason why some people are hesitant to accept GM technology. Since for most people nanotechnology foods will not be perceived as tampering with nature, few people will be opposed to nanotechnology on moral grounds [72].

12.4.2 Scientists and Industrial Perspective

Scientists in the field of nanotechnology are in general more optimistic about the potential benefits and less concerned about the risks of this technology than the public [64, 75]. The study by Scheufele and colleagues [75] suggests that most experts expect that nanotechnology may lead to a better treatment of human diseases and improved ways to clean up the environment. Scientists were more concerned than the public that nanotechnology may lead to more pollution and environmental contamination and new health problems.

Lay people and experts assess the risk associated with nanotechnology differently [64]. Lay people tend to perceive higher risks associated with nanotechnology applications than experts. Another study also found that, for most risks associated with nanotechnology, lay people perceive more risks than experts [75]. However, regarding the risk of more pollution, experts expressed more concern than lay people. This result fits well with the outcomes of the studies discussed in the previous section. It seems that lay people are not especially concerned about a possible impact of nanotechnology on the environment.

Results of an industry survey in Switzerland and in Germany raises some doubts whether all companies properly address possible risks associated with nanotechnology [23]. The way lay people perceive nanotechnology food applications, in conjunction with an industry that may not address the risks associated with a technology as expected by the public, may lead to a social amplification process [76]. Applications in the food or health domains are associated with a high level of dread and distrust [64]. As a result, such nanotechnology applications are most likely to become controversial topics.

12.5 Conclusions

The release of nanoparticles can occur at every stage of the life cycle of a product. Analysis of the life cycles and research about possible effects of nanoparticle products on the environment are still in their infancy. Therefore, it is still unknown in which stages of the life cycle of a product it is most likely that nanoparticles enter the environment. Humans may be confronted with nanoparticles in food through various pathways. Based on the results of available studies, it is still unclear if nanoparticles are problematic for human health and the environment and if nano-food should be treated separately.

Most lay people know very little about nanotechnology, and most people do not have strong attitudes toward this new technology. This poses a problem for studies dealing with lay people's risk perception. People may not be able to answer questions without receiving further information. However, providing additional information may influence people's attitudes in a certain direction. It is difficult to forecast how the public will react to nanotechnology in the future. Based on available risk perception research, it seems that lay people are less concerned about environmental problems associated with nanotechnology, but mainly with nanoparticles that are consumed with foodstuff.

Nanotechnology is an enabling technology, and it is used for a heterogeneous set of applications like ammunition, car paint or foodstuffs. Incidents in one domain may have spill-over effects on other domains. A problem in one field of applications may be extrapolated to other applications because the same label "nanotechnology" is used. This is similar to what has been labeled as "guilt by association" [77]. The industry may be well advised, therefore, not to emphasize nanotechnology in marketing their products.

References

1 Sanguansri, P., and Augustin, M.A. (2006) Nanoscale materials development – a food industry perspective. *Trends Food Sci. Technol.*, **17**, 547–556.

2 Kuzma, J., and VerHage, P. (2006) Nanotechnology in agriculture and food production: anticipated applications. Project on Emerging Nanotechnologies, Woodrow Wilson International Center for Scholars, Washington, DC. Available from: http://www.nanotechproject.org/process/assets/files/2706/94_pen4_agfood.pdf (accessed 8 November 2010).

3 Allianz & OECD (2005) Opportunities and Risks of Nanotechnology, Allianz, Munich.

4 Borm, P.J.A., and Berube, D. (2008) A tale of opportunities, uncertainties, and risks. *Nano Today*, **3**, 56–59.

5 Bauer, C., Buchgeister, J., Hischier, R., Oganietz, W.R., Schebek, L., and Warsen, J. (2008) Towards a framework for life cycle thinking in the assessment of nanotechnology. *J. Clean. Prod.*, **16**, 910–926.

6 Chatterjee, R. (2008) The challenge of regulating nanomaterials. *Environ. Sci. Technol.*, **42**, 339–343.

7 Chau, C.-F., Wu, S.-H., and Yen, G.-C. (2007) The development of regulations for food nanotechnology. *Trends Food Sci. Technol.*, **18**, 269–280.

8 Pusztai, A., and Bardocz, S. (2006) The future of nanotechnology in food science and nutrition: can science predict its safety? in *Nanotechnology: Risk, Ethics and Law* (eds G. Hunt and M.D. Mehta), Earthscan, London, pp. 167–179.

9 Nel, A., Xia, T., Mädler, L., and Li, N. (2006) Toxic potential of materials at the nanolevel. *Science*, **311**, 622–627.

10 Oberdörster, G., Oberdörster, E., and Oberdörster, J. (2005) Nanotoxicology: an emerging discipline evolving from studies of ultrafine particles. *Environ. Health Perspect.*, **113**, 823–839.

11 Köhler, A.R., Som, C., Helland, A., and Gottschalk, F. (2008) Studying the potential release of carbon nanotubes throughout the application life cycle. *J. Clean. Prod.*, **16**, 927–937.

12 Kreyling, W.G., Semmler-Behnke, M., and Möller, W. (2006) Health implications of nanoparticles. *J. Nanopart. Res.*, **8**, 543–562.

13 Jandt, K.D. (2006) Probing the future in functional soft drinks on the nanometre scale – towards tooth friendly soft drinks. *Trends Food Sci. Technol.*, **17**, 263–271.

14 Weiss, J., Takhistov, P., and McClements, J.D. (2006) Functional materials in food nanotechnology. *J. Food Sci.*, **71**, R107–R116.
15 Sorrentino, A., Gorrasi, G., and Vittoria, V. (2007) Potential perspectives of bio-nanocomposites for food packaging applications. *Trends Food Sci. Technol.*, **18**, 84–95.
16 Würth, B. (2007) Emissions of engineered and unintentionally produced nanoparticles to the soil. Diploma thesis, Department of Environmental Sciences, ETH Zürich.
17 NanoSys GmbH (2005) Nano-Argentum 10, Technisches Merkblatt. Available at: http://www.nanosys.ch/tedablas/d/tm_nanoargentum041207.pdf (accessed 10 November 2010).
18 Joseph, T., and Morrison, M. (2006) Nanotechnology in agriculture and food, Nanoforum Report, European Nanotechnology Gateway. Available at: http://www.nanoforum.org/ (accessed 10 November 2010).
19 Hsu, L.Y., and Chein, H.M. (2007) Evaluation of nanoparticle emission for TiO_2 nanopowder coating materials. *J. Nanopart. Res.*, **9** (1), 157–163.
20 Müller, N.C., and Nowack, B. (2008) Exposure modeling of engineered nanoparticles in the environment. *Environ. Sci. Technol.*, **42**, 4447–4453.
21 Benn, T.M., and Westerhoff, P. (2008) Nanoparticle silver released into water from commercially available sock fabrics. *Environ. Sci. Technol.*, **42**, 4133–4139.
22 Blaser, S.A., Scheringer, M., MacLeod, M., and Hungerbuhler, K. (2008) Estimation of cumulative aquatic exposure and risk due to silver: contribution of nano-functionalized plastics and textiles. *Sci. Total Environ.*, **390** (2-3), 396–409.
23 Helland, A., Scheringer, M., Siegrist, M., Kastenholz, H.G., Wiek, A., and Scholz, R.W. (2008) Risk assessment of engineered nanomaterials–survey of industrial approaches. *Environ. Sci. Technol.*, **42**, 640–646.
24 Schmid, K., and Riediker, M. (2008) Use of nanoparticles in Swiss industry: a targeted survey. *Environ. Sci. Technol.*, **42**, 2253–2260.
25 Colvin, V.L. (2003) The potential environmental impact of engineered nanomaterials. *Nat. Biotechnol.*, **21**, 1166–1170.
26 Boxall, A.B.A., Chaudhry, Q., Sinclair, C., Jones, A.D., Aitken, R., Jefferson, B., *et al.* (2007) Current and Future Predicted Environmental Exposure to Engineered Nanoparticles, Central Science Laboratory, Sand Hutton, UK.
27 Nowack, B., and Bucheli, T.D. (2007) Occurrence, behavior and effects of nanoparticles in the environment. *Environ. Pollut.*, **150**, 5–22.
28 Omelia, C.R. (1980) Aquasols–the behavior of small particles in aquatic systems. *Environ. Sci. Technol.*, **14** (9), 1052–1060.
29 Mackay, C.E., Johns, M., Salatas, J.H., Bessinger, B., and Perri, M. (2006) Stochastic probability modeling to predict the environmental stability of nanoparticles in aqueous suspension. *Integr. Environ. Assess. Manage.*, **2** (3), 293–298.
30 Fukushi, K., and Sato, T. (2005) Using a surface complexation model to predict the nature and stability of nanoparticles. *Environ. Sci. Technol.*, **39**, 1250–1256.
31 Kallay, N., and Zalac, S. (2002) Stability of nanodispersions: a model for kinetics of aggregation of nanoparticles. *J. Colloid Interface Sci.*, **253** (1), 70–76.
32 Kallay, N., and Zalac, S. (2001) Introduction of the surface complexation model into the theory of colloid stability. *Croat. Chem. Acta*, **74** (3), 479–497.
33 Grolimund, D., Elimelech, M., Borkovec, M., Barmettler, K., Kretzschmar, R., and Sticher, H. (1998) Transport of *in situ* mobilized colloidal particles in packed soil columns. *Environ. Sci. Technol.*, **32** (22), 3562–3569.
34 Sen, T.K., and Khilar, K.C. (2006) Review on subsurface colloids and colloid-associated contaminant transport in saturated porous media. *Adv. Colloid Interface Sci.*, **119** (2-3), 71–96.
35 Dunphy Guzman, K.A., Finnegan, D.L., and Banfield, J.F. (2006) Influence of surface potential on aggregation and transport of titania nanoparticles. *Environ. Sci. Technol.*, **40**, 7688–7693.

36 Lecoanet, H.F., Bottero, J.Y., and Wiesner, M.R. (2004) Laboratory assessment of the mobility of nanomaterials in porous media. *Environ. Sci. Technol.*, **38**, 5164–5169.
37 Lecoanet, H.F., and Wiesner, M.R. (2004) Velocity effects on fullerene and oxide nanoparticle deposition in porous media. *Environ. Sci. Technol.*, **38**, 4377–4382.
38 Rao, G.P., Lu, C., and Su, F. (2007) Sorption of divalent metal ions from aqueous solution by carbon nanotubes: a review. *Separ. Purif. Technol.*, **58**, 224–231.
39 Waychunas, G.A., Kim, C.S., and Banfield, J.F. (2005) Nanoparticulate iron oxide minerals in soils and sediments: unique properties and contaminant scavenging mechanisms. *J. Nanopart. Res.*, **7**, 409–433.
40 Chen, W., Duan, L., and Zhu, D.Q. (2007) Adsorption of polar and nonpolar organic chemicals to carbon nanotubes. *Environ. Sci. Technol.*, **41** (24), 8295–8300.
41 Nowack, B. (2008) Pollution prevention and treatment using nanotechnology, in *Nanotechnology*, Vol. 2, *Environmental Aspects* (ed. H.F. Krug), Springer, Berlin, pp. 1–15.
42 Yang, K., Zhu, L., and Xing, B. (2006) Adsorption of polycyclic aromatic hydrocarbons by carbon nanomaterials. *Environ. Sci. Technol.*, **40**, 1855–1861.
43 Hochella, M.F., and Madden, A.S. (2005) Earth's nano-compartment for toxic metals. *Elements*, **1** (4), 199–203.
44 Maynard, A.D. (2006) Nanotechnology: assessing the risks. *Nano Today*, **1**, 22–33.
45 Brayner, R. (2008) The toxicological impact of nanoparticles. *Nano Today*, **3**, 48–55.
46 Helland, A., Kastenholz, H., Thidell, A., Arnfalk, P., and Deppert, K. (2006) Nanoparticulate materials and regulatory policy in Europe: an analysis of stakeholder perspectives. *J. Nanopart. Res.*, **8**, 709–719.
47 Donaldson, K., Aitken, R., Tran, L., Stone, V., Duffin, R., Forrest, G., *et al.* (2006) Carbon nanotubes: a review of their properties in relation to pulmonary toxicology and workplace safety. *Toxicol. Sci.*, **92** (1), 5–22.
48 Lewinski, N., Colvin, V., and Drezek, R. (2008) Cytotoxicity of nanoparticles. *Small*, **4** (1), 26–49.
49 Lynch, I., Dawson, K.A., and Linse, S. (2006) Detecting cryptic epitopes created by nanoparticles. *Sci. STKE*, **2006** (327), pe14. doi: 10.1126/stke.3272006pe14.
50 Rothen-Rutishauser, B.M., Schürch, S., Haenni, B., Kapp, N., and Gehr, P. (2006) Interaction of fine particles and nanoparticles with red blood cells visualized with advanced microscope techniques. *Environ. Sci. Technol.*, **40**, 4353–4359.
51 Smart, S.K., Cassady, A.I., Lu, G.Q., and Martin, D.J. (2006) The biocompatibility of carbon nanotubes. *Carbon*, **44** (6), 1034–1047.
52 Chithrani, B.D., Ghazani, A.A., and Chan, W.C.W. (2006) Determining the size and shape dependence of gold nanoparticle uptake into mammalian cells. *Nano Lett.*, **6** (4), 662–668.
53 Limbach, L.K., Li, Y., Grass, R.N., Brunner, T.J., Hintermann, M.A., Muller, M., *et al.* (2005) Oxide nanoparticle uptake in human lung fibroblasts: effects of particle size, agglomeration, and diffusion at low concentrations. *Environ. Sci. Technol.*, **39**, 9370–9376.
54 Chaudhry, Q., Scotter, M., Blackburn, J., Ross, B., Boxall, A.B.A., Castle, L., *et al.* (2008) Applications and implications of nanotechnologies for the food sector. *Food Addit. Contam.*, **25** (3), 241–258.
55 des Rieux, A., Fievez, V., Garinot, M., Schneider, Y.J., and Préat, V. (2006) Nanoparticles as potential oral delivery systems of proteins and vaccines: a mechanistic approach. *J. Control. Release*, **116**, 1–27.
56 Renn, O., and Roco, M.C. (2006) Nanotechnology and the need for risk governance. *J. Nanopart. Res.*, **8**, 153–191.
57 Balbus, J.B., Florini, K., Denison, R.A., and Walsh, S.A. (2007) Protecting workers and the environment: an environmental NGO's perspective on nanotechnology. *J. Nanopart. Res.*, **9**, 11–22.
58 IRGC (2007) Nanotechnology Risk Governance: Recommendations for a Global, Coordinated Approach to the

Governance of Potential Risks, International Risk Governance Council, Geneva.

59 Marchant, G.E., Sylvester, D.J., and Abbott, K.W. (2008) Risk management principles for nanotechnology. *Nanoethics*, **2**, 43–60.

60 Tyshenko, M.G., and Krewski, D. (2008) A risk management framework for the regulation of nanomaterials. *Int. J. Nanotechnol.*, **5** (1), 143–160.

61 Cobb, M.D., and Macoubrie, J. (2004) Public perceptions about nanotechnology: risks, benefits and trust. *J. Nanopart. Res.*, **6**, 395–405.

62 Gaskell, G., Ten Eyck, T., Jackson, J., and Veltri, G. (2005) Imaging nanotechnology: cultural support for technological innovation in Europe and the United States. *Public Underst. Sci.*, **14**, 81–90.

63 Waldron, A.M., Spencer, D., and Batt, C.A. (2006) The current state of public understanding of nanotechnology. *J. Nanopart. Res.*, **8**, 569–575.

64 Siegrist, M., Keller, C., Kastenholz, H., Frey, S., and Wiek, A. (2007) Laypeople's and experts' perception of nanotechnology hazards. *Risk Anal.*, **27**, 59–69.

65 Durant, J., Bauer, M.W., and Gaskell, G. (eds) (1998) *Biotechnology in the Public Sphere*, Science Museum, London.

66 Siegrist, M., Cousin, M.-E., Kastenholz, H., and Wiek, A. (2007) Public acceptance of nanotechnology foods and food packaging: the influence of affect and trust. *Appetite*, **49**, 459–466.

67 Siegrist, M., Stampfli, N., Kastenholz, H., and Keller, C. (2008) Perceived risks and perceived benefits of different nanotechnology foods and nanotechnology food packaging. *Appetite*, **51**, 283–290.

68 Burri, R.V., and Bellucci, S. (2008) Public perception of nanotechnology. *J. Nanopart. Res.*, **10**, 387–391.

69 Pidgeon, N., and Rogers-Hayden, T. (2007) Opening up nanotechnology dialogue with the publics: risk communication or "upstream engagement"? *Health Risk Soc.*, **9**, 191–210.

70 Visschers, V.H.M., Meertens, R.M., Passchier, W.F., and de Vries, N.K. (2007) How does the general public evaluate risk information? The impact of associations with other risks. *Risk Anal.*, **27**, 715–727.

71 Batt, C.A., Waldron, A.M., and Broadwater, N. (2008) Numbers, scale and symbols: the public understanding of nanotechnology. *J. Nanopart. Res.*, **10**, 1141–1148.

72 Siegrist, M. (2008) Factors influencing public acceptance of innovative food technologies and products. *Trends Food Sci. Technol.*, **19** (11), 603–608.

73 Rozin, P. (2006) Naturalness judgments by lay Americans: process dominates content in judgments of food or water acceptability and naturalness. *Judgm. Decis. Mak.*, **1** (2), 91–97.

74 Sjöberg, L. (2000) Perceived risk and tampering with nature. *J. Risk Res.*, **3**, 353–367.

75 Scheufele, D.A., Corley, E.A., Dunwoody, S., Shih, T.-J., Hillback, E., and Guston, D.H. (2007) Scientists worry about some risks more than the public. *Nat. Nanotechnol.*, **2**, 732–734.

76 Pidgeon, N., Kasperson, R.E., and Slovic, P. (2003) *The Social Amplification of Risk*, Cambridge University Press, Cambridge.

77 Keller, K.H. (2007) Nanotechnology and society. *J. Nanopart. Res.*, **9**, 5–10.

13
Nanotechnology and Food Allergy

E.N. Clare Mills, Yuri Aleexev, and Alan R. Mackie

13.1
Introduction

Food allergies are one of several different types of reproducible adverse reactions to foods that have been described, which also include enzyme deficiencies, such as lactose intolerance, and pharmacological reactions to foods rich in compounds such as histamine. Food allergies share a common characteristic, namely an immunological basis, and so far two different forms have been recognized [1]. One of them involves the humoral arm of the immune system with the development of food-specific immunoglobulin E (IgE) responses that can trigger a host of reactions usually classified as type I hypersensitivity reactions. The other type involves activation of immune cells in the gut, and is manifested as the gluten intolerance syndrome known as celiac disease.

With regard to IgE-mediated food allergies, during normal healthy functioning, the immune system produces a type of immunoglobulin known as IgE, the role of which is to defend the body from parasitic infections, such as malaria. For reasons not fully understood, some individuals begin to make IgE in response to various environmental agents, including dust, pollens, and foods, which can lead to the development of allergic reactions. Such IgE-mediated allergies develop in two phases: (i) sensitization when IgE production is stimulated, and (ii) elicitation when an individual experiences an adverse reaction, mediated by IgE, upon re-exposure to an allergen. Both stages are triggered by allergens, which are almost always proteins. In an allergic reaction, allergen is recognized by IgE bound to the surface of histamine containing mast cells, cross-linking the IgE in the process and triggering the release of inflammatory mediators such as histamine. These mediators cause the acute inflammatory reactions that become manifested as respiratory (asthma, rhinitis), cutaneous (eczema, urticaria) or gastrointestinal (vomiting, diarrhea) symptoms, which may occur alone or in combination in an allergic reaction. A rare but very severe reaction is anaphylactic shock characterized by respiratory symptoms, fainting, itching, urticaria, swelling of the throat or other mucous membranes, and a dramatic loss of blood pressure.

Nanotechnology in the Agri-Food Sector: Implications for the Future, First Edition. Edited by Lynn J. Frewer, Willem Norde, Arnout Fischer, Frans Kampers.

In contrast to the rapid onset characteristic of type I hypersensitivity reactions, the gluten intolerance syndrome, celiac disease, can take between hours and days to manifest itself after consumption of gluten-containing food. It is thought that around 1% of the population suffers from this gluten intolerance syndrome, and it seems to affect women more than men. It is caused by the recognition of gluten peptides that result from digestion, which have first been deamidated by the action of gut mucosal transglutaminase. These deamidated peptides can bind to receptors, known as class II human histocompatibility leukocyte antigen receptors, of certain types, namely DQ2 and DQ8. This then triggers an abnormal cell-mediated immune response, which results in an inflammatory reaction in the gastrointestinal mucosa, which causes the loss of the normal villous architecture that is characteristic of celiac disease [2].

There is no proactive treatment available for either IgE-mediated food allergies or celiac disease. Consequently, individuals who suffer from these conditions have to practice food avoidance and, in the case of IgE-mediated allergies, are provided with medication (such as adrenalin pens) to be used in case of accidental consumption of a problem food. In practice, it can be difficult to avoid some problem foods, especially widely used ingredients such as wheat, cows' milk or hens' egg. It is generally held that the vast majority of food allergies are caused by a limited number of foods [3], although a large number of foods have been documented as causing food allergies, reflecting the diversity of food species that humans consume. In order to help allergic consumers avoid problem foods, legislation has been brought in around the world that makes it mandatory to label certain allergenic foods and derived ingredients, irrespective of the level to which they are added to a foodstuff [4].

13.2
Molecules in Foods Involved in Triggering Allergies

The molecules that trigger both types of immunological reactions to foods are known as allergens and to date those responsible for almost all food allergies are proteins. Those involved in triggering celiac disease are confined to the prolamin seed storage proteins of cereals (wheat, rye, and barley). In contrast, the proteins involved in triggering IgE-mediated food allergies are still being identified and characterized, but they originate from a diverse range of foods of both plant and animal origin. Food allergens triggering IgE-mediated reactions appear to be restricted to certain structural types or protein families [4], and this has led to a classification based on protein family membership [5]. Thus, an analysis of plant food allergen families has shown that they belong to only 27 protein families [6], with four protein families (prolamin, cupin, Bet v 1, and profilin families) accounting for more than 65% of all plant food allergens. The distribution of animal food allergens was similar [7], with three protein families (tropomyosin, parvalbumin, and casein families) dominating. These observations suggest that conserved structures and biological activities play a role in determining or promoting allergenic properties of proteins, and are in part explained by the conservation of surface structures in certain families, such as the Bet v 1 and parvalbumin superfamilies, which promote IgE cross-reactivity [6, 7].

The characteristics of major allergen families are summarized below. More detailed reviews of food allergen structure and properties, including both major and minor allergen families, can be found in references [8] and [9].

13.2.1
Plant Food Allergens

13.2.1.1 Prolamin Superfamily

The prolamin superfamily comprises the seed storage prolamins of cereals, 2S albumins, non-specific lipid transfer proteins, and α-amylase/trypsin inhibitors of cereals [4, 5]. Apart from the seed storage prolamins, these are all low-molecular-weight cysteine-rich proteins that share the same three-dimensional fold, are rich in α-helices, and are generally stable to thermal processing and proteolysis. The 2S albumins are a major group of storage proteins present in many dicotyledonous plants. They include major allergens from tree nuts and seeds such as Brazil nut, walnut, sesame, and mustard. The non-specific lipid transfer proteins play an important role in plant defense against fungi and bacteria. They are found in a diverse range of plant foods, including fruits, nuts, seeds, and vegetables, and are an important group of allergens in the Mediterranean area [10]. The family of cereal α-amylase and protease inhibitors mediates a certain degree of resistance to insect pests that feed on plant tissues, allergenic members having been identified in wheat, barley, rice, and corn. Like the 2S albumins and the non-specific lipid transfer proteins, these allergens are able to sensitize susceptible individuals through either ingestion or inhalation.

The prolamin seed storage proteins appear to have evolved through insertion of a highly repetitive domain within the cysteine skeleton. They are involved in triggering some IgE-mediated allergies and, more importantly, trigger the gluten intolerance syndrome, celiac disease, involving the homologous proteins from wheat, rye, and barley. Many celiacs can tolerate oats, which contain much lower levels of prolamin storage proteins (known as avenins). But while they have a slightly different structure compared to the prolamins from the Triticeae, there are still some concerns about the safety of oats for celiacs in general.

It appears that, while all prolamin fractions appear to trigger the condition, the most potent appears to be α-gliadin, which can trigger more severe reactions [11]. The key feature appears to involve recognition of prolamin-derived peptides by receptors such as histocompatibility leukocyte antigen DQ2 (and some to DQ8), which results in stimulation of T-cell responses that initiate inflammatory reactions. One particular peptide that appears to stimulate the majority of T-cells in untreated celiacs corresponds to a 33-amino-acid peptide derived from α-gliadin. This peptide is not completely digested by enzymes in the gastrointestinal tract lumen or the brush border enzymes of the mucosa and includes epitopes corresponding to amino acid sequences such as Pro-Phe-Pro-Gln-Pro-Gln-Leu-Pro-Tyr, Pro-Gln-Pro-Gln-Leu-Pro-Tyr-Pro-Gln, and Pro-Tyr-Pro-Gln-Pro-Gln-Leu-Pro-Tyr [12].

13.2.1.2 **Cupin Superfamily**

The cupins are a functionally diverse superfamily of proteins that share a β-barrel structural core domain to which the term "cupin" (Latin *cupa*, meaning "barrel") was given. The cupin superfamily comprises the major globulin storage proteins mainly from legumes and nuts. The globulins are divided into the 7S vicilin-like globulins and the 11S legumin-like globulins. Globulins have been found to be highly relevant allergens in plant foods including peanuts, soybean, lentils, walnut, hazelnut, and sesame [4, 5]. Despite having very low levels of sequence identity, members of the cupin superfamily have highly conserved structures. In contrast to the Bet v 1 family of plant food allergens, there is little evidence of IgE cross-reactivity between cupin allergens, with an overall sequence identity of less than 40%. This results in very limited cross-reactivity between cupins from even closely related species such as peanut and pea [13].

13.2.1.3 **Bet v 1 Family**

Individuals with pollen allergy frequently suffer from allergic symptoms after eating certain plant foods. The majority of these reactions are caused by allergens of Rosaceae fruits like apple, peach, cherry, and apricot, and certain vegetables such as celery root (celeriac) and carrot, which cross-react with allergens that are present in birch pollen, particularly the major birch pollen allergen Bet v 1 [5]. Bet v 1 was the first of many allergens published that showed homology to family 10 of the pathogenesis-related proteins. Bet v 1-type allergens are rather unstable to heating and digestion. Consequently, symptoms are mostly restricted to the oral cavity. In general, Bet v 1 from birch pollen is thought to act as the primary sensitizing agent, with allergies to foods developing subsequently [14]. The overall high levels of conserved surface residues between the members of the Bet v 1 family plays an important role in conservation of IgE binding sites and underlies the fruit–vegetable–pollen cross-reactive syndromes [6].

13.2.1.4 **Profilins**

Being cytosolic proteins, profilins are ubiquitous proteins found in all eukaryotic cells, which are thought to play a role in regulating the polymerization and depolymerization of actin during a variety of cellular processes including cell movement [15]. Like members of the Bet v 1 family, profilins are involved in cross-reactive allergies, where sensitization to pollens results in IgE responses toward homologs found in fresh fruits and vegetables [16]. However, the clinical relevance of plant food profilin-specific IgE is still under debate [17].

13.2.2
Animal Food Allergens

13.2.2.1 **Tropomyosins**

Tropomyosins are cytoskeletal proteins and, together with other contractile proteins, such as actin and myosin, play a key role in regulation of muscle contraction [18]. Together with actin and myosin, tropomyosins play a key regulatory role in muscle contraction. Being two-stranded α-helical coiled proteins, tropomyosins form head-to-tail polymers along the actin filaments. Tropomyosins have been described as allergens in Crustacea (such as shrimps, crab, and lobster) and Molluscs (such as abalone) and are recognized as invertebrate pan-allergens [19]. The proteins are heat-stable and, because of the extensive homologies between invertebrates, tend to show IgE cross-reactivity between crustacean and molluscs [20, 21].

13.2.2.2 Parvalbumins

The second largest animal food allergen family are the fish β-parvalbumins, a calcium-binding protein found in the white muscle with a characteristic EF-hand structure [22]. They have been characterized as allergens in many different fish species and are considered as pan-allergens in fish [23], their conservations of surface structures explaining the IgE cross-reactivity that is frequently observed between fish species [7]. The proteins show considerable thermal stability when calcium is bound [24], but changes in conformation when calcium is lost is associated with a loss of IgE reactivity [25, 26].

13.2.2.3 Caseins

The major proteins found in milk, caseins are structurally mobile proteins that bind calcium through clusters of phosphoserine and/or phosphothreonine residues. The casein fraction of cows' milk comprises α_{s1}-, α_{s2}- and β-caseins, which assemble into micelles stabilized by κ-casein [27]. They are the major food allergens in cows' milk allergy, which is primarily an allergy of infancy. There is considerable sequence similarity between caseins from different species, with sequence identities of over 90% between cows' milk and goats' milk caseins, explaining the cross-reactive allergies between cows' milk and goats' milk [28].

13.3 Food Structure, Processing, and Food Allergy

While we are gaining an extensive knowledge of the molecules in foods that trigger food allergies, they are not consumed as individual purified molecules, but rather as part of foods. Indeed, many allergens are abundant in foods and make an important contribution to forming the food structure itself. During food processing procedures, allergens may undergo complex physical and chemical changes, altering their three-dimensional structure, and promoting interactions (both covalent and non-covalent) with other food constituents, including proteins, lipids, and sugars. These changes, coupled with the effects of the food matrix itself, may affect the release and stability of allergens, and all have the potential to either reduce or enhance the allergenic potential of food allergens by modifying the way in which they are presented to the immune system. Such effects may be mediated at both a molecular and a macroscopic level.

13.3.1
Molecular Effects of Food Processing on Allergenicity

The impact of processing on allergenic potential, especially in terms of eliciting an allergic reaction, can be considered in terms of the effect that processing-induced changes in allergen structures have on IgE binding, particularly in relation to IgE epitopes. One type of epitope comprises linear stretches of contiguous amino acids and is generally known as a linear or continuous epitope. IgE binding to such epitopes is usually unaffected by the folded state of the protein. A second type of epitope (often known as a conformational epitope) is formed from different segments of a polypeptide chain that are brought together in space as a consequence of the way in which a protein is folded. Such epitopes can be disrupted as a consequence of protein denaturation, reducing or even abolishing antibody recognition [29]. It has been suggested that an antibody developed toward a highly disordered state (such as a denatured protein) is able to recognize the more highly ordered states found in native, folded proteins possibly because it recognizes linear epitopes [30]. This is especially so if an epitope is located on the surface of the folded form of the protein. However, antibodies directed toward a folded protein (conformational epitopes) tend to be directed to conformational epitopes and hence often recognize denatured forms poorly, if at all.

Lastly, small molecules (such as sugars) attached to proteins can form part of an epitope. Such molecules are unable to elicit an immune response alone but can stimulate humoral immune responses when linked to a carrier molecule, such as a protein. Known as haptens, such small molecular substituents on proteins have been shown to stimulate IgE responses, particularly in relation to cross-reactive carbohydrate determinants. However, for cross-reactive carbohydrate determinants at least, such haptens are often unable to stimulate a biological response, possibly because their sparse distribution on a protein limits their ability to cross-link IgE on mast cells and hence stimulate histamine release [31, 32].

Thus, food processing has the ability to destroy IgE epitopes, but may also introduce novel (neo)epitopes into a protein, either by changing protein conformation, by linking proteins together in aggregated states, or as a consequence of introducing novel haptens, through Maillard modification, by conjugation with lipid oxidation products, or via a host of other chemical changes that may result from the thermal treatments frequently used in food production. The effect of thermal processing on allergen structure depends on many factors such as time–temperature combinations, protein concentration, water activity, and whether a protein is heated alone or in combination with other proteins or food ingredients such as sugars. These types of interaction can be illustrated by processing-induced changes that have been described for the major whey proteins (see Figure 13.1), β-lactoglobulin (β-Lg) and α-lactalbumin (α-La), which have been structurally well defined using techniques such as nuclear magnetic resonance spectroscopy and X-ray diffraction, and are both important cows' milk allergens [33]. Other food allergens that have been shown to undergo similar types of interactions, unfolding and forming aggregates, are the allergenic 7S and 11S seed storage globulins from

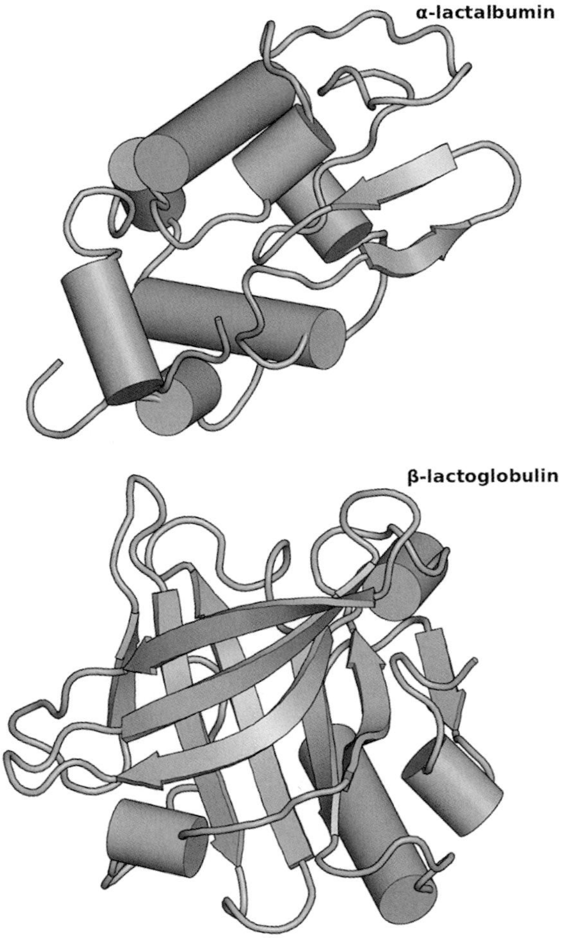

Figure 13.1 Structures of the important whey proteins in cows' milk: α-lactalbumin (Protein Data Bank number 1HFX) and β-lactoglobulin (Protein Data Bank number 1BSY) at pH 7.1. The α-helices are shown as cylinders. The β-pleated sheet and loops are shown as broad flat arrows and as strings/wires, respectively. (The pictures were generated using the open-source molecular visualization system PyMOL.)

foods such as peanut and soybean [34] and the allergenic potato tuber protein patatin [35].

A 18 400 dalton retinol binding protein, β-Lg is a β-barrel protein belonging to the lipocalin superfamily and is stabilized by two intramolecular disulfide bonds (Cys^{106}–Cys^{119} and Cys^{66}–Cys^{160}), together with a single free cysteine residue (Cys^{121}) [36]. It is present as a mixture of monomers and dimers at neutral pH, dissociating on heating to 70 °C [37] and appears to adopt a partially folded state following thermal denaturation [38], before forming thread-like aggregates around 50 nm in diameter [39]. Heat-induced unfolding of β-Lg reveals the buried Cys^{121}, which is then able to catalyze disulfide interchange with other disulfides in β-Lg to form a non-native monomer in which Cys^{119} is exposed [40].

Also being a low-molecular-weight 14200 dalton disulfide-bond-stabilized calcium binding protein, α-La has a role in regulating lactose synthase [41]. Primarily an α-helical protein, it exists at low pH or moderately elevated temperatures in a partially folded or "molten globule" state [42]. Thus α-La expands on formation of the low-pH-induced molten globule from 19.4 to 21.6 nm, while after heating it is further expanded to 23.8 nm. The heat-induced partially folded form is kinetically trapped as a consequence of intermolecular disulfide interchange and retains much of the secondary structure of the native protein.

While thermally induced changes in the structure of these important whey allergens are well defined, little is known about their impact on IgE reactivity. Some limited studies have shown that the IgE binding capacity of β-Lg (variants A and B) is reduced following thermal treatments able to denature the protein, some trace of IgE binding remaining [43]. Such studies are also consistent with clinical observations that the allergenicity of extensively baked milk products (muffins, heated to 180 °C for 30 min) is substantially reduced, especially in children whose cows' milk allergy is beginning to resolve [44]. Investigations into the effect of processing on sensitization potential are more difficult to undertake and rely on the use of animal models. Nevertheless there are indications that heat-induced aggregation of whey proteins affected the path of uptake across the mucosal barrier and that soluble proteins were endocytosed by epithelial cells. But after pasteurization the resulting aggregates were preferentially taken up by the Peyer's patches and this was associated with a shift toward a Th2-associated antibody and cytokine pattern. However, the soluble proteins were much more effective in triggering an anaphylactic reaction [45].

13.3.2 Macroscopic Effects of Food Processing on Allergenicity

Both naturally occurring food structures, and those formed in fabricated foods, may act to trap allergens, preventing their becoming solubilized in fluids such as saliva, gastric or duodenal secretions, and possibly protecting them from degradation by intestinal proteases. This can affect their allergenic potential, in terms of both sensitization and elicitation.

13.3.2.1 Natural Cellular Structures

Allergens are contained within the natural cellular and tissue structures of fruits, vegetables, and seeds, and in the cellular and fibrous structures of meats. For example, the non-specific lipid transfer protein allergens of fruits are largely confined to the skins of fruits such as apple and peach, reflecting the greater allergenicity of peel with respect to flesh for patients suffering from fruit allergies involving non-specific lipid transfer proteins [46]. In contrast, Bet v 1 allergens are largely confined to the flesh [47]. It may be that differences in the structure and components in different fruits and vegetables account for the different allergenic properties of homologous allergens. For example, the allergenic Bet v 1 homolog of celery root (celeriac) has been shown to be stable to processing, retaining its ability to elicit an allergic reaction after cooking [48], but the Mal d 1 homolog found in apple is lost after processing [49].

Natural structures, in particular the plant cell walls found in a particular plant tissue, may affect the stability, release, and presentation of allergenic molecules to the immune system. For example, the mechanical break-up of plant tissues, either during food processing (such as cooking, or preparation of fruit purée) or during chewing, is determined by the plant cell wall properties and will both affect release of allergens into solution and generate a range of particulate structures made up of fragments of the original plant tissue structure. The cell wall structure and composition will also determine how intact cells, clusters of cells or larger fragments of plant tissue structures respond to the environment of the upper gastrointestinal tract, and hence may alter the ingress of degradative enzymes and biosurfactants, as well as the release of allergens into the gut lumen. Similarly, it may be that the cellular and fibrous structures found in the flesh of animals, such as fish, crustacean, and molluscs, may affect the way in which fish and shellfish allergens are released from cooked flesh.

13.3.2.2 Processed Food Structures

Structures formed in complex processed foodstuffs may also affect the stability and release of allergenic molecules. Many foods are in the form of dispersions, with one phase (such as oil, starch granules or other particulates) dispersed in a second immiscible phase in the form of droplets (like oil droplets in water found in sauces such as mayonnaise), air bubbles (like the air bubbles found in bread dough) or particulates (like starch granules in a sauce made using corn starch). These dispersions include the following.

- **Gels** These can be like either the low-pH-set gels of milk-based yogurts or the heat-set gels formed when boiling an egg.
- **Foams** In this group fall the whipped egg whites in meringues and mousse-style desserts. In some cases the foams become set by cooking, with either the protein or starch forming a solid network, which usually needs to rupture following baking to form a sponge network such as is found in cakes.
- **Emulsions** Either oil-in-water (salad dressings or cream) or water-in-oil (spreads and margarines) emulsions, these are unstable unless a surface-active agent is added, such as a protein or a low-molecular-weight surfactant such as lecithin.

In many cases, food structures are formed from the allergenic proteins – gels may be formed from milk or egg proteins, or set foams formed by gluten proteins in bread and cakes. Additionally, other allergens such as whey proteins may be used as emulsifiers. However, there is an almost complete lack of knowledge on how such classical food structures may affect the allergenic potential of foods. This is partly because many clinical investigations have utilized soluble extracts of foods and processed food systems rather than investigating the allergenic activity of the insoluble matrix because of the technical difficulties in studying such insoluble systems. One of the few clinical studies undertaken in this difficult area of research showed that enhancing the fat content of a chocolate matrix containing peanuts affected the kinetics of allergen release and potentiated severe allergic reactions [50]. Such studies that have been published have often been restricted to investigations on the ability of processed foods to elicit reactions in individuals already suffering from a food allergy. We currently lack effective animal models for investigating the potential for allergens or foods to sensitize, and hence our knowledge base in this topic is almost non-existent.

13.3.3
Molecular and Macroscopic Effects of Processing on Allergenicity of Foods

The complex interplay between molecular and macroscopic effects of food processing in relation to allergenicity of foods can be illustrated by a couple of well-characterized allergen families, the Bet v 1 and prolamin superfamilies. One type of food allergy where the IgE binding is dominated by conformational epitopes is the pollen–fruit allergy syndrome involving the birch pollen allergen Bet v 1. In this condition, individuals become sensitized to native Bet v 1 through inhalation of birch pollen, and consequently the main IgE binding sites are primarily directed toward conformational epitopes on the native protein [51, 52]. Thus, it might be expected that processing could disrupt these conformational IgE epitopes, reducing the allergenicity of a cooked, compared with a fresh, food. However, the extent to which this happens will be determined by the inherent thermostability of the protein. Bet v 1 itself is relatively thermostable, the protein irreversibly unfolding only at temperatures above 68 °C [53], and in some foods, such as celeriac, this is expressed in the stability of the allergenic Bet v 1 homologue, Api g 1, to processing [10]. Similarly the Bet v 1 homolog from soybean, Gly m 4, retains its allergenicity even in a processed soya-based food supplement [54, 55].

However, this is not so for all foods involved in the birch pollen–fruit allergy syndrome, and especially for fruits such as apple [49], while roasting hazelnuts reduced but did not abolish their allergenic properties in a group of patients with birch-pollen-associated allergy to hazelnuts [56], as has been shown more recently by others [57]. Therefore, it appears that other factors, such as the food matrix itself, as well as the inherent thermostability of a protein and the type of processing procedures employed, may be responsible for the apparent lability of Bet v 1 homologs in foods such as apple compared with celery root.

Another family of allergens that are inherently thermostable are the various members of the prolamin superfamily. With the exception of the prolamin seed storage proteins of cereals, the large number of intramolecular disulfide bonds present in these proteins play an important role in determining their thermostability. Both the 2S albumin allergens, such as Ber e 1 from Brazil nut and Ses i 1 from sesame seeds, have secondary structures that are almost unaltered by heating [58, 59], as well as the allergenic non-specific lipid transfer proteins from a variety of fruits such as apple [60]. However, despite such inherent thermostability, in some instances the allergenicity of specific lipid transfer proteins is retained even after the extensive thermal treatments and fermentation involved in brewing and wine-making [61], as is the IgE binding capacity of wheat α-amylase inhibitors when a model cooking procedure involving preparation of a flour gel comparable to a porridge was used in a study of wheat allergy [62]. However, in the same study of wheat allergy, some patients lost their IgE binding capacity toward wheat-specific lipid transfer protein [62], while in a study of specific lipid transfer protein-mediated rice allergy, boiling abolished IgE binding [63].

In such complex food systems, there is an interplay between the stability of individual allergens, coupled with interactions with other components in the food matrix that could render proteins insoluble and hence no longer accessible and able to trigger a reaction. The ability of wheat prolamins to form disulfide bonds could alter the allergenic properties of other ingredients in baked goods, and it has been shown that the egg white allergen ovomucoid becomes disulfide-linked to the gluten proteins during baking, rendering it insoluble and hence reducing the allergenic activity of soluble extracts made from such baked goods [64]. Alternatively, this loss of IgE reactivity might be due to leaching of the allergen into the cooking water, as has been observed for another prolamin superfamily member, the peanut allergen Ara h 2 [65].

As well as physical changes induced in protein structure through denaturation and aggregation, processing may introduce the formation of complexes with other food components that may also alter protein stability and bioaccessibility. Thus, the plant polyphenol epigallocatechin has been shown to cause compaction of cows' milk caseins, with the casein molecules wrapping around the polyphenol, forming a complex held together by hydrophobic interactions [66]. Modification of peanut allergens Ara h 1 and Ara h 2 with phytic acid showed that this compound reduced both their solubility and their IgE reactivity, an effect mirrored by treatment of peanut butter with phytic acid [67].

13.4
Impact of Nanoscale Structures on Allergenic Potential of Foods

Our lack of knowledge about the impact that food processing and structure have on the allergenicity of foods makes it difficult to assess the potential impact that novel processes, including the use of nanoscale structures in foods, will have on allergenicity. However, as described above, the formation of protein aggregates and networks, complexes with lipids, and other food components, results in the formation of nanoscale structures, which we have been consuming probably ever since mankind began using heat to preserve and cook foods. There are no published data on the impact of nanotechnology in relation to food allergy, and studies in relation to the impact of fabricated nanoscale structures on allergy in general are in their infancy, particularly regarding their use in drug delivery. Thus, delivery of a deoxyribonucleic acid (DNA) vector expressing transforming growth factor beta (TGFβ) in chitosan nanoparticles via the gastrointestinal tract was able to ameliorate the symptoms of food allergy in an animal model, using the egg allergen, ovalbumin, as a model food allergen [68]. Similar beneficial effects have been observed in using chitosan particles to deliver mite allergens for immunotherapy in the treatment of mite allergy [69] and with biodegradable poly(D,L-lactic-*co*-glycolic acid) nanospheres used to deliver Bet v 1 in immunotherapy [70].

Other studies have focused on adverse effects of inorganic nanoparticles, such as titanium dioxide, on allergic reactions to personal products such as cosmetics, using conditions such as atopic dermatitis, a symptom often associated with food allergies, especially in infants, as a model system. Using a dust mite model system in mice, titanium dioxide nanoparticles irrespective of size (15, 50 or 100 nm in size) were found to aggravate immunological markers of the atopic dermatitis-like skin lesions in this model system, including serum IgE [71]. In contrast, nanocrystalline silver had beneficial effects in reducing inflammation in a guinea pig model of contact dermatitis to a similar extent as topical steroids [72].

The impact of nanoparticles, especially those resulting from atmospheric pollution and found in for example, diesel exhaust, has also been studied in relation to allergic disease, and relates to their ability to have a pro-inflammatory effect on the respiratory epithelium. The concerns are that they might have adjuvant effects on allergic sensitization, and this is reflected in reports of studies undertaken in animal models using ultra-fine carbon particles, which increased inflammation by increasing oxidative stress [73–76]. In contrast to such adverse effects, there are indications that novel carbon structures, such as fullerenes, may have beneficial effects, reducing mediator release involved in elicitation of allergic reactions, including a model of anaphylaxis [77].

The efficacy of such nanoscale structures for delivery of therapeutics is explained in part by the observations that biodegradable nanoparticles, such as poly(D,L-lactic-*co*-glycolic acid), can be taken up by cellular models of the respiratory and gut epithelium [78], although the cell models do not include the mucus layer, an important biophysical barrier that particulates must traverse before contacting the underlying epithelium. However, there is evidence that combinations of dextran and chitosan nanoparticles 500 nm in size were muco-adhesive, and were effective at rendering insulin bioavailable through the oral route [79]. Such effectiveness at overcoming the gut barrier has clear benefits in terms of delivery of bioactive molecules via the oral route. However, there is almost nothing known about the potential impact on allergenicity, especially of nanoparticles that may be included in foods to enhance delivery of important health-promoting micronutrients [80–82].

13.5 Conclusions

Food allergy is an emerging problem, and while our documentation of the molecules responsible for triggering allergic reactions is extensive, the way in which these molecules are altered by food processing conditions and how food structures may alter their presentation to the immune system is very incomplete. Of particular relevance to considering the potential impact of nanotechnology, undoubtedly one of the most important aspects is the use of nanoparticles to deliver therapeutic agents, such as those involved in immunotherapy. Utilization of nanoparticles for oral delivery of other important therapeutics, such as insulin, is showing promise, and it is likely, as therapies are developed for food allergy, that such technology may play an important role in providing the effective cure for food allergy that is currently lacking. Such a therapy would undoubtedly improve the quality of life for food-allergic consumers, which can be acute [83, 84].

The broader utilization of nanoparticles in foods will, as for other types of novel technology, need to undergo an allergenicity risk assessment [85], although there can be difficulties in undertaking such assessments for other types of novel process or novel foods, including genetically modified organisms, partly because of our lack of effective animal models for food allergy. Biologically derived nanoparticles are probably produced during the digestion of foods, and nanoscale structures have been described in conventionally processed foods for many years. Any nanoparticle-containing ingredients derived from allergenic food that it is mandatory to label will need to be declared, and in this way allergic consumers will be able to avoid their consumption. However, novel types of bionanoparticles and inorganic nanoparticles based on carbon or silver for example, may have unintended effects, but there are no clear agreed experimental approaches or frameworks to develop data on which to base an effective risk assessment. Further research is required to address these gaps in our knowledge and hence ensure that the considerable benefits that may arise from this new technology are realized while minimizing the risks of potentiating existing allergic conditions or introducing new ones.

Acknowledgments

The authors acknowledge the support of the UK Biological and Biotechnological Sciences Research Council through the competitive strategic grant to the Institute of Food Research.

References

1 Johansson, S.G.O., Hourihane, J.O'B., Bousquet, J., Bruijnzeel-Kooman, C., Drejborg, S., Haahtela, T., Kowlaski, M.L., Mygind, N., Ring, J., van Cauwenberge, P., van Hage-Hemsten, M., and Wüthrich, B. (2001) A revised nomenclature for allergy. An EACCI position statement from the EACCI nomenclature task force. *Allergy*, **56**, 813–824.

2 Hischenhuber, C., Crevel, R., Jarry, B., Maki, M., Moneret-Vautrin, D.A., Romano, A., Troncone, R., and Ward, R. (2006) Review article: Safe amounts of gluten for patients with wheat allergy or coeliac disease. *Aliment. Pharmacol. Ther.*, **23**, 559–575.

3 Bush, R.K., and Hefle, S.L. (1996) Food allergens. *CRC Crit. Rev. Food Sci. Nutr.*, **36**, S119–S163.

4 Mills, E.N.C., Jenkins, J.A., Alcocer, M.J., and Shewry, P.R. (2004) Structural, biological, and evolutionary relationships of plant food allergens sensitizing via the gastrointestinal tract. *Crit. Rev. Food Sci. Nutr.*, **44**, 379–407.
5 Breiteneder, H., and Radauer, C. (2004) A classification of plant food allergens. *J. Allergy Clin. Immunol.*, **113**, 821–830.
6 Jenkins, J.A., Griffiths-Jones, S., Shewry, P.R., Breiteneder, H., and Mills, E.N.C. (2005) Structural relatedness of plant food allergens with specific reference to cross-reactive allergens: an *in silico* analysis. *J. Allergy Clin. Immunol.*, **115**, 163–170.
7 Jenkins, J.A., Breiteneder, H., and Mills, E.N.C. (2007) Evolutionary distance from human homologs reflects allergenicity of animal food proteins. *J. Allergy Clin. Immunol.*, **120**, 1399–1405.
8 Breiteneder, H., and Mills, E.N.C. (2008) Food allergens: molecular and immunological characteristics, in *Food Allergy: Adverse Reactions to Foods and Food Additives*, 4th edn (eds D.D. Metcalfe, H.A. Sampson, and R.A. Simon), Blackwell Publishing, Malden, MA, pp. 43–61.
9 Mills E.N.C., Johnson, P., Alexeev, Y., and Breiteneder, H. (2009) Identification and characterisation of food allergens, in *Management of Food Allergens* (eds J. Coutts and R. Fielder), Blackwell Publishing, Oxford, pp. 42–69.
10 Fernandez-Rivas, M., Bolhaar, S., Gonzalez-Mancebo, E., Asero, R., van Leeuwen, A., Bohle, B., Ma, Y., Ebner, C., Rigby, N., Sancho, A.I., Miles, S., Zuidmeer, L., Knulst, A., Breiteneder, H., Mills, E.N.C., Hoffmann-Sommergruber, K., and van Ree, R. (2006) Apple allergy across Europe: how allergen sensitization profiles determine clinical expression of plant food allergies. *J. Allergy Clin. Immunol.*, **18**, 481–488.
11 Howdle, P.D., Ciclitira, P.J., Simpson, F.G., and Losowsky, M.S. (1984) Are all gliadins toxic in coeliac disease? An *in vitro* study of α, β, γ and ω-gliadins. *Scand. J. Gastroenterol.*, **19**, 41–47.
12 Shan, L., Molberg, O., Parrot, I., Hausch, F., Filiz, F., Gray, G.M., Sollid, L.M., and Khosla, C. (2002) Structural basis for gluten intolerance in celiac sprue. *Science*, **297**, 2275–2279.
13 Wensing, M., Knulst, A.C., Piersma, S., O'Kane, F., Knol, E.F., and Koppelman, S.J. (2003) Patients with anaphylaxis to pea can have peanut allergy caused by cross-reactive IgE to vicilin (Ara h 1). *J. Allergy Clin. Immunol.*, **111**, 420–424.
14 Vieths, S., Scheurer, S., and Ballmer-Weber, B. (2002) Current understanding of cross-reactivity of food allergens and pollen. *Ann. N.Y. Acad. Sci.*, **964**, 47–68.
15 Witke, W. (2004) The role of profilin complexes in cell motility and other cellular processes. *Trends Cell Biol.*, **14**, 461–469.
16 Radauer, C., and Hoffmann-Sommergrube, K. (2004) Profilins, in *Plant Food Allergens* (eds E.N.C. Mills and P.R. Shewry), Blackwell Publishing, Oxford.
17 Wensing, M., Akkerdaas, J.H., van Leeuwen, W.A., Stapel, S.O., Bruijnzeel-Koomen, C.A., Aalberse, R.C., Bast, B.J., Knulst, A.C., and van Ree, R. (2002) IgE to Bet v 1 and profilin: cross-reactivity patterns and clinical relevance. *J. Allergy Clin. Immunol.*, **110**, 435–442.
18 MacLeod, A.R. (1987) Genetic origin of diversity of human cytoskeletal tropomyosins. *Bioessays*, **6**, 208–212.
19 Reese, G., Ayuso, R., and Lehrer, S.B. (1999) Tropomyosin: an invertebrate pan-allergen. *Int. Arch. Allergy Immunol.*, **119**, 247–258.
20 Motoyama, K., Ishizaki, S., Nagashima, Y., and Shiomi, K. (2006) Cephalopod tropomyosins: identification as major allergens and molecular cloning. *Food Chem. Toxicol.*, **44**, 1997–2002.
21 Lehrer, S.B., Ibanez, M.D., McCants, M.L., Daul, C.B., and Morgan, J.E. (1990) Characterization of water-soluble shrimp allergens released during boiling. *J. Allergy Clin. Immunol.*, **85**, 1005–1013.
22 Lewit-Bentley, A., and Rety, S. (2000) EF-hand calcium-binding proteins. *Curr. Opin. Struct. Biol.*, **10**, 637–643.
23 Bernhisel-Broadbent, J., Scanlon, S.M., and Sampson, H.A. (1992) Fish hypersensitivity. I. *In vitro* and oral challenge results in fish-allergic patients. *J. Allergy Clin. Immunol.*, **89**, 730–737.

24 Filimonov, V.V., Pfeil, W., Tsalkova, T.N., and Privalov, P.L. (1978) Thermodynamic investigations of proteins. IV. Calcium binding protein parvalbumin. *Biophys. Chem.*, **8**, 117–122.
25 Bugajska-Schretter, A., Elfman, L., Fuchs, T., Kapiotis, S., Rumpold, H., Valenta, R., and Spitzauer, S. (1998) Parvalbumin, a cross-reactive fish allergen, contains IgE-binding epitopes sensitive to periodate treatment and Ca^{2+} depletion. *J. Allergy Clin. Immunol.*, **101**, 67–74.
26 Bugajska-Schretter, A., Grote, M., Vangelista, L., Valent, P., Sperr, W.R., Rumpold, H., Valenta, R., Pastore, A., Reichelt, R., Valenta, R., and Spitzauer, S. (2000) Purification, biochemical, and immunological characterisation of a major food allergen: different immunoglobulin E recognition of the apo- and calcium-bound forms of carp parvalbumin. *Gut*, **46**, 661–669.
27 Tuinier, R., and de Kruif, C.G. (2002) Stability of casein micelles in milk. *J. Chem. Phys.*, **117**, 1290–1295.
28 Bellioni-Businco, B., Paganelli, R., Lucenti, P., Giampietro, P.G., Perborn, H., and Businco, L. (1999) Allergenicity of goat's milk in children with cow's milk allergy. *J. Allergy Clin. Immunol.*, **103**, 1191–1194.
29 Van Regenmortel, M.H.V. (1992) Molecular dissection of protein antigens, in *Structure of Antigens*, vol. 1 (ed. M.H.V. Van Regenmortel), CRC Press, Boca Raton, FL, pp. 1–28.
30 Dyson, H.J., Jeng, M.F., Tennant, L.L., Slaby, I., Lindell, M., Cui, D.S., Kuprin, S., and Holmgren, A. (1997) Effects of buried charged groups on cysteine thiol ionization and reactivity in *Escherichia coli* thioredoxin: structural and functional characterization of mutants of Asp 26 and Lys 57. *Biochemistry*, **36**, 2622–2636.
31 Aalberse, R.C. (1998) Clinical relevance of carbohydrate allergen epitopes. *Allergy*, **53**, 54–57Koppelman, S.J., van Koningsveld, G.A., Knulst, A.C., Gruppen, H., Pigmans, I.G., and de Jongh, H.H. (2002) Effect of heat-induced aggregation on the IgE binding of patatin (Sol t 1) is dominated by other potato proteins. *J. Agric. Food Chem.*, **50**, 1562–1568.
32 Foetisch, K., Westphal, S., Lauer, I., Retzek, M., Altmann, F., Kolarich, D., Scheurer, S., and Vieths, S. (2003) Biological activity of IgE specific for cross-reactive carbohydrate determinants. *J. Allergy Clin. Immunol.*, **111**, 889–896.
33 Wal, J.-M. (2002) Cow's milk proteins/ allergens. *Ann. Allergy Asthma Immunol.*, **89**, 3–10.
34 Mills, E.N.C., Jenkins, J., Marigheto, N., Belton, P.S., Gunning, A.P., and Morris, V.J. (2002) Allergens of the cupin superfamily. *Biochem. Soc. Trans.*, **30**, 925–929.
35 Koppelman, S.J., Bruijnzeel-Koomen, C.A., Hessing, M., and de Jongh, H.H. (1999) Heat-induced conformational changes of Ara h 1, a major peanut allergen, do not affect its allergenic properties. *J. Biol. Chem.*, **274**, 4770–4777.
36 Brownlow, S., Cabral, J.H.M., Cooper, R., Flower, D.R., Yewdall, S.J., Polikarpov, I., North, A.C.T., and Sawyer, L. (1997) Bovine beta-lactoglobulin at 1.8 angstrom resolution – still an enigmatic lipocalin. *Structure*, **5**, 481–495.
37 Aymard, P., Durand, D., and Nicolai, T. (1996) The effect of temperature and ionic strength on the dimerisation of beta-lactoglobulin. *Int. J. Biol. Macromol.*, **19**, 213–221.
38 Casal, H.L., Kohler, U., and Mantsch, H.H. (1988) Structural and conformational changes of beta-lactoglobulin-B – an infrared spectroscopic study of the effect of pH and temperature. *Biochim. Biophys. Acta*, **957**, 11–20.
39 Carrotta, R., Bauer, R., Waninge, R., and Rischel, C. (2001) Conformational characterization of oligomeric intermediates and aggregates in beta-lactoglobulin heat aggregation. *Protein Sci.*, **10**, 1312–1318.
40 Croguennec, T., Bouhallab, S., Molle, D., O'Kennedy, B.T., and Mehra, R. (2003) Stable monomeric intermediate with exposed Cys-119 is formed during heat denaturation of beta-lactoglobulin. *Biochem. Biophys. Res. Commun.*, **301**, 465–471.

41 Brew, K., and Grobler, J.A. (1992) Alpha-Lactalbumin, in *Advances in Dairy Chemistry*, vol. 1, *Proteins* (ed. P.F. Fox), Elsevier Applied Science, New York, pp. 191–229.
42 Baum, J., Dobson, C.M., Evans, P.A., and Hanley, C. (1989) Characterization of a partly folded protein by NMR methods – studies on the molten globule state of guinea-pig alpha-lactalbumin. *Biochemistry*, **28**, 7–13.
43 Ehn, B.M., Allmere, T., Telemo, E., Bengtsson, U., and Ekstrand, B. (2005) Modification of IgE binding to beta-lactoglobulin by fermentation and proteolysis of cow's milk. *J. Agric. Food Chem.*, **53**, 3743–3748.
44 Nowak-Wegrzyn, A., Bloom, K.A., Sicherer, S.H., Shreffler, W.G., Noone, S., Wanich, N., and Sampson, H.A. (2008) Tolerance to extensively heated milk in children with cow's milk allergy. *J. Allergy Clin. Immunol.*, **122**, 342–347.
45 Roth-Walter, F., Berin, M.C., Arnaboldi, P., Escalante, C.R., Dahan, S., Rauch, J., Jensen-Jarolim, E., and Mayer, L. (2008) Pasteurization of milk proteins promotes allergic sensitization by enhancing uptake through Peyer's patches. *Allergy*, **63**, 882–890.
46 Fernandez-Rivas, M., and Cuevas, M. (1999) Peels of Rosaceae fruits have a higher allergenicity than pulps. *Clin. Exp. Allergy*, **29**, 1239–1247.
47 Sancho, A.I., Foxall, R., Browne, T., Dey, R., Zuidmeer, L., Marzban, G., Waldron, K.W., van Ree, R., Hoffmann-Sommergruber, K., Laimer, M., and Mills, E.N.C. (2006) Effect of postharvest storage on the expression of the apple allergen Mal d 1. *J. Agric. Food Chem.*, **54**, 5917–5923
48 Ballmer-Weber, B.K., Hoffmann, A., Wüthrich, B., Lüttkopf, D., Pompei, C., Wangorsch, A., Kästner, M., and Vieths, S. (2002) Influence of food processing on the allergenicity of celery: DBPCFC with celery spice and cooked celery in patients with celery allergy. *Allergy*, **57**, 228–235.
49 Asero, R., Mistrello, G., Roncarolo, D., and Amato, S. (2006) PT with heat-processed apple peel extract to detect LTP hypersensitivity. *Allerg. Immunol. (Paris)*, **38**, 351–354.
50 Grimshaw, K.E.C., King, R.M., Nordlee, J.A., Hefle, S.L., Warner, J.O., and Hourihane, J.O'B. (2003) Presentation of allergen in different food preparations affects the nature of the allergic reaction – a case series. *Clin. Exp. Allergy*, **33**, 1581–1585.
51 Gajhede, M., Osmark, P., Poulsen, F.M., Ipsen, H., Larsen, J.N., van Neerven, R.J.J., Schou, C., Lowenstein, H., and Spangfort, M.D. (1996) X-ray and NMR structure of Bet v 1, the origin of birch pollen allergy. *Nat. Struct. Biol.*, **3**, 1040–1045.
52 Neudecker, P., Schweimer, K., Nerkamp, J., Scheurer, S., Vieths, S., Sticht, H., and Rosch, P. (2001) Allergic cross-reactivity made visible: solution structure of the major cherry allergen Pru av 1. *J. Biol. Chem.*, **276**, 22756–22763.
53 Mogensen, J.E., Ferreras, M., Wimmer, R., Petersen, S.V., Enghild, J.J., and Otzen, D.E. (2007) The major allergen from birch tree pollen, Bet v 1, binds and permeabilizes membranes. *Biochemistry*, **46**, 3356–3365.
54 Mittag, D., Vieths, S., Vogel, L., Becker, W.M., Rihs, H.P., Helbling, A., Wuthrich, B., and Ballmer-Weber, B.K. (2004) Soybean allergy in patients allergic to birch pollen: clinical investigation and molecular characterization of allergens. *J. Allergy Clin. Immunol.*, **113**, 148–154.
55 Kleine-Tebbe, J., Vogel, L., Crowell, D.N., Haustein, U.F., and Vieths, S. (2002) Severe oral allergy syndrome and anaphylactic reactions caused by a Bet v 1-related PR-10 protein in soybean, SAM22. *J. Allergy Clin. Immunol.*, **110**, 797–804.
56 Hansen, K.S., Ballmer-Weber, B.K., Luttkopf, D., Skov, P.S., Wuthrich, B., Bindslev-Jensen, C., Vieths, S., and Poulsen, L.K. (2003) Roasted hazelnuts – allergenic activity evaluated by double-blind, placebo-controlled food challenge. *Allergy*, **58**, 132–138.
57 Worm, M., Hompes, S., Fiedler, E.M., Illner, A.K., Zuberbier, T., and Vieths, S. (2009) Impact of native, heat-processed and encapsulated hazelnuts on the allergic response in hazelnut-allergic patients. *Clin. Exp. Allergy*, **39**, 159–166.

58 Moreno, F.J., Mellon, F.A., Wickham, M.S.J., Bottrill, A.R., and Mills, E.N.C. (2005) Stability of the major allergen Brazil nut 2S albumin (Ber e 1) to physiologically relevant *in vitro* gastrointestinal digestion. *FEBS J.*, **272**, 341–352.

59 Moreno, F.J., Maldonado, B.M., Wellner, N., and Mills, E.N.C. (2005) Thermostability and *in vitro* digestibility of a purified major allergen 2S albumin (Ses i 1) from white sesame seeds (*Sesamum indicum* L.). *Biochim. Biophys. Acta Proteins Proteomics*, **1752**, 142–153.

60 Sancho, A.I., Rigby, N.M., Zuidmeer, L., Asero, R., Mistrello, G., Amato, S., Gonzalez-Mancebo, E., Fernandez-Rivas, M., Ree, R., and Mills, E.N.C. (2005) The effect of thermal processing on the IgE reactivity of the non-specific lipid transfer protein from apple, Mal d 3. *Allergy*, **60**, 1262–1268.

61 Schad, S.G., Trcka, J., Vieths, S., Scheurer, S., Conti, A., Brocker, E.B., and Trautmann, A. (2005) Wine anaphylaxis in a German patient: IgE-mediated allergy against a lipid transfer protein of grapes. *Int. Arch. Allergy Immunol.*, **136**, 159–164.

62 Pastorello, E.A., Farioli, L., Conti, A., Pravettoni, V., Bonomi, S., Iametti, S., Fortunato, D., Scibilia, J., Bindslev-Jensen, C., Ballmer-Weber, B., Robino, A.M., and Ortolani, C. (2007) Wheat IgE-mediated food allergy in European patients: alpha-amylase inhibitors, lipid transfer proteins and low-molecular-weight glutenins. Allergenic molecules recognized by double-blind, placebo-controlled food challenge. *Int. Arch. Allergy Immunol.*, **144**, 10–22.

63 Asero, R., Amato, S., Alfieri, B., Folloni, S., and Mistrello, G. (2007) Rice: another potential cause of food allergy in patients sensitized to lipid transfer protein. *Int. Arch. Allergy Immunol.*, **143**, 69–74.

64 Kato, Y., Oozawa, E., and Matsuda, T. (2001) Decrease in antigenic and allergenic potentials of ovomucoid by heating in the presence of wheat flour: dependence on wheat variety and intermolecular disulfide bridges. *J. Agric. Food Chem.*, **49**, 3661–3665.

65 Mondoulet, L., Paty, E., Drumare, M.F., Ah-Leung, S., Scheinmann, P., Willemot, R.M., Wal, J.M., and Bernard, H. (2005) Influence of thermal processing on the allergenicity of peanut proteins. *J. Agric. Food Chem.*, **53**, 4547–4553.

66 Jöbstl, E., Howse, J.R., Fairclough, J.P., and Williamson, M.P. (2006) Noncovalent cross-linking of casein by epigallocatechin gallate characterized by single molecule force microscopy. *J. Agric. Food Chem.*, **54**, 4077–4081.

67 Chung, S.Y., and Champagne, E.T. (2007) Effects of phytic acid on peanut allergens and allergenic properties of extracts. *J. Agric. Food Chem.*, **55**, 9054–9058.

68 Li, F., Wang, L., Jin, X.M., Yan, C.H., Jiang, S., and Shen, X.M. (2009) The immunologic effect of TGF-beta1 chitosan nanoparticle plasmids on ovalbumin-induced allergic BALB/c mice. *Immunobiology*, **214**, 87–99.

69 Li, J., Liu, Z., Wu, Y., Wu, H., and Ran, P. (2008) Chitosan microparticles loaded with mite group 2 allergen Der f 2 alleviate asthma in mice. *J. Invest. Allergol. Clin. Immunol.*, **18**, 454–460.

70 Schöll, I., Weissenböck, A., Förster-Waldl, E., Untersmayr, E., Walter, F., Willheim, M., Boltz-Nitulescu, G., Scheiner, O., Gabor, F., and Jensen-Jarolim, E. (2004) Allergen-loaded biodegradable poly(D,L-lactic-*co*-glycolic) acid nanoparticles down-regulate an ongoing Th2 response in the BALB/c mouse model. *Clin. Exp. Allergy*, **34**, 315–321.

71 Yanagisawa, R., Takano, H., Inoue, K., Koike, E., Kamachi, T., Sadakane, K., and Ichinose, T. (2009) Titanium dioxide nanoparticles aggravate atopic dermatitis-like skin lesions in NC/Nga mice. *Exp. Biol. Med.*, **234**, 314–322.

72 Bhol, K.C., Alroy, J., and Schechter, P.J. (2004) Anti-inflammatory effect of topical nanocrystalline silver cream on allergic contact dermatitis in a guinea pig model. *Clin. Exp. Dermatol.*, **29**, 282–287.

73 Inoue, K., Takano, H., Yanagisawa, R., Sakurai, M., Abe, S., Yoshino, S., Yamaki, K., and Yoshikawa, T. (2007) Effects of nanoparticles on lung physiology in the presence or absence of

antigen. *Int. J. Immunopathol. Pharmacol.*, **20**, 737–744.

74 Li, N., Xia, T., and Nel, A.E. (2008) The role of oxidative stress in ambient particulate matter-induced lung diseases and its implications in the toxicity of engineered nanoparticles. *Free Radic. Biol. Med.*, **44**, 1689–1699.

75 Alessandrini, F., Beck-Speier, I., Krappmann, D., Weichenmeier, I., Takenaka, S., Karg, E., Kloo, B., Schulz, H., Jakob, T., Mempel, M., and Behrendt, H. (2009) Role of oxidative stress in ultrafine particle-induced exacerbation of allergic lung inflammation. *Am. J. Respir. Crit. Care Med.*, **179**, 984–991.

76 Nygaard, U.C., Hansen, J.S., Samuelsen, M., Alberg, T., Marioara, C.D., and Løvik, M. (2009) Single-walled and multi-walled carbon nanotubes promote allergic immune responses in mice. *Toxicol. Sci.*, **109**, 113–123.

77 Ryan, J.J., Bateman, H.R., Stover, A., Gomez, G., Norton, S.K., Zhao, W., Schwartz, L.B., Lenk, R., and Kepley, C.L. (2007) Fullerene nanomaterials inhibit the allergic response. *J. Immunol.*, **179**, 665–672.

78 Cartiera, M.S., Johnson, K.M., Rajendran, V., Caplan, M.J., and Saltzman, W.M. (2009) The uptake and intracellular fate of PLGA nanoparticles in epithelial cells. *Biomaterials*, **30**, 2790–2798.

79 Sarmento, B., Ribeiro, A., Veiga, F., Ferreira, D., and Neufeld, R. (2007) Oral bioavailability of insulin contained in polysaccharide nanoparticles. *Biomacromolecules*, **8**, 3054–3060.

80 Jones, O.G., Decker, E.A., and McClements, D.J. (2009) Formation of biopolymer particles by thermal treatment of beta-lactoglobulin–pectin complexes. *Food Hydrocoll.*, **23**, 1312–1321.

81 Weiss, J., Decker, E.A., McClements, D.J., Kristbergsson, K., Helgason, T., and Awad, T. (2008) Solid lipid nanoparticles as delivery systems for bioactive food components. *Food Biophys.*, **3**, 146–154.

82 McClements, D.J., Decker, E.A., Park, Y., and Weiss, J. (2008) Designing food structure to control stability, digestion, release and absorption of lipophilic food components. *Food Biophys.*, **3**, 219–228.

83 Flokstra-de Blok, B.M., Dunn Galvin, A., Vlieg-Boerstra, B.J., Oude Elberink, J.N., Duiverman, E.J., Hourihane, J.O'B., and Dubois, A.E. (2008) Development and validation of the self-administered Food Allergy Quality of Life Questionnaire for adolescents. *J. Allergy Clin. Immunol.*, **122**, 139–144.

84 Dunn Galvin, A., Flokstra-de Blok, B.M., Burks, A.W., Dubois, A.E., and Hourihane, J.O'B. (2008) Food allergy QoL questionnaire for children aged 0–12 years: content, construct, and cross-cultural validity. *Clin. Exp. Allergy*, **38**, 977–986.

85 EFSA (2009) The potential risks arising from nanoscience and nanotechnologies on food and feed safety (Scientific Opinion). *EFSA J.*, **7**, 958 . Available at: http://www.efsa.europa.eu/en/efsajournal/doc/958.pdf (accessed 9 November 2010).

14
Communication of Risks and Benefits of Nanotechnology: the Issue of Societal Acceptance of Emerging Technologies

Lynn J. Frewer, Arnout R.H. Fischer, and J.(Hans)C.M. van Trijp

14.1
Introduction

The successful development, implementation, and commercialization of novel technologies is contingent on societal acceptance of these same technologies and their specific applications. New technologies associated with risks, and risks perceived by the public, have not been successfully commercialized in the past. Thus the introduction of a new technology will be contingent on being perceived to be of acceptable risk (for example, in terms of its potential impact on human health and the environment), as well as filling a need, or providing a putative benefit to the end-user, even if the benefit has not yet been recognized as important by society. Failure to deliver desirable and tangible consumer benefits would, at best, lead to public indifference to the new technology and its applications.

An individual's motivation to adopt the applications and products of the new technology would also be low under circumstances where risks are perceived to be high and benefits low. If the new technology does not align with the values and preferences of the public, public indifference might easily turn into societal rejection of the technology. Therefore public response toward the risks and benefits of new technologies should be understood, as public response may determine success or failure of the new technology. Some technologies have been described as transformative, inasmuch as their impact extends to other areas of society beyond that originally intended during their development.

Transformative technologies can be broadly defined as technologies with applications or impacts on society and the economy, which also have the potential for long-term effects on values, power structures, and ideas within society as a whole [1]. Nanotechnology, including its application to food production, and across the agri-food sector more generally, may represent such a transformative technology, inasmuch as it will result in changes to the way society organizes and regulates itself.

Technological innovations in food production have been singled out as a special case, notable because food-related issues and innovations co-evolved with human

Nanotechnology in the Agri-Food Sector: Implications for the Future, First Edition. Edited by Lynn J. Frewer, Willem Norde, Arnout Fischer, Frans Kampers.

civilization. Human "hunter-gatherers" mastered the art of agriculture, introducing new varieties of edible crops and ways to grow and process these into safer and longer-lasting foodstuffs, thus improving food security and food quality. The rise of agricultural technology shifted the dominant societal structure from a nomadic hunter-gatherer society to one focused on stationary communities, with time available to develop more complex technologies, artistic expression, and cultural structures. As part of this, food consumption was not only integral to community survival, but also had culturally symbolic associations, and implications for employment and societal organization. Agriculture rapidly became a dominant occupation for many human beings, which further led to the evolution of agriculturally based economies. Thus the invention of agriculture can be described as transformative inasmuch as it showed the capacity to transform society by introducing completely new ways of living. What once started as simple means to control the growth of natural food sources to facilitate food security now involves application of state-of-the-art technology, involving highly mechanized tools and technological innovations for food production and food processing.

A more recent example, the "green revolution", resulted in a significant increase in agricultural productivity resulting from the introduction of high-yield varieties of grains, the use of pesticides, and improved management techniques. This enabled an increase in agricultural yields, in line with the needs of the increasing global population. An additional effect was the change in farming practices from small agrarian units to industrial-scale farming, with associated changes in the structure of rural communities and increased migration from the countryside to urban environments. The growth of cities was in turn supported by the reduced labor-intensive food production methods and the much lower risk of failed harvests, facilitated by the same "green revolution" [2].

In the context of the global market and international trade, food is treated as a commodity, and its import and export enables transactions within the global "food market". Surplus food or food scarcity determines, to a certain extent, the overall well-being of the global community. Applications of new technologies to food production may therefore be regarded as a decisive element of product and process innovation [3], and a key driver of the globalization of food production, trade, and associated food preference and culture.

Transformative technologies are not limited to the agri-food sector. In another example, developments in information and communication technology (ICT) have revolutionized not only how people communicate and deal with information, but also how they structure their working and leisure activities. At the same time, the transmission and storage of electronic information raises new issues regarding governance and privacy [4]. It is of interest to note that one aspect of ICT, the application of radiofrequency identification (RFID) technology, has now also been applied in the agri-food sector to facilitate tracking and tracing of foods and ingredients, and thus food safety and food quality. RFID technology represents an example of a technology associated with some societal concerns (for example, regarding privacy) but which has been widely accepted in the agri-food sector by various food-chain actors. At the same time, public awareness appears to be rather low, or at least reflects low levels of concern on the part of broader society.

It is likely that nanotechnology will represent an example of such a transformative technology, not only in the area of food production, food processing, and food storage, but also in other societal domains (paints, clothing, medicine, reinforced plastics for use in tennis rackets, neural enhancement, etc.). There is already discussion of the need to revise existing legislative frameworks regarding the protection of human health and the environment. In addition, the potential impact of the technology as a whole on society may be regarded as being greater than the benefits resulting from specific applications by both the scientific community and key stakeholders in industry and the policy arena.

Societal responses to emerging technological developments, particularly those which have potential to be transformative, may be ambiguous. Some technologies are accepted by society, with low levels of societal debate and controversy. Even within technologies, some applications are more readily accepted than other applications of the same technology. Most medical applications of technology are perceived by citizens to be very high in benefit, and tend to be more acceptable compared to many other applications of new technologies, even when they are simultaneously associated with risks. Exceptions seem to occur when the perceived benefits of a specific medical innovation appear to be very low for the individual receiving it, or when an individual is exposed to the medical innovation on what is perceived to be an involuntary basis – as was the case with, for example, the measles–mumps–rubella (MMR) vaccine or fluoridation of the water supply.

In the agri-food sector, public acceptance of food technologies has included, for example, the application of high-pressure processing to improve food safety and food quality, or the fortification of foods with micronutrients. In other examples of food-related technology, public negativity and consumer rejection have resulted in the failure of these technologies to deliver societal benefits, as consumers would not accept their application to foods and ingredients. Other examples of food technologies associated with public negativity can easily be identified. Food irradiation is the process of exposing food to ionizing radiation to destroy microorganisms, bacteria, viruses, or insects that might be present in the food, or to delay ripening or improve rehydration. In the 1980s, when food irradiation technology was ready for application, the technology was regarded by scientific experts as a major advance in the area of food safety and quality. However, the negative response on the part of consumers toward "irradiated" foods has resulted in low levels of uptake of food, although medical applications (for example, related to the sterilization of dressings) are widely accepted. Within the food domain, the exception appears to be irradiation of herbs and spices, possibly because public awareness of this food irradiation application is low, or perhaps because the alternative process, exposure to nitrous oxide, is viewed more negatively by consumers.

The application of technologies to food production may be particularly sensitive inasmuch as their acceptance by consumers is concerned [5, 6]. Foods (and their preparation) are associated with many traditional socio-cultural preferences and practices. In addition, global changes in food supply may be ubiquitous, and affect many consumers, particularly in a globalizing world economy where food chains (and food webs) are internationalizing. One reason why food is particularly sensitive is because food is ingested and "taken in" to the body, with the potential to influence health, and indeed one's conception of "self", negatively. It is perhaps not surprising that public negativity associated with the introduction of novel technologies has focused on applications in the agri-food sector, as opposed to the medical or materials sectors, as has been the case with genetically modified foods and ingredients [7]

Predicting public response regarding the application of nanotechnology to food will require timely introduction of information about end-user requirements, demands, and values into the development of new agri-food technologies. In the past, it has often been assumed that the public essentially consisted of uneducated people, who would react positively to technological innovation if they could be educated to accept the underpinning science. It is now widely recognized that public opinion, whether derived from perceptions or other non-technical concerns, is valuable, albeit representing a different perspective from that expressed by experts. The perception and opinion of different sectors of the public need to be considered as part of the policy process [8].

It is thus important to address societal and consumer preferences and demands as part of the process of technology development and implementation. In the present context, this may be particularly relevant in terms of the application of nanotechnology to food production.

14.2
Science and Society: Lessons for Nanotechnology Applied to Food Production

Food is vital for human survival, and concern about its production and preparation is widespread [9]. In this context, new technologies are increasingly being applied to ensure food security, as well as to provide additional consumer benefits related to human health and food quality. For example, limiting the health burden caused by vitamin A deficiency, particularly in developing countries, resulted in the development of "golden rice", which has been genetically modified to increase the β-carotene content of the diet for those consumers who are deficient in this particular micronutrient. However, the problems associated with delivering the health benefits of golden rice to the populations who needed them go beyond the technological ability to develop the product.

First, there is a problem of distributing the product to consumer populations who would benefit from consuming it (the "end-of-pipe" problem). If the distribution problem could be overcome, then it is arguable that the same distribution process should be able to supply vitamin supplements, or facilitate the distribution of existing foods rich in β-carotene, as another relevant way to solve to the public health problem under consideration.

Second, consumers may be reluctant to consume the rice, which may appear discolored (yellow) compared to traditionally used variants of the same food, because of the increased β-carotene content. This increased health benefit was achieved at the cost of reduced "quality" perception.

In addition, it cannot be assumed that all consumers will accept the underpinning food technology, in this case genetic modification, even if there is a benefit associated with the end-product. Perceived benefits must outweigh perceived risks for consumer acceptance to occur. Developing an informed understanding of consumer responses to emerging technologies and their applications is key to optimizing the strategic development of science and technology in the future, as well as developing and refining commercialization strategies associated with specific products [10].

To date, societal responses to the application of different technologies in the agri-food sector has been a focus of greater societal concern compared to, for example, medical applications of the same technologies [11]. This is, in part, because many of the technologies in the agri-food sector have been developed without reference to potential consumer acceptance of different applications in the agri-food sector *per se* [3], and the situation is contextualized by an increasingly internationalized market. Although many new agri-food technologies promise to deliver profound benefits to society, they may also be associated with substantial risks in terms of both environmental and human health impacts and consumer perceptions of risk. As is the case for other transformative technologies, the impact of nanotechnology on society can be very broad and potentially unintended by the developers of the technology.

Historically, research into the determinants of public responses associated with emerging technologies (in general, and those specifically in the agri-food sector) has tended to occur subsequent to public rejection of the application of technologies. Research needs to be directed toward exploring how such products will be received before they have been developed, allowing consumer and public demands and preferences to be integrated into the design of products and technology implementation from the start.

Producers, processors, and wholesalers of food, as well as retailers, may usefully negotiate the development of new products with relevant stakeholders across regional and even cross-continental borders. Successful development of new products is ultimately dependent on society-wide consumer acceptance and purchasing of produced foods.

In parallel with developments in risk perception and communication, some authors have noted a decline in public confidence in risk analysis practices [12]. While this decline in public confidence in science underpinning technology commercialization has been evident in many technology sectors, the recent example of European negativity toward genetically modified organisms, particularly as applied to agri-food, is the example frequently cited as being of greatest relevance to societal adoption of nanotechnology, particularly in the context of its commercialization (for example, see references [13, 14]).

Despite this observation, the direct comparisons between potential societal responses to nanotechnology and other technologies may be limited. For example, nanotechnology is not as "contained" in terms of the range of potential applications as is genetic modification. The term "nanotechnology" essentially covers a broader range of sub-technologies with different applications. In addition, various combinations and convergences of emerging technologies (for example, nanotechnology, biotechnology, information and communication technology, cognitive science, and engineering in the case of human enhancement) may also raise ethical issues that include, but go beyond, those associated with genetic modification. These larger issues may influence the societal response to nanotechnology as a whole, even when specific applications in themselves do not raise such negative response.

For example, the use of nano-sieves (or nanofilters) in food processing may be acceptable in itself, but may become unacceptable in the context of widespread rejection of nanotechnology as a whole. At the same time, there is societal demand to "revisit" current risk assessment approaches, as they may not adequately address safety issues associated with nanoparticles, which may be fueled by heightened consumer risk perceptions and distrust in industry and regulatory institutions. In the absence of a clear trajectory regarding the development of public responses, it seems that companies are currently downplaying the use of nanotechnology in their products for fear of a negative consumer response and triggering distrust in nanotechnology. At this stage, it is relevant to consider how such consumer risk (and benefit) perceptions are formed, and what their consequences are in terms of consumer and citizen behaviors.

14.3
A Short Introduction to the Psychology of Risk–Benefit Perception

Much public negativity associated with the way in which risks are managed and regulated has been the result of risk managers, assessors, and other key actors in the process of risk analysis failing to take account of the actual concerns of the public when assessing, managing, and communicating information about risks. Risk assessment and management were traditionally performed without involving the public. This has (subsequently) had a negative impact on public perceptions regarding the motives of regulators, science, and industry in taking decisions or actions in relation to risk assessment priorities, resource allocation, and risk mitigation activities [15]. This may be partly the result of risk communication being implemented as a one-way transmission of the outcomes of scientific risk assessments, and the failure of responsible institutions to incorporate public concerns, values, and fears into the broader societal debate. (The interested reader should see reference [16], for example, but also Marvin *et al.* in Chapter 17 in this volume for a recently developed alternative to integrate public consultation into risk management.) Communication that is based on technical risk assessment, but does not explicitly address public concerns, is likely to have a limited role in reassuring the public. Hence it is necessary to know the actual public concerns to decide which risks should be assessed. Understanding the rationale behind public concerns thus becomes of great importance for anyone involved with new technologies.

Research into the various processes by which risk analysis practices interface with society has been evolving since the 1970s. The research of Paul Slovic and colleagues, which indicated that lay people incorporate psychological factors into their personal assessment of the acceptability of different hazards, was initially assumed to be evidence of public irrationality by different actors in the regulatory and industrial communities, and appeared to explain why risk management decisions acceptable to expert communities (for example, regulators and scientists) were not acceptable to some members of the public. The decision by some regulatory and industry stakeholders to continue to implement technologies despite negative consumer perceptions not only resulted in high levels of consumer distrust in the motives of regulators and industry, but prevented the successful commercialization of some technologies, notably in the area of genetically modified foods.

Consequently, risk communication activities at this time focused on changing public views on risk to become aligned with expert views, with emphasis on communication directed toward risk acceptance, in particular in the area of emerging technologies. The process has been described as the "deficit model" [17], whereby expert and elite organizations and institutions assumed that the various sectors of the public are in some way deficient in their understanding of risk. As a consequence, it was reasoned that the acceptance of emerging technologies and other hazards was contingent on public trust in institutions with responsibility for regulating the associated risks, rather than on the public understanding the technical assessment of the risk.

The literature suggests that public distrust resulted from the failure of these institutions to take public concerns into account. The underpinning rationale appeared to promote the notion that increased public trust in regulatory bodies with responsibility for consumer protection, industry, and science would increase technology acceptance. It was assumed that an increase in trust could be achieved by a greater emphasis on increased transparency in the process of risk analysis, in particularly risk assessment and risk management. While there is some limited evidence to suggest that increased transparency is a precondition for trust in institutional activities to develop, increased transparency in itself is not a trust-increasing event [18]. Lack of transparency may result in decreased trust, but trust *per se* is a result of citizen perceptions of institutional honesty, concern for public welfare, and competence.

A second approach to developing trust focused on greater public inclusion in the process of policy development, specifically focusing on the argument that more extensive public consultation and participation in risk management and other science and technology issues would restore public confidence in institutions with responsibility for public and consumer protection (see, for example, reference [19]). At the time of writing, increased public consultation appears to play a limited role in increasing public confidence, because there is little evidence that the output of the consultation exercise influences the policy process (see Chapter 15 in this volume, and references [20, 21]), and because there is scant evidence that institutional responses to broader consultations are able to tolerate lack of consensus in public opinion.

While there seems to be institutional and governmental motivation to conduct public consultations regarding the future of nanotechnology, it is not clear what will be done with the outputs, how lack of consensus will be handled, nor what will be the concrete ambition of the consultation. There is also some evidence that, for example, in countries with a long history of consultation, the public are suffering from "consultation overload", and the original ambition, increasing public confidence in science and technology, is more recently construed by the public as a route to technology acceptance – in other words, as a way to implement the technology regardless of public concern or demand for the technology.

More recent approaches to nanotechnology development focus on combining social science and policy research with natural science and engineering processes into "real-time technology assessment", which proposes an ongoing interaction between technology and society, permitting an iterative embedding of societal values in the emerging framework containing technology. It is argued that such an approach does not run the same risks of creating public negativity as traditional public participation exercises, which may destroy societal trust if public rejection of specific activities is not internalized into the regulatory framework [22], or block specific innovations that would be appreciated by consumers because of a societal rejection of a type of technological application at an early point in time. The effectiveness of such approaches are contingent on developing an understanding of the formation of public perceptions of nanotechnology.

14.4
How do People Form Perceptions of New Technologies

At present, there is little research regarding public perceptions of, and attitudes toward, nanotechnology. In part, this reflects low levels of public exposure to different applications under development (perhaps because of economic interests reflecting concerns about a public backlash toward different developments). Although there is a literature demanding that research into societal issues be conducted (for example, see reference [23]), in practice contemporary empirical analysis into the science and society issues of nanotechnology remains somewhat scarce.

This has parallels to research on public attitudes to biotechnology in the 1990s, where the need to understand public attitudes, and how these were forming, was identified, but research into the process was not being conducted [24]. On the one hand, there is frequent reference in the literature to the potential impact of science fiction and film/literary references to nanotechnology as being influential in terms of crystallizing public views. For example, reference is made to understand the impact of Michael Crichton's novel "Prey" as an irresponsible piece of fiction portraying nanotechnology as "out of control" and uncertain [13]. A similar discussion precluded the release of the film "Jurassic Park" (also based on a novel by Michael Crichton) and its impact on perceptions of genetic engineering. On the other hand, however, there is no evidence to suggest that public attitudes were influenced by what are clearly works of fiction (and perceived to be such). Of greater concern is the negative backlash following "overselling" of a particular technology by those scientists developing it (for example, as has been the case for human metabolomics).

At this point, it is of interest to review the existing literature on consumer attitudes to nanotechnology, with the caveat that public awareness regarding nanotechnology is not, at present, extensive [16, 25]. There is some evidence to suggest that food-related applications of nanotechnology may also result in a more negative consumer response compared to other nanotechnology applications [26]. Nevertheless, some examples of nanotechnology in foods or in food contact materials are already on the market (for example, nano-silica has long been added to non-dairy coffee creamer). Nanomaterials in food packaging are beginning to enter the market. "Smart packaging", where active components in the package control the atmosphere surrounding fresh food products, where labels respond to molecules in the atmosphere to indicate the condition of the packaged product, or where ultraviolet blockers on plastic wine bottles preserve product quality, are already possible.

Although these products may enter the market, the extent to which consumers are actually aware that nanotechnology is being applied within the agri-food sector is debatable. It is reasonable to assume that attitudes toward nanotechnology are likely to start developing in the near future, and will be formed by direct experience with the technology and its applications [27], or be driven by an affective or emotional response to the issue or application [28–30]. In the case of nanotechnology, consumers have, to date, little (conscious) experience with nanotechnology products [31], implying that information provided by external sources will probably play a dominant role in the current, early stage of public opinion formation [32].

Fischer *et al.* [33] have demonstrated that simultaneous exposure to risk and benefit information does not necessarily result in positive or negative public attitudes toward nanotechnology. After receiving balanced risk and benefit information, some individuals develop positive or negative attitudes toward nanotechnology following the provision of combined risk and benefit information. Other individuals remain neutral and do not develop strong attitudes toward nanotechnology. These results suggest that individuals develop different attitudes despite receiving the same information about nanotechnology.

Similarly, Kahan *et al.* [34] report that members of the public readily form opinions on whether the potential risks of nanotechnology outweigh its potential benefits. These are largely driven by affective or emotional responses, as well as other attitudes held by the individuals receiving the information. For example, attitudes toward environmental risks generally explain more of the differences in individuals' perceptions of nanotechnology's risks and benefits than do the other attitudes held by these individuals. The authors report that these views are amenable to influence by the provision of additional information, but that individuals exposed to balanced information polarize along cultural and political lines.

14.5 Nanotechnology Communication in the Business Context

Commercialization, aimed at the generation of willingness to buy and, potentially, willingness to pay, would require the communication of nanotechnology applications in terms of a unique benefit proposition to the consumer, primarily in terms of usefulness and ease of use [35]. In other words, what specific consumer needs are addressed by the nanotechnology application that would put it at a competitive advantage?

Cost–benefit considerations on the part of the consumer have traditionally been operationalized in terms of functional benefits of improved product performance (e.g., better taste, higher convenience, safety, etc.). However, increasingly, consumers consider not only "what the product delivers" in terms of personal benefits, but also "how the product is brought about" in terms of social and environmental impact. This is why perceived risk and uncertainty with the new technology is an essential part of the positioning challenge, as already discussed in the previous sections. Consumer behavior research [36, 37] reveals an important asymmetry, with negative societal perceptions outweighing positive contributions as a determinant of consumer acceptance. This is why consideration of public attitudes toward nanotechnology needs to be an integral part of commercial communication, particularly so, as commercial stakeholders are likely to be judged with a considerable level of skepticism and distrust.

Nanotechnology applications may occur at almost all stages of the agricultural value chain, with different potential benefits for various chain actors, all the way through from primary producers to end consumers. For effective communication of nanotechnology, it is important to identify *a priori* the business model underlying the nanotechnology application. Such a business model would include identification of "where value is being added", "to whose benefit", and "with what positioning". Following on this, the strategic management literature [38] suggests three dominant bases for realizing competitive advantage from nanotechnology applications: cost leadership, differentiation, and focus strategy. Increasingly, corporate social responsibility has come to the forefront in the business environment as a strategic orientation for competitive advantage to which nanotechnology can add considerably.

Cost leadership ("doing the same thing at lower cost") would imply that food with nanotechnology applications performs at parity with competing products (i.e., delivering existing benefit and benefit performance) in the marketplace, but that the nanotech applications add value by reducing costs anywhere along the total supply chain. This may occur, for example, through more efficient processing, and logistics with fewer losses, or more effective sourcing from increased productivity in primary production. Cost leadership approaches would not be actively communicated in terms of consumer benefit (as it delivers an established benefit), but (some of) the cost reduction may be delivered to the consumer in terms of lower price.

Nanotechnology applications also have the potential to deliver improved (e.g., step changes in taste, texture, and/or health quality of the product) or even completely new benefits to chain partners, in particular the end consumer (e.g., packaging from which freshness can be inferred). Such added consumer benefits from nanotechnology may contribute to differentiation strategies whereby the rationale for applying nanotechnology is described, together with justification for increased pricing where appropriate. Specific nanotechnology applications also have the potential for being associated with a focus strategy ("beneficial for a specific subgroup") in which the benefits brought forward by the technology would be targeted at one or two specific groups of consumers. A possible application would be the use of nanotechnology to produce a non-allergenic product, thereby opening much broader access to products for this specific segment.

Corporate social responsibility strategies, based on nanotechnology applications, would communicate the added value for society at large (e.g., in reducing environmental and social benefits) rather than the end benefits to the consumers more specifically. A key example here would be nanotechnology applications to increase the effectiveness of pesticides (and thereby reducing their use) or the development of new plant varieties that are resistant to unfavorable production conditions (such as drought and saline areas). As discussed in previous chapters, the potential for the introduction of new risks may entail the development of novel risk assessment paradigms. These also need to address consumer concerns and priorities.

Nanotechnology applications can occur at the level of process innovation and product innovation, and effective communication depends on the successful integration between the two. This is particularly important, as, in their perception and valuation of technology-based food innovations [35], consumers reply on their personal cost–benefit considerations ("what is in it for me in terms of improved product performance?") as well as risk and uncertainty, which is also largely related to process innovation ("how has this product been developed and with what consequences to whom"). For effective communication strategy, in terms of communication objectives, target audience, and message content, it is important to distinguish between different combinations of process and product innovation. This is illustrated in Table 14.1. The table distinguishes between communication strategies depending on whether the benefits arise for the chain actors (process innovation) and/or to tangible consumer benefits (product innovation).

For the net benefits delivered by the product innovation to the consumer, the end consumer is the target audience, and the marketing objective (contained in the message content) will focus on communication of those benefits to consumers and justifying the potential price premium being charged. For the benefits related to the chain actors and society at large, communication is more complicated, as, due to the complexity of the issue, the value and benefit distribution across the chain cannot easily be verified by the consumer. Consumer perception of chain

Table 14.1 A topology of societal (profit, corporate social responsibility) and consumer benefit and risk associated with different market introductions of a new technology.

Process innovation	Products from technological innovation		
Net benefit to chain actors	**Net benefits to end consumer**		
	Negative	**Neutral**	**Positive**
Negative			Consumer benefit at "societal cost"
Neutral			Consumer benefit at no societal cost
Positive		Societal benefits at no "consumer cost"	Synergetic quality delivery

and societal benefits are much more based on *indirect* communication through stakeholders of the value chain, most notably commercial stakeholders, non-governmental organizations, consumer organizations, media, scientists, and governments. As commercial stakeholders are not necessarily seen as an independent and trustworthy source of information, communication on chain benefits will to a large degree be delivered indirectly via these stakeholder groups. Table 14.1 highlights the importance of the aligning chain actors and social benefits of nanotechnology with actual product benefits to the consumer, as a win–win proposition between public and consumer–private interest.

14.6 Conclusion

To date, societal acceptance of new technologies has often been studied after these technologies, and their applications, have been introduced. Nanotechnology provides a unique opportunity to examine theoretical models of public opinion formation under circumstances in which consumers are only just beginning to make sense of the potential perceived risks, costs, and benefits associated with technological innovation. Consumer perceptions of risk, benefit, and cost are unlikely to be stable over time, but may have some predictable properties, which should be considered and implemented in an early stage of technology development, application, and commercialization. The success or failure of new technologies depends both on societal responses, which may create legal and governmental obstacles, and on end-user uptake, which may create or prevent the cash flow needed for further development. Careful positioning of the technology and adopting the relevant associated communication is an essential precondition to prevent adverse reactions from society.

References

1 Dufour, P., and Hassan, M.H.A. (2005) Nanotechnology for development. *Issues Sci. Technol.*, **22** (1), 15–16.
2 Borlaug, N.E. (2000) Ending world hunger. The promise of biotechnology and the threat of antiscience zealotry. *Plant Physiol.*, **124** (2), 487–490.
3 Henson, S., Annou, M., Cranfield, J., and Ryks, J. (2008) Understanding consumer attitudes toward food technologies in Canada. *Risk Anal.*, **28** (6), 1601–1617.
4 Van Kleef, E., Fischer, A.R.H., Khan, M., and Frewer, L.J. (2010) Risk and benefit perceptions of mobile phone and base station technology in Bangladesh. *Risk Anal.*, **30** (6), 1002–1015.
5 Rozin, P. (2007) Food choice: an introduction, in *Understanding Consumers of Food Products* (eds L.J. Frewer and H. van Trijp), Woodhead, Cambridge, pp. 3–29.
6 Rozin, P., Haidt, J., and McCauley, C.R. (2000) Disgust, in *Handbook of Emotions*, 2nd edn (eds M. Lewis and J.M. Haviland-Jones), Guilford, New York, pp. 637–653.
7 Frewer, L.J., and Shepherd, R. (1995) Ethical concerns and risk perceptions associated with different applications of genetic engineering: interrelationships with the perceived need for regulation of the technology. *Agric. Human Values*, **12** (1), 48–57.
8 Hansen, J., Holm, L., Frewer, L.J., Robinson, P., and Sandoe, P. (2003) Beyond the knowledge deficit: recent research into lay and expert attitudes to food risks. *Appetite*, **41** (2), 111–121.
9 Hohl, K., and Gaskell, G. (2008) European public perceptions of food risk: cross-national and methodological comparisons. *Risk Anal.*, **28** (2), 311–324.
10 Frewer, L.J., Howard, C., and Shepherd, R. (1997) Public concerns about general and specific applications of genetic engineering: risk, benefit and ethics. *Sci. Technol. Hum. Values*, **22**, 98–124.
11 Bredahl, L. (2001) Determinants of consumer attitudes and purchase intentions with regard to genetically modified food – results of a cross-national survey. *J. Consum. Policy*, **24** (1), 23.
12 Jensen, K.K., and Sandoe, P. (2002) Food safety and ethics: the interplay between science and values. *J. Agric. Environ. Ethics*, **15**, 245–253.
13 Bainbridge, W.S. (2004) Religion and science. *Futures*, **36** (9), 1009–1023.
14 Einsiedel, E.F., and Goldenberg, L. (2004) Dwarfing the social? Nanotechnology lessons from the biotechnology front. *Bull. Sci. Technol. Soc.*, **24** (1), 28–33.
15 Wentholt, M.T.A., Rowe, G., König, A., Marvin, H.J.P., and Frewer, L.J. (2009) The views of key stakeholders on an evolving food risk governance framework: results from a Delphi study. *Food Policy*, **34** (6), 539–548.
16 Cobb, M.D., and Macoubrie, J. (2004) Public perceptions about nanotechnology: risks, benefits and trust. *J. Nanopart. Res.*, **6** (4), 395–405.
17 Hilgartner, S. (1990) The dominant view of popularisation: conceptual problems, political issues. *Soc. Stud. Sci.*, **20**, 519–539.
18 Frewer, L.J., Howard, C., Hedderley, D., and Shepherd, R. (1996) What determines trust in information about food-related risks? Underlying psychological constructs. *Risk Anal.*, **16** (4), 473–486.
19 Renn, O., Webler, T., and Widermann, P. (1995) *Fairness and Competence in Citizen Participation*, Kluwer Academic, Dordrecht, The Netherlands.
20 Rowe, G., and Frewer, L.J. (2004) Evaluating public participation exercises: a research agenda. *Sci. Technol. Hum. Values*, **29** (4), 512–556.
21 Rowe, G., and Frewer, L.J. (2000) Public participation methods: a framework for evaluation. *Sci. Technol. Hum. Values*, **25**, 3–29.
22 Guston, D.H., and Sarewitz, D. (2002) Real-time technology assessment. *Technol. Soc.*, **24** (1-2), 93–109.
23 Pilarski, L.M., Adamia, S., Pilarski, P.M., Prakash, R., Lauzon, J., and Backhouse, C.J. (2004) Improved diagnosis and

monitoring of cancer using portable microfluidics platforms. Paper presented at the Proceedings – 2004 International Conference on MEMS, NANO and Smart Systems, ICMENS 2004.

24 Frewer, L.J., Shepherd, R., and Sparks, P. (1994) The interrelationship between perceived knowledge, control and risk associated with a range of food-related hazards targeted at the individual, other people and society. *J. Food Safety*, **14** (1), 19–40.

25 Pardo, R., Midden, C., and Miller, J.D. (2002) Attitudes toward biotechnology in the European Union. *J. Biotechnol.*, **98** (1), 9–24.

26 Siegrist, M., Wiek, A., Helland, A., and Kastenholz, H. (2007) Risks and nanotechnology: the public is more concerned than experts and industry. *Nat. Nanotechnol.*, **2** (2), 67.

27 Fischer, A R.H., and De Vries, P.W. (2008) Everyday behaviour and everyday risk: an exploration how people respond to frequently encountered risks. *Health Risk Soc.*, **10** (4), 385–397.

28 Alhakami, A.S., and Slovic, P. (1994) A psychological study of the inverse relationship between perceived risk and perceived benefit. *Risk Anal.*, **14** (6), 1085–1096.

29 Finucane, M.L., Alhakami, A.S., Slovic, P., and Johnson, S.M. (2000) The affect heuristic in judgments of risks and benefits. *J. Behav. Decis. Mak.*, **13** (1), 1–17.

30 Slovic, P., Finucane, M.L., Peters, E., and MacGregor, D.G. (2004) Risk as analysis and risk as feelings: some thoughts about affect, reason, risk, and rationality. *Risk Anal.*, **24** (2), 311–322.

31 Waldron, A.M., Spencer, D., and Batt, C.A. (2006) The current state of public understanding of nanotechnology. *J. Nanopart. Res.*, **8** (5), 569–575.

32 Fischer, A.R.H., and Frewer, L.J. (2009) Consumer familiarity with foods and the perception of risks and benefits. *Food Qual. Prefer.*, **20** (8), 576–585.

33 Fischer, A.R.H., Van Dijk, H., De Jonge, J., Rowe, G., and Frewer, L.J. The impact of information on attitudinal ambivalence: the case of nanotechnology in food production (submitted).

34 Kahan, D.M., Braman, D., Slovic, P., Gastil, J., and Cohen, G.L. (2007) Cultural cognition of the risks and benefits of nanotechnology. *Nat. Nanotechnol.*, **4** (2), 87–90.

35 Ronteltap, A., van Trijp, J.C.M., Renes, R.J., and Frewer, L.J. (2007) Consumer acceptance of technology-based food innovations: lessons for the future of nutrigenomics. *Appetite*, **49** (1), 1–17.

36 Klein, J., and Dawar, N. (2004) Corporate social responsibility and consumers' attributions and brand evaluations in a product-harm crisis. *Int. J. Res. Mark.*, **21** (3), 203–217.

37 Sen, S., and Bhattacharya, C.B. (2001) Does doing good always lead to doing better? Consumer reactions to corporate social responsibility. *J. Mark. Res.*, **38** (2), 225–243.

38 Porter, M.E. (1985) *Competitive Advantage*, Free Press, New York.

15
Public Engagement with Emerging Issues in Agri-Food Nanotechnology

Lynn J. Frewer, Arnout R.H. Fischer, and Gene Rowe

15.1
Introduction

Internationally, there is increasing investment in both the private and public sectors in *technologically driven* innovations, such as foods developed with novel technologies, improved food safety and food quality initiatives, and improved sustainability in terms of food production, processing, delivery and consumption. As has been discussed in other chapters in this volume, however, societal responses to agri-food nanotechnology may not necessarily be positive. Successful implementation and commercialization of emerging technologies is contingent on societal acceptance of the technology overall, as well as on consumer acceptance of specific applications, particularly at a time when public confidence in technological innovation is generally low – this is particularly true within the agri-food sector, which has been beset by past societal controversies. As a consequence, it is important to identify societal concerns regarding new developments in the area of emerging technologies, possibly to allow the timely opportunity to pre-empt or ameliorate these [1], or to change the trajectory of technology development in line with societal preferences.

In the case of agri-food nanotechnologies, there is potential for profound societal and consumer *benefit* to be associated with its application. Indeed, signs of consumer negativity and distrust in the motives of both industry and regulatory institutions – regarding this and other technologies – are already appearing. At the present time, societal attitudes toward nanotechnology have not yet fully crystallized [2, 3], but this may change as societal debate about the risks and benefits of nanotechnology intensifies and products begin to be made available to consumers. Developing effective public engagement is therefore key to understanding consumer–citizen priorities and preferences for future food production systems and their products. The aim of this chapter is to review different mechanisms for public engagement, to consider their application to the nanotechnologies issues, and to comment on their likely limitations and effectiveness – particularly with regard to the appropriate criteria for their subsequent acceptance by those involved in the process and society more generally.

Nanotechnology in the Agri-Food Sector: Implications for the Future, First Edition. Edited by Lynn J. Frewer, Willem Norde, Arnout Fischer, Frans Kampers.

It is argued that it is important for event sponsors (in particular) to recognize the different aims of consultation and engagement mechanisms, and choose a mechanism according to their specific consultative aims [4]. Furthermore, given that an appropriately selected mechanism (for a given situation) may still be inappropriately applied, it is important that the exercise itself should be independently *evaluated*. That is, the purpose and aims of such exercises should be made explicit by the sponsors, and any exercise conducted should be of appropriate rigor regarding independent evaluation of both process and policy impact (or at least the use of the outcomes) to enable scrutiny by the interested end-user communities, and to facilitate comparisons in time and across different geographic regions and population groups. Public engagement associated with technology development or other technological or societal initiatives (including agri-food nanotechnologies) should recognize that such initiatives affect society over and above product development and commercialization.

15.2
What Is "Public Engagement"?

Perceptions of risk and benefit associated with different applications of food technology, "sustainable" production, or other future foods issues, may have a direct influence on consumer acceptance of specific products. Societal acceptance of emerging technologies is also likely to be contingent on "societal trust" in regulators, regulatory institutions, industry, and other actors in the technology sector under consideration [5]. Citizens' trust in industrial and regulatory actors with responsibility for consumer protection and optimization of the economic success of the applications of emerging food technologies may influence perceptions of risk and benefit. The occurrence of various food safety incidents associated with existing and emerging technologies, many of which have had international and national consequences for quality of life and economic functioning, has highlighted the need to develop and maintain public confidence in the management of the food supply [6]. A case in point is the negative public reactions to genetic modification in the agri-food sectors, which have had important consequences for commercialization of the technology as well as for international food risk governance [7, 8].

A key element in developing societal trust in the motives of actors involved with developing the products of novel technologies is to ensure that these actors take account of the concerns of interested stakeholders (including the general public), and address these concerns in the process of research and policy formulation, as well as in considering potential commercialization strategies [9]. Activities geared toward involving the public have, in the agri-food sector, ranged from traditional consultations (e.g., quantitative surveys of nationally representative populations) and focus groups conducted in order to evaluate citizen views, through to exercises specifically focusing on citizen involvement in the decision-making process. In other words, there has been a growing trend within many societies to involve a wider range of stakeholders in policy decisions than has traditionally been the case in the past. Such stakeholders may include the public, or, at least, representatives of the public, as well as consumer, environmental, and industry interest groups, and other interested end-users and stakeholders.

Historically, public policy regarding science and technology has been determined by politicians and policy-makers, often with the aid of expert advice. Decisions resulting from this process have then generally been communicated to the public, under the assumption that the communication recipients will understand and believe the information, and think and behave appropriately in response – that is, as the policy-makers have predicted and think appropriate [10, 11]. In contemporary societies, this model has, to some degree, failed: a loss of trust in decision-makers and their advisors has led to public and stakeholder skepticism in the motives of actors in the policy, regulatory, and industry communities [12], including the potential for societal mobilization against policy directives and associated activities (e.g., the introduction of innovative but controversial food technologies and their specific applications [13–16]).

As a consequence, many actors in the policy community and beyond have shifted to a position where it is assumed that public trust and confidence may be gained and maintained if decision-makers are perceived by society more generally to be actively obtaining broader views associated with policy development and regulatory activities. This activity has been referred to as public (or citizen or stakeholder) engagement, involvement, or participation [17, 18]. Various reasons have been put forward to explain the recent increased popularity of stakeholder involvement [19–21]. Among the assumed benefits of engagement are:

- regaining of societal trust in policy-makers and policy decisions;
- public acquisition of political efficacy;
- enhancement of democracy;
- societal acceptance of decisions associated with policy development and implementation;
- improvement of policy decisions.

It is of interest to note that, despite the assumed existence of these benefits, empirical tests of the societal impact of public engagement are scarce [22].

In order to achieve the goal of "greater participation", various methodologies have been developed to facilitate public and stakeholder involvement in the policy process. Three broad categories of approach can be identified, as shown in Table 15.1 (adapted from [18]). The first can be described as "public communication", entailing a unidirectional flow of information from the expert community to the public. Hence public opinion on the matter under consideration cannot influence the policy process. Specific activities included in the communication category are beyond the scope of the present chapter, and are not considered further. The second approach can be described as "public consultation". This group of approaches utilizes more traditional opinion elicitation methods, where the opinion of the public is gathered and used in policy development or implementation with little or no interaction with sponsors, policy-makers or expert communi-

Table 15.1 An overview of methodologies applied to facilitate public and stakeholder involvement (adapted from [18]).

Consultation	Participation	Communication
Citizens' panel	Action planning workshop	Cable TV
Consultation document	Citizens' jury	Drop-in centers
Electronic consultation	Consensus conference	Hotline
Focus group	Deliberative opinion poll	Information broadcasts
Opinion poll	Negotiated rule-making	Internet information
Referendum	Planning cell	
Survey	Technology assessment	
Telepolling		
Delphi		
Public hearing		
Expert group		

ties. Information about public opinion is provided to the expert community, after which there is no further public involvement. The information flow is "from the public to the experts" and is often, but not always, confined to a single period of data sampling. A third group of methodologies consists of more innovative methods that actively aim to engage participants in an ongoing dialog with sponsors – "public engagement" or "public participation". These methods aim to be truly interactive.

Both the consultation and participation methods may potentially involve one, or a combination, of "citizens", "the public", "consumers", "stakeholders" or "experts". Descriptions of each of the terms included in the table are provided in the glossary given in the Appendix at the end of this chapter.

The issue then arises as to which class of approaches is most useful for engaging with the public under specific circumstances – with recent arguments (as previously discussed) suggesting that more participatory approaches are more appropriate in many current social–political–technological contexts. This is particularly the case with nanotechnology, where increased participation (particularly early in the process, or "upstream") has been prescribed by a major inquiry into nanotechnologies in the UK conducted by the Royal Society and Royal Academy of Engineering (see reference [23] for a description), and has been encapsulated in the major National Nanotechnology Initiative (NNI) in the USA [9].

15.3
Evaluating the Effectiveness of Public and Stakeholder Engagement

Rowe and Frewer [22] reviewed the literature on public engagement and found little empirical evidence for the existence of the generally assumed benefits – with only a few studies having critically examined engagement exercises (in any context, over and above agri-food nanotechnologies) to see what they have achieved and whether they have delivered the benefits proposed. Of course, this does not *necessarily* mean that such benefits have not been obtained, but this does indicate that this issue has not been the focus of empirical investigation.

A major problem in collating evidence to support the assumed benefits associated with public and stakeholder engagement is a lack of theoretical perspectives about *how* to evaluate engagement exercises. In addition, it should be recognized that there are practical difficulties in attempting to conduct such evaluations, whether these are of expert consultations or public engagement types [18, 21, 24]. A key element in doing this effectively concerns defining what is meant by an "effective" exercise – a complex issue that is open to debate and dispute. In spite of this, some consensus regarding best practice in evaluation is beginning to emerge in the literature. For example, two recent evaluation frameworks [25, 26] essentially agree upon a dichotomy of effectiveness requirements. These are that an engagement exercise should be perceived as being *fair* by those participating (as well as outside observers), and that the process itself should be competently enacted in a manner allowing appropriate interactions and exchanges of knowledge and/or information between those involved.

In addition to identifying and using appropriate evaluation criteria, it is generally recognized that evaluation of public and stakeholder engagement should be done independently, particularly as the underpinning assumptions of the associated benefits of applying such exercises may be erroneous, and those conducting engagement often have a positive expectation of the process they have promoted and run. Indeed, it has been argued that, under some circumstances, engagement may not result in the assumed benefits for the policy process or society in general [27–30]. Caution may particularly be called for where engagement involves the public, and under circumstances where the policy topic is highly complex [31], as is the case with emerging technologies applied in the agri-food sector.

Less-than-rigorous application of method and (in the case of public engagement) lack of independent evaluation may act as a barrier to the uptake of public consultation conclusions in policy measures and subsequent debates – possibly correctly so, if the outputs are somehow flawed (e.g., have been derived through a faulty process). Rigorous evaluation of participation exercises is thus important to enable other researchers and end-users to have access to credible research findings, to enable scrutiny by interested end-user communities, to facilitate comparisons in time and space as well as across different cultural and geographical regions [32], and to avoid repetition and redundancy in research activities.

Initiatives must be deployed to collect the data on public acceptance and knowledge following more classical methods to facilitate opinion sampling over multiple countries and participant groups, to be able to make any claims of the generalizability of results from small group involvement (participation being characterized by limited participant numbers) to the larger public in affected regions. An important issue in the global or international context (the level at which innovations in food technology may apply) is the acquisition of data that can distinguish between those attitudinal factors which are invariant over cultures and people, and those which are prone to be influenced by cultural factors and historical contexts. Policy, regulation, and strategic development of technology innovation can then be harmonized internationally, or "fine-tuned" to the needs of specific countries or regions. At the present time, stakeholder and public participation in the governance and strategic development of agri-food (nano)technology is fragmented. It is essential that ongoing activities are reviewed and harmonized in order to avoid duplication of effort and inappropriate allocation of resources. The peer-review process is important in this regard – while also being useful in increasing transparency of policy impacts, as well as in identifying gaps in ongoing activities and developing best practice in stakeholder and societal dialog regarding existing and emerging societal concerns.

The one-way stream of information *from* the public *to* the expert community is still a more frequently utilized method compared to "genuine" public engagement, despite frequent references to the *need* to engage the public in the development of, for example, agri-food nanotechnology or other emerging food technologies. While the notion of public engagement has become salient in the minds of researchers, policy-makers, and decision-makers in industry, there is little concerted or considered use of the various available methodologies. It may be that conducting and publishing evaluations of engagement exercises may aid in convincing those using these traditional approaches to try novel methods.

Finally, the value of public engagement in terms of its assumed advantages also needs to be assessed. Key questions include the following.

- Does public engagement increase trust in policy-makers and industry, and in political and regulatory processes more generally?
- Is public engagement justifiable in its own right, as a means of enhancing the democratic process?
- Are the public more likely to accept decisions associated with policy development and implementation following public engagement or public consultation, and how does lack of consensus across different groups of those consulted influence this process?
- How and in what way are policy decisions improved following public consultation or engagement.
- Is there already adequate information available regarding public opinion, concerns, and values associated with either nanotechnology or sustainable chemistry?

15.4
Public Engagement Examples

As noted, nanotechnology has been viewed through a similar lens to genetic modification [1], with one of the lessons emerging from prior debate being that of the difficulty of *communicating* technology benefits to a cautious (and indeed skeptical) public. Furthermore, it has been noted that using *consultative* approaches, such as surveys, are of little use in revealing the ways in which people will interpret and understand novel and complex technologies when the issue of interest is one about which people know little [33]–as has been shown to be the case with nanotechnologies. Proponents of nanotechnology, as well as interested social scientists, have thus sought more interactive public engagements using *participative* approaches.

There have, consequently, been a growing number of participative exercises used on nanotechnology issues, though relatively few have made it into the academic press. One of these was "Nanojury UK", the first citizens' jury on nanotechnology in the UK (see the Appendix at the end of this chapter for a description of this general method). This has been described by Pidgeon and Rogers-Hayden [3]. Burri [34] reports on a similar approach (a "citizens' panel") held in Switzerland, while Evers and D'Silva [35] discuss another citizens' panel held as part of a Flemish technology assessment project (albeit on nano-applications in the medical, not agri-food, domain).

Meanwhile, Pidgeon *et al.* [33] have described the first comparative US–UK public participation experiment, which comprised four concurrent half-day public *workshops* debating energy and health nanotechnologies. One interesting result from this study was that participants focused on benefits rather than risks and, in general, had a high regard for science and technology. Rather than the country in which the exercise was run, it was application *context* that was the most significant source of attitudinal differences, with energy applications viewed in a substantially more positive light than applications in human health and enhancement in both countries (where agri-food application would fit in this analysis is an empirical question). The authors also reported that more subtle differences were present in views about the equitable distribution of benefits, corporate and governmental trustworthiness, the risks to realizing benefits, and in consumerist attitudes. It is arguable that more profound cross-cultural differences may be observed under circumstances where cultural or economic differences are more profound, but this has not been subject to empirical test as far as can be ascertained.

What is notable is the general lack of rigorous evaluation of these and other initiatives. However, some research has attempted to consider the qualities of such participative processes and what they achieve. Powell and Kleinman [36], for example, drew on in-depth interviews with participants of a *consensus conference* (see the Appendix at the end of this chapter) in the USA on nanotechnology to consider how citizen participants felt the consensus conference experience had affected their *knowledge and efficacy* related to participation in nanotechnology issues; which aspects of the conference they thought shaped their knowledge and efficacy; and whether they felt motivated to engage in future participatory mechanisms related to nanotechnology issues. They concluded that, even if consensus conferences have little or no *influence* on policy or policy-makers, they may empower citizens by improving their perceived abilities to participate meaningfully in technoscientific issues.

Similarly, Besley *et al.* [37] explored interpersonal discussion following participation in a novel program of citizen engagement about nanotechnology (using a process they called "citizen schools" – a program of lessons and engagement between scientists and public lasting several weeks, and therefore considerably more prolonged than the majority of participative approaches). Participants answered closed- and open-ended questions about their discursive behavior in a post-engagement survey. Respondents reported moderate levels of post-engagement discussion and appeared to say positive things about both nanotechnology and the experts who contributed to the engagement program. Respondents also reported primarily talking about nanotechnology in terms of scientific progress while using a range of fairness and competence frames to discuss experts and the program.

On the negative side, Hamlett and Cobb [38] collected data from a small set of public deliberations on nanotechnology to test the concern that group deliberations may bias toward the original majority preferences because of cognitive and affective errors in decision-making, such as deference to the numerical majority opinion held within a group, and they found some evidence for this polarization hypothesis. Indeed, Rogers-Hayden and Pidgeon [39] have also emphasized that the kinds of conversations that emerge from public engagement and other approaches to understanding public attitudes may not lead to harmonious development of nanotechnologies (as perhaps hoped by proponents), but may open up differences in visions – although they point out that this is necessary if public participation is to move "upstream", beyond mere consultation to encompass a "co-creation of nanotechnology for sustainability" ([39] p. 1010).

15.5 Recommendations for Conducting Public Engagement and Public Consultation Exercises

In this chapter different ways have been discussed in which the public may be engaged in the current debate about agri-food nanotechnologies. The multitude of methods available have been briefly reviewed, and the different intentions and characteristics of these noted (see the Appendix at the end of this chapter). The trend toward more "public participation" has been described, and some examples drawn from different areas of application of nanotechnology. In particular, the issue of "evaluation" has been raised, highlighting the need for researchers to seriously consider the processes and consequences of their engagement methods. From this, a number of recommendations for conducting such approaches can be identified.

1) The goals of the exercise must be clearly defined at the outset. Often, participative processes are framed in such a way that encourages participants to believe that their views will have real impact upon an issue, when this is not truly the sponsors' intent, and once participants realize this, they can feel aggrieved – as can external observers and stakeholders. If an exercise is simply

being conducted to collect views, then a simple consultative approach–such as a survey or focus groups–would be better. If the intent of an exercise is more than this–to give other stakeholders (which may include the public) some real say and ownership of the problem issue, then participative approaches seem suitable. (Evers and D'Silva [35] note that it is often unclear how regulatory actors can actually proceed from the output of such exercises, and perhaps this needs to be clearly specified at the outset.) Being clear on the aims of a process can also help to inform the appropriate criteria for evaluating that process (see point 3).

2) There are many different methodologies available to facilitate consultative and participatory processes, which are of direct utility in understanding science and society issues. From the available literature (both refereed and non-refereed), the approach favored by academics working in this area tends to be the *survey* and *focus group* for consultation, and the *consensus conference, citizens' jury* and (in the area of ethical impact) *technology assessment* in the context of deliberative processes and public engagement. This may, to some extent, reflect the fact that these methods are established and accepted within the academic and policy communities, rather than because they are inherently better or more suited to the purposes to which they are put.

3) It is important to consider the timing of a public engagement event. There has been a view developing that engagement with the public (in particular, public *participation*) should occur early in the process of developing novel technologies–so-called "upstream engagement". Rogers-Hayden and Pidgeon [23] discussed some of the promise and perils of moving public debate upstream, however, concluding that there is a risk of merely replacing the perceived deficit in public understanding of science with a perceived deficit in public engagement with science–so caution is needed.

4) It is essential that public engagement exercises are systematically evaluated against appropriate criteria by independent evaluators (i.e., evaluators who are not in any way involved in the development of the exercise or the sponsoring organization). An example of such evaluation criteria has been provided by Rowe and Frewer [26].

5) In order to ensure that the results of a public consultation or public engagement exercise meet the rigorous standards required by peer review, publication in a peer-reviewed journals is desirable. This also ensures that the results of the research are available to other researchers, and helps to prevent duplication of effort and allocation of resources to similar activities.

It is important to note that, while a public engagement activity may provide indicators of emerging public concerns and preferences, it will not provide data that will facilitate systematic comparison of emerging concerns and preferences across different segments of the population (e.g., demographic groups, regional groups, or groups of individuals with specific preferences or attitudes to technology or food production). The results of the public consultation need to be taken together with existing knowledge about determinants of individuals' perceptions of risk–benefit already available in the published literature, to validate and explain identified patterns. Ultimately, such qualitative processes can be useful in informing the development of a survey instrument for collecting quantitative data regarding attitudes toward (for example) nanotechnology.

Appendix

Glossary (from [18])

Action planning workshop An action planning workshop is an intensive workshop of one or two full days in which representatives of the involved parties join together to review the current status of the issues, set the future goals and write a detailed action plan for the next steps to be taken.

Citizens' jury A citizens' jury is a mechanism of participatory action research that draws on the symbolism, and some of the practices, of a legal trial by jury. The "jury" is made up of people who are usually selected "at random" from a local or national population. The jurors cross question experts they have called to provide different perspectives on the topic. The jury then collectively produces a summary of their conclusions, typically in a short report.

Citizens' panel Citizens' panels involve a broadly representative sample of the local population, who have agreed to take part in consultation activity. They can involve between 500 and 3000 people. Panel members are then asked to complete surveys on a regular basis. This can involve the whole panel, or particular target groups within the total panel.

Consensus conference A consensus conference is a chaired public hearing with an audience drawn from the public and with active participation of 10–15 lay people and a corresponding number of different experts. The experts may be from different disciplines and/or from different schools within a discipline. The conference may last several days, plus the time for preparation. The purpose is to produce an informed debate on a limited subject presented in the form of six or seven main questions.

Consultation document This is a document that identifies an issue and proposes one or several ways to deal with the issue. These proposals are then offered as concrete ideas to the view from the public and are open for comments and adjustments. If multiple proposals are offered, the preference of the public for one of the proposals can be asked for.

Deliberative opinion poll Deliberative polling combines small-group discussions involving a small numbers of participants with random sampling of public opinion. Citizens are invited to take part at random, so that a large enough participant group will provide a relatively accurate, scientific representation of public opinion.

Delphi The Delphi method is a systematic interactive forecasting method for obtaining forecasts from a panel of independent experts. The carefully selected experts answer questionnaires in two or more rounds. After each round, a facilitator provides an anonymous summary of the experts' forecasts and reasoning from the previous round. In subsequent rounds, participants are encouraged to revise their earlier answers in light of the replies of other members of the group, which is assumed to facilitate consensus-building. The process is stopped after a predefined stop criterion (e.g., number of rounds, achievement of consensus, stability of results).

Drop-in centers Drop-in centers are places in the community to which the interested public can go without prior appointment. The public receive information on the status and future directions of different issues (e.g., town planning), but may also comment and provide feedback on plans in such a center.

Electronic consultation This refers to an exchange between government and citizens using the Internet. Electronic consultation represents a specific form of online deliberation. Online consultation consists in using the Internet to ask a group of people their opinion on one or more specific topics, allowing for trade-offs and dialog between participants. Generally, electronic consultation is used to identify or access options, or to evaluate ongoing activities. This enables governments to draft more citizen-centered policy.

Focus group A focus group is a form of qualitative research in which a group of people are asked about their attitude toward a product, service, concept, advertisement, idea, or packaging. Questions are asked in an interactive group setting, where participants are free to talk with other group members. Focus groups can be used for gaining access to various cultural and social groups, selecting sites to study, sampling of such sites, and raising unexpected issues for exploration.

Hotline This is a direct telephone number that people can ring to ask questions and give comments, or to put forward views on a specific issue.

Negotiated rule-making Negotiated rule-making is a process in which an advisory committee made up of disparate interest groups negotiates the terms of a rule or issue between each other.

Opinion poll An opinion poll is a survey of opinion from a particular sample. Opinion polls are usually designed to represent the opinions of a population by asking a small number of people a series of questions and then extrapolating the answers to the larger group within confidence intervals.

Planning cell Planning cells might be defined as a non-partisan, *ad hoc*, randomly selected, single-issue, short-term micro-parliament. The planning cell is presented with an issue, discusses it, and drafts recommendations and the assessments.

Public hearings/inquiry A public hearing or inquiry is an official review of issues ordered by the government. A public inquiry differs from more general inquiries or reviews in that evidence submitted to the inquiry is heard in a public environment. Interested members of the public and organizations may not only make (written) evidential submissions, as is the case with most inquiries, but also listen to oral evidence given by other parties.

Referendum A referendum is a direct vote in which an entire electorate is asked to either accept or reject a particular proposal.

Survey Surveys are used to collect quantitative information about items in a population on a certain issue. A survey may focus on opinions or factual information depending on its purpose. Most surveys involve administering questions to individuals. When the questions are administered by a researcher, the survey is called a structured interview or a researcher-administered survey. When the questions are administered by the respondent, the survey is referred to as a questionnaire or a self-administered survey.

Technology assessment Technology assessment is the study and evaluation of new technologies. It is based on the conviction that new developments within, and discoveries by, the scientific community are relevant for the world at large rather than just for the scientific experts themselves, and that technological progress can never be free of ethical implications. Technology assessment explicitly recognizes the fact that scientists normally are not trained ethicists themselves and accordingly ought to be very careful when passing ethical judgment on new findings, projects, or work in progress.

Telepolling Telepolling is a way of administering an opinion poll (see definition of opinion poll) by means of telephone interviews.

References

1 Macoubrie, J. (2006) Nanotechnology: public concerns, reasoning and trust in government. *Public Underst. Sci.*, **15** (2), 221–241.

2 Burri, R.V., and Bellucci, S. (2008) Public perception of nanotechnology. *J. Nanopart. Res.*, **10** (3), 387–391.

3 Pidgeon, N., and Rogers-Hayden, T. (2007) Opening up nanotechnology dialogue with the publics: risk communication or "upstream engagement"? *Health Risk Soc.*, **9** (2), 191–210.

4 Powell, M.C., and Colin, M. (2008) Meaningful citizen engagement in science and technology–what would it really take? *Sci. Commun.*, **30** (1), 126–136.

5 Fischer, A.R.H., and Frewer, L.J. (2007) Public acceptance of new technologies in food products and production, in *Risk and the Public Acceptability of New Technologies* (eds R. Flynn and P. Bellaby), Palgrave Macmillan, Basingstoke, UK, pp. 6[illegible]–85.

6 Houghton, J.R., Rowe, G., Frewer, L.J., Van Kleef, E., Chryssochoidis, G., Kehagia, O., Korzen-Bohr, S., Lassen, J., Pfenning, U., and Strada, A. (2008) The quality of food risk management in Europe: perspectives and priorities. *J. Food Policy*, **33**, 13–26.

7 Gaskell, G., Allum, N., and Stares, S. (2003) Europeans and Biotechnology in 2002: Eurobarometer 58.0, Methodology Institute, London School of Economics, London, UK.

8 Horlick-Jones, T., Walls, J., Rowe, G., Pidgeon, N., Poortinga, W., Murdock, G., and O'Riordan, T. (200[illegible]) *The GM Debate:*

Risk, Politics and Public Engagement, Routledge, London.
9 Sandler, R., and Kay, W.D. (2006) The National Nanotechnology Initiative and the social good. *J. Law Med. Ethics*, **34** (4), 675–681.
10 Hilgartner, S. (1990) The dominant view of popularisation: conceptual problems, political issues. *Soc. Stud. Sci.*, **20**, 519–539.
11 Irwin, A. (2006) The politics of talk: coming to terms with the "new" scientific governance. *Soc. Stud. Sci.*, **36** (2), 299–320.
12 Laird, F.N. (1989) The decline of deference: the political context of risk communication. *Risk Anal.*, **9** (4), 545–550.
13 Flynn, R., and Bellaby, P. (eds) (2007) *Risk and the Public Acceptability of New Technologies*, Palgrave Macmillan, Basingstoke, UK.
14 Frewer, L.J., Howard, C., and Shepherd, R. (1998) Understanding public attitudes to technology. *J. Risk Res.*, **1**, 221–237.
15 Frewer, L.J., Scholderer, J., and Bredahl, L. (2003) Communicating about the risks and benefits of genetically modified foods: effects of different information strategies. *Risk Anal.*, **23** (6), 1117–1133.
16 Levidow, L., and Marris, C. (2001) Science and governance in Europe: lessons from the case of agricultural biotechnology. *Sci. Public Policy*, **28** (5), 345–360.
17 Fischhoff, B. (1995) Risk perception and communication: twenty years of process. *Risk Anal.*, **15**, 137–145.
18 Rowe, G., and Frewer, L.J. (2005) A typology of public engagement mechanisms. *Sci. Technol. Human Values*, **30** (2), 251–290.
19 Fishkin, J.S. (1991) *Democracy and Deliberation: New Directions for Democratic Reform*, Yale University Press, New Haven, CT.
20 Wynne, B. (2006) Public engagement as a means of restoring public trust in science–hitting the notes, but missing the music? *Community Genet.*, **9** (3), 211–220.
21 Walls, J., Rowe, G., and Frewer, L.J. (2010) Stakeholder engagement in food risk management: evaluation of an iterated workshop approach. *Public Underst. Sci.*, in press. doi: 10.1177/0963662509354543
22 Rowe, G., and Frewer, L.J. (2004) Evaluating public participation exercises: a research agenda. *Sci. Technol. Human Values*, **29** (4), 512–556.
23 Rogers-Hayden, T., and Pidgeon, N. (2007) Moving engagement "upstream"? Nanotechnologies and the Royal Society and Royal Academy of Engineering's inquiry. *Public Underst. Sci.*, **16** (3), 345–364.
24 Wentholt, M., Rowe, G., Konig, A., Marvin, H., and Frewer, L. (2009) The views of key stakeholders on an evolving food risk governance framework: results from a Delphi study. *Food Policy*, **34** (6), 539–548.
25 Webler, T. (1995) "Right" discourse in citizen participation: an evaluative yardstick, in *Fairness and Competence in Citizen Participation: Evaluating Models for Environmental Discourse* (eds O. Renn, T. Webler, and P. Wiedemann), Kluwer Academic, Dordrecht, The Netherlands, pp. 35–86.
26 Rowe, G., and Frewer, L.J. (2000) Public participation methods: a framework for evaluation. *Sci. Technol. Human Values*, **25** (1), 3–29.
27 Delli Carpini, M.X., Cook, F.L., and Jacobs, L.R. (2004) Public deliberation, discursive participation, and citizen engagement: a review of the empirical literature. *Annu. Rev. Polit. Sci.*, **7**, 315–344.
28 Mendelberg, T. (2002) The deliberative citizen: theory and evidence, in *Research in Micropolitics: Political Decision-Making, Deliberation and Participation* (eds M.X. Delli Carpini, L. Huddy, and R. Shapiro), JAI Press, Greenwich, CT, pp. 151–193.
29 Ryfe, D.M. (2005) Does deliberative democracy work? *Annu. Rev. Polit. Sci.*, **8**, 49–71.
30 Sunstein, C.R. (2002) The law of group polarization. *J. Polit. Philos.*, **10** (2), 175–195.
31 Rowe, G. (2007) Public engagement in food policy, in *Understanding Consumers of Food Products* (eds L.J. Frewer and J.C.M. van Trijp), Woodhead Publishing, Cambridge, UK, pp. 592–609.

32 Fischer, A.R.H., and Frewer, L.J. (2008) Stakeholder and Societal Consultation about Emerging Issues in the Chemical Sector. A Review for the Suschem Technology Platform, Wageningen University, The Netherlands.

33 Pidgeon, N., Harthorn, B.H., Bryant, K., and Rogers-Hayden, T. (2009) Deliberating the risks of nanotechnologies for energy and health applications in the United States and United Kingdom. *Nat. Nanotechnol.*, **4** (2), 95–98.

34 Burri, R.V. (2009) Coping with uncertainty: assessing nanotechnologies in a citizen panel in Switzerland. *Public Underst. Sci.*, **18** (5), 498–511.

35 Evers, J., and D'Silva, J. (2009) Knowledge transfer from citizens' panels to regulatory bodies in the domain of nano-enabled medical applications. *Innov. Eur. J. Soc. Sci. Res.*, **22** (1), 125–142.

36 Powell, M., and Kleinman, D.L. (2008) Building citizen capacities for participation in nanotechnology decision-making: the democratic credentials of the consensus conference model. *Public Underst. Sci.*, **17** (3), 329–348.

37 Besley, J.C., Kramer, V.L., Yao, Q.J., and Toumey, C. (2008) Interpersonal discussion following citizen engagement about nanotechnology: what, if anything, do they say? *Sci. Commun.*, **30** (2), 209–235.

38 Hamlett, P.W., and Cobb, M.D. (2006) Potential solutions to public deliberation problems: structured deliberations and polarization cascades. *Policy Stud. J.*, **34** (4), 629–648.

39 Rogers-Hayden, T., and Pidgeon, N. (2008) Developments in nanotechnology public engagement in the UK: "upstream" towards sustainability? *J. Clean. Prod.*, **16** (8–9), 1010–1013.

16
Nano-Ethics

Roger Strand

16.1
Introduction: Historical Background

Some readers might wonder if the title of this chapter is a joke. What is "nano-ethics"? Does it exist? The answer is yes. "Nano-ethics", the study of the ethical impacts, issues, and aspects of nanoscience and nanotechnology, emerged with the new (twenty-first) century and is currently consolidating as a research field, a field of expertise, and a set of practices in the regulation and governance of nanoscience and nanotechnology. For instance, in 2007 the academic publisher Springer launched a new journal called *Nano-ethics.* Several research anthologies [1, 2] and governmental and non-governmental reports on the ethics of nanotechnology have been published [3–6]. Often, for the ethical aspects/impacts of nanoscience and nanotechnology, the broader terms ethical, legal, and societal/social aspects (ELSA) and ethical, legal, and societal/social impacts (ELSI) are used; others prefer the term social and ethical interactions with nano (SEIN) [7]. We will return to some of the underlying reasons for such differences. For the sake of simplicity, I will use the term "nano-ethics" throughout this chapter, because it makes little sense to distinguish sharply between an "ethical" and a "societal" issue.

One may easily identify debates on the ethics of science and technology in other technological fields and at earlier times. Notably, many physicists participated in what we would now call ethics debates in the decades following World War II, when the full implications of nuclear technology, including the hydrogen bomb, became evident. It was only with the advent of biotechnology, however, that the ethics of science and technology became a focus of attention in its own right, with dedicated research projects, university programs, academic journals, ethics committees, and even specific regulations and laws being implemented in many countries. Of particular importance was molecular biologist and Nobel laureate James Watson's initiative in the 1980s to set aside a certain percentage (3%) of the total budget of the Human Genome Project for ELSI research. This decision implied a vast increase in the funding for bioethics research and was copied in many European countries and what is now known as the European Union. This policy has largely been continued as public nanotechnology

Nanotechnology in the Agri-Food Sector: Implications for the Future, First Edition. Edited by Lynn J. Frewer, Willem Norde, Arnout Fischer, Frans Kampers.

funding programs have been set up. For instance, it has been claimed that more than 40 million US dollars are spent each year on nano-ethics/ELSA in the USA alone[1)].

What do nano-ethicists discuss, then? The question is simple, but the answer is complicated, and most of this chapter will be devoted to providing an introductory guide to nano-ethics. A first observation to be made is the contrast between the ethics of nuclear technology and that of biotechnology. With respect to nuclear technology, one might somewhat disrespectfully say that the bomb arrived first and the ethical qualms arrived a few years later. In the case of biotechnology, much of the ethical debate occurred simultaneously with the development of the technology itself, and often the ethical problems appeared to be quite self-evident and pressing. For instance, it was seen as obvious that the handling of, and experimentation with, human embryos demanded careful ethical evaluation, at least in the Judeo-Christian cultural sphere. "Simultaneous ethics" means, however, ethics that ameliorates the effects of undesirable technologies, products or possibilities, rather than preventing the problems from emerging.

With the advent of nanotechnology, it was therefore argued by many ethicists that society should "seize the day" and take the "historic opportunity" to install ethics *in advance of* the technological development itself. Nano-ethics should not only deal with existing technologies, but also prepare for future technology, and foresee and prevent ethical problems. Furthermore, even many strong proponents of nanotechnology have called for ethics to be included at an early stage. Often, reference has been made to the political controversies over genetically modified food in Europe as an example of an unwanted situation. Ethics has been conceived both by nano-optimists and nano-skeptics as a way to avoid massive expenditure on the development of products that, at the end of the day, are found to be unwanted by citizens and consumers.

The diversity of nanotechnology ranges from well-established production methods for nanostructured materials to, say, basic research on hypothesized functional couplings between computers and animal (or human) brains. From this, one may appreciate how diverse nano-ethics debates may be. The nano-ethics literature discusses a myriad of existing and non-existing technologies; with observed or suspected or postulated impacts; and the impacts may be controversial, trivial, and difficult to identify. As might be expected by academics under such uncertain conditions, there is also a lot of discussion about what are the appropriate topics and methods for nano-ethics, and how this field should develop [8, 9]. We may concur with Kjølberg & Wickson [7] that the nano-ethics field in a certain sense is *immature*.

Immaturity does not imply fault or uselessness, though. On the contrary, in what follows, I will draw upon the current diversity of opinions of what nano-ethics is and ought to be in order to explain how one may ask ethical questions about a

1) In the absence of an authoritative reference, the blog of the "Editors of The American Journal of Bioethics" has been consulted: http://blog.bioethics.net/2006/01/nanoethics-the-elsi-of-21st-century-bioethics/ (accessed 16 November 2010).

certain technology in different ways and at different levels; they will be called three different ethical "gazes", or ways of looking at the system of research, development, and production of technology. This "system" is difficult to define in the general case, as it may or may not include this or that form of directed or applied research, various activities associated with technology transfer, and so forth, in the concrete case. However, I believe a general and rather undifferentiated concept of the "system" to be useful. From now on, I shall use the words "gaze" and "system" in this sense. This may all sound abstract, but we shall see that the implications are highly practical and policy-relevant.

16.2 Identifying and Avoiding Unethical Nanotechnological Products

Which nanotechnological products and processes are, or could be, unethical, and in what respect? In my subjective experience, this is the intuitive nano-ethics question for many journalists, policy-makers, and scientists. Many ethicists, perhaps striving to be useful in the eyes of those who pay their salaries, apply this kind of ethical gaze at the system, looking for potential harm, injustice, inequity, threats to human self-determination and dignity, and so on.

The first question to be asked is, of course, if the nano-product is or may be harmful, a question that belongs as much to risk–hazard assessment and management as to ethics. For instance, it is an open question as to whether free nanoparticles may travel through the body or the ecosystem and give rise to novel human health risks or environmental effects. The desirable properties of nanoparticles are due to their small size, which gives them a higher surface-to-volume ratio and different chemical properties. It is by no means unthinkable that these same characteristics may cause unforeseen effects that could be harmful. Such questions must be studied by the appropriate scientific disciplines, such as toxicology and ecotoxicology, and can of course not be decided upon by ethicists. No conclusion about harm can be made in the general case; it will depend upon the stability, mobility, and reactivity of the type of particles, their use, the adequacy and reliability of safety measures, and so on. Ethics may still be useful by debating the right thing to do given the certainty or uncertainty about positive and negative effects. Typically, a designed and certain benefit has to be weighed against uncertain or even unidentified harm.

One important discussion that follows is whether ordinary risk assessment and management procedures should be used, whether some version of the precautionary principle should be invoked, or if even more cautious measures are required. The ETC Group [10] has called for a moratorium to be applied to the environmental release of free nanoparticles on these grounds. Another non-governmental organization, Friends of the Earth, has argued that definitions of nanoparticles should be reworked to be more precautionary (including particle size up to 300 nm) [11, 12]. As nanotechnology gradually enters human medicine and food industries, one should expect ever more focus on the aspect of potential harm.

It is also important to bear in mind that the question of harm is not only one of undesired secondary effects. One may easily imagine nano-terrorism and other malevolent uses of nanotechnology, for instance by designing highly reactive particles that may penetrate the body or foodstuffs. Furthermore, it has been pointed out that nanotechnologies, because of their small size, might more easily evade detection, especially if one does not know what to look for. Again, this would be relevant for industries that could be the victim of sabotage. As for military research on nanotechnology, this is not within the expertise of the present author; and, indeed, sources of reliable public information do not abound.

Along the same lines, questions of autonomy and dignity of humans have been discussed. In the case of nanotechnology, one could imagine a further miniaturization of tracer technologies, for instance to improve logistics or knowledge of origin of products, which might also be used in surveillance of unaware subjects (see also Chapter 4 in this volume). Even with subjects aware of their use, nanotechnology-based medical technologies might constitute a complexity of "inner surveillance" and precision control over physiological parameters through directed medication within the body, to the extent that the subject is hardly informed and in control any more. It has been argued that such benevolent and medically beneficial technologies may also be a threat to perceived personal autonomy and integrity. Taking it to the extreme, the envisaged coupling of biotechnology, information and communication technology (ICT), and cognitive science at the nano level – so-called nano-bio-info-cogno (NBIC) – has raised discussions over potential technologies for human enhancement, that is, technologies that improve human senses and capacities, either for the individual or even for (part of) the human species. On this issue, the North Atlantic Ocean appears to be a sharp line of division: while arguments in favor of human enhancement and *transhumanism* are utterly politically incorrect in Europe, transhumanist visions have actually been put forth by central nanotechnology proponents in the USA [13, 14]. If one consults the web page of the MIT Institute of Soldier Nanotechnologies,[2] one may furthermore be convinced that the issue is not purely one of science fiction. I shall return below to the European response to US transhumanism.

Finally, the ethics debate has discussed the so-called *nano-divide*, or how the organization of research and development of nanotechnology might increase global injustice. It is true that nanoscience and nanotechnology are dominated by wealthy and developed countries: North America, Western Europe, Japan, and then South Korea and China. The nano-divide is not just focused on the question of who develops and owns nanotechnologies, but also whether these technologies get built into more production systems so that the lagging behind of poorer economies will become an increasingly large disadvantage in the global market.

Summing up, the ethical gaze I have described is one that screens and scrutinizes the properties of a given nanotechnological product, process or activity, and that investigates its actual or potential effects. Normally, this ethical focus is *negative* in the sense that there is no need for ethics if there is no harm or threat to

2) See http://web.mit.edu/isn/ (accessed 16 November 2010).

anyone. The ethics consists in identifying problems and then figuring out what to do with them: if something should be discouraged or prohibited, or if special antagonistic measures should be taken. Exactly for this reason, this ethical gaze is closely related to legal and regulatory institutions and procedures. What I have described here is the kind of ethics that is a central part of what goes on in ethics reviews and ethics committees, as well as governmental reports.

16.3 Ensuring Ethical Nanotechnological Research, Innovation, and Production

A distinctly different, but equally important, ethical gaze is that which looks at the *actions* that lead to nanotechnological products. The question is then no longer if the product is unethical, but if the researchers, developers, and producers have behaved in ethically justifiable ways, and if their institutions and companies are organized in an accountable and responsible manner that allows, encourages, and ensures ethical behavior. This is important for all fields of science and technology, and not less important for nanoscience and nanotechnology, for two reasons. First, there is big money involved, with high expectations of profit. Second, as already mentioned, nanotechnologies may involve particular challenges with respect to detection, controllability, and unknown harmful effects. In other words, in particular in terms of sins of omission, there is what a television series crime investigator might call both *motive* and *opportunity*.

At the same time, the practices and institutions of research have changed vastly. Until World War II, science was a lifestyle choice and involved a small elite. Since 1945, the *gentlemen* have become vastly outnumbered by the *players*, and research is now ordinary work, not even particularly well paid or highly esteemed, at least not for the majority of the research workforce. In the natural sciences, many researchers do not enjoy the freedom to develop their own research questions, but rather work as "super-technicians" within large research teams. Many senior researchers have vested interests in the products of their own research. Without exaggerating the sense of vocation and ethical virtues of the scientists in the past, it is not difficult to understand that ideas of new public management and quality assurance found their ways into a research world with big expenditures, big workforce, and big safety challenges. Hence, to avoid fraud and corruption, researchers nowadays are required to store data in prescribed ways and to disclose their personal economic interests. Universities and research institutions produce ethical guidelines and demand that their employees and students comply with them; ethics courses are offered or even required; and there are ever more national and international research guidelines. The author's home country, Norway, passed its Research Ethics Act in 2006, actually making breaches of research ethics illegal.

To a large extent, this has been a matter of codifying and enforcing ideals and norms of research ethics that already existed. More than 60 years ago, Robert K. Merton [15] formulated his "ethos of science", arguing that efficient knowledge production depended upon open access to others' work, a disinterested attitude (i.e., only interested in truth), and a methodically critical attitude ("organized skepticism"), and so on. Seen with this ethical gaze, good ethics is a prerequisite of good science; indeed, they are almost the same thing. Likewise, one may argue that there can be no functional economic market in a society where everybody is prone to lie, cheat or steal.

Nonetheless, the development of the institutional ethical gaze goes beyond the classical norms of the ethos of science, business ethics, and common morality. The clearest example of this is the high-level expert report to the European Commission called *Converging Technologies for the European Knowledge Society* (CTEKS for short) [16]. The CTEKS report acknowledges that ordinary honesty combined with the ethical gaze at products is not enough to avoid ethical problems with nanotechnologies. It is fair to see the report as a response to the US NBIC report [14] that to a large extent advocated human enhancement and only envisaged a *post hoc*, corrective role for ethics. The CTEKS question was accordingly: How can we ensure that nanotechnology development does not take a harmful, unethical, and dangerous direction? Formulated in the usual self-content European jargon: How do we ensure that the technology development is in accordance with European values?

One should appreciate how radical the reflection provided by the CTEKS report actually is. So far in this chapter, I have only discussed ethical gazes that look for anomalies – faults or sins – within a system that is never questioned *per se*. The CTEKS report, however, acknowledges the fact that researchers and developers with good intentions, complying with every ethics guideline there is, may still produce something dangerous or unethical. Of course, it may then be identified as such by an ethics committee – but then it may be too late. The world may already have changed, because something is introduced and dispersed into our ecosystem, or our bodies, or our space of possible ill-intended actions. Again, at the heart of the issue we find the power and the smallness of nanotechnology, potentially eluding detection and retraction.

The CTEKS report tries to solve this challenge by demanding that research shall be planned in accordance with European values. The convergence of sciences and technologies at the nanoscale does not happen arbitrarily and by itself, they argue, it requires that technical goals are set. The answer is therefore to organize broad political processes to define the social purposes to which these goals are to correspond. For instance, they mention reduction of obesity as a health problem, as a purpose that is in accordance with European values, while human enhancement is not. This value choice must then be translated into ethically responsible research policies.

16.4
Nano-Ethics as the Question of the Good Nanotechnology Society

The CTEKS report is a suitable departure for explaining a third ethical gaze, namely that which asks about the good nanotechnology society [17]. First, it is clear that CTEKS aspired to provide a road to that society. It is even possible that it could do so in certain domains of nanotechnology development, perhaps also in the agri-food sector. The more applied and the less "fundamental" character of the research and technology involved, the more relevant CTEKS appears to be. It would be exciting to see attempts at democratic involvement of citizens in the design of novel foods, rather than treating the same people only as consumers whose behavior is predicted through focus group methodologies. A number of so-called upstream engagement exercises have been devised over the latter years, in particular with respect to nanotechnology (see also Chapter 15 in this volume). It remains to see whether such exercises have had significant impact and that this impact has resulted in a better nanotechnology society (see e.g. [18]). Moreover, when the research and development activities take on a more "basic", fundamental character, it is hard to see that a solution such as CTEKS could work at all. There is no one-to-one correspondence between the preset goal of a basic research project and its results; on the contrary, open-endedness is a defining character of science [19, 20].

What CTEKS clearly showed, however, is how close a relationship there is between the ethical and the political. This was explained in full by another European expert group in their report *Taking the European Knowledge Society Seriously* [21], which talked about *the unpolitics of ethics*. Ethics – in particular, in the shape of expert ethicist committees and reports – effectively serves to remove attention from and to depoliticize politically controversial issues: "Don't worry, we have a group of ethics experts working on it!" The original, broadly defined political issue, perhaps vaguely or just implicitly expressed as "But do we really need this novel food? Do we, as a society, really *want* it?", is transformed by the above-described myopic ethical gazes into questions of health risk, religious qualms about tampering with nature, or new "ethical accounting" practices in research and development. Accordingly, the political issue is reduced to a set of so-called ethical issues that are of a technical nature and have a technical solution, and the public can be reassured as everything is under the control of the ethical experts. The big question is, of course, whether the people *really* are reassured, and for how long, by such procedures.

In the introduction, I posed the question of what nano-ethicists discuss, and replied that the answer is complicated. By now the reader will know why. The choice of ethical gaze is in itself an ethical and political choice, and this is as true for the technologist and producer as it is for the ethicist. In my view, there is still a lot to learn from careful reflection upon the controversies surrounding genetically modified food. According to the European expert group cited above, ethics contributes to depoliticize controversial issues, in particular if the ethics is narrowly construed as expert deliberation upon limited questions of a more ethical–technical nature. This may be true, though the genetic modification controversies also show that the involvement of ethics and ethicists does not eliminate or preempt the political potential. In other words: Ethics projects, ethics groups, and ethicist advice do not make the real problems go away in cases where the public really has an opinion. A narrow approach to ethics accordingly runs the risk of failing to predict, prevent or prepare for a big controversy at a later stage.

In this respect, it is interesting to note the development of "codes of conduct" for nanotechnology. In February 2008, the European Commission published their

Code of Conduct for Responsible Nanosciences and Nanotechnologies Research [22].[3) The concept of ethics does not in itself play a prominent role in the document, although it is of course said that research should be in accordance with ethical principles and comply with ethics guidelines and ethical review. There is no doubt, however, that the entire code is consistent with–and perhaps informed by–the broader view on ethics and politics that I have described as the third ethical gaze. Indeed, the first principle of the code is called "meaning", and reads as follows [22]:

> ***Meaning*** N&N [nanoscience and nanotechnology] research activities should be comprehensible to the public. They should respect fundamental rights and be conducted in the interest of the well-being of individuals and society in their design, implementation, dissemination and use.

Furthermore, the code recommends an inclusive approach to governance [22]:

> Good governance of N&N research should take into account the need and desire of all stakeholders to be aware of the specific challenges and opportunities raised by N&N. A general culture of responsibility should be created in view of challenges and opportunities that may be raised in the future and that we cannot at present foresee.

"All stakeholders" is understood as "Member States, employers, research funders, researchers and more generally all individuals and civil society organizations engaged, involved or interested in N&N research" [22]. One may of course discuss how realistic such aspirations are, and to what extent soft regulation such as this recommendation by the European Commission will have any implications. Entering into the general discussion on soft regulation will go beyond the scope of this chapter; however, it should be recalled that the communication on the precautionary principle [23] and the White Paper on governance, two definitely influential texts from the European Commission [24], were both "mere" recommendations. The effect of such recommendations, guidelines, and codes depends on the creative work of facilitating (or averting) their implementation and use.

16.5
Conclusion: The Ethical Challenge Ahead for the Nano-Agri-Food Sector

In this chapter, I have described three nano-ethical gazes that ask the following type of questions:

1) What ethical problems (harm, injustice, inequity, threats to human self-determination and dignity, etc.) are raised by the nanotechnological *product* or process under scrutiny?

3) A UK non-governmental initiative along the same lines can be found at: http://www.responsiblenanocode.org (accessed 16 November 2010).

2) Are the *actions* of researchers, developers, and producers organized in an ethically responsible (benevolent, honest, accountable) way?

3) What would constitute a *good* society with nanotechnology, and what path leads to this society?

The multiplicity of these gazes corresponds to the eternal diversity of ethics, being concerned with *the morally right and wrong* as well as *the good life*. All three questions are important and, I would claim, necessary, and they are related to each other. I have argued that academic and applied ethics have directed too much attention to the two first-mentioned questions, while the third type of question appears to be on the rise, not always under the label of ethics, but also as the (political) issue of *governance of nanotechnology*.

Just as little as any other argument, the argument of this chapter cannot be politically neutral. Indeed, to insist on the relevance of the third gaze and third type of question, is to say that the objective of current innovation policies is not self-evidently good in the moral sense. There is a long and strong tradition, in particular in the industrialized world, to see scientific and technological progress as something inherently and unquestionably good. This is why ethics has been relegated to the minor role of avoiding what we could call moral adverse effects. The lesson from the advent of the nuclear bomb is that progress is inherently two-sided. The lesson from the genetic modification controversies, if not before, is that people know about the ambiguity of progress, and may actually say "no thanks" to new and technically speaking better products. Accordingly, developers and producers are left with two options.

The first option is to accept that judging the quality of new products – quality in the broadest sense, technical, ethical, political – is, and should be, a collective, societal task. If so, all three ethical gazes are necessary, and the industry needs slow and sincere dialog with the public.

The other option is not to accept this claim, and instead to develop more sophisticated knowledge of consumer behavior together with more effective means of persuasion, so that the public will not resist the introduction of what scientists, technologists, and industrialists believe to be rational technologies and better products. This option violates most of ethics' general principles, such as the respect for the self-determination and dignity of others. In other words, the ethical challenge for the sector is in one sense simple: to be ethical or not to be ethical, that is the question.

Acknowledgments

This chapter builds upon the countless discussions within the Nanoethics Group at the Centre for the Study of the Sciences and the Humanities, University of Bergen, in particular with Fern Wickson and Kamilla Lein Kjølberg, as well as the fruitful collaboration with Rune Nydal, Program for Applied Ethics, the Norwegian University of Science and Technology. The financial support from the Research Council of Norway to the Nanoethics Group is gratefully acknowledged.

References

1 Hunt, G., and Metha, M. (eds) (2006) *Nanotechnology, Risk, Ethics and Law*, Earthscan, London.

2 Allhof, F., Lin, P., Moor, J., and Weckert, J. (eds) (2007) *NanoEthics. The Ethical and Social Implication of Nanotechnology*, John Wiley & Sons, Inc., Hoboken, NJ.

3 Royal Society and Royal Academy of Engineering (2004) Nanoscience and Nanotechnologies: Opportunities and Uncertainties. The Royal Society, London, UK. Available at: http://www.nanotec.org.uk/finalReport.htm (accessed 9 November 2010).

4 Tegart, G. (2006) Environmental, social, legal and ethical aspects of the development of nanotechnologies in Australia. A report from the National Academies Forum for the National Nanotechnology Strategy Taskforce, Department of Industry, Tourism and Resources, National Academies Forum, Parkville, Victoria, Australia.

5 National Science Foundation (2001) Societal implications of nanoscience and nanotechnology, Report from the *Workshop held at the National Science Foundation, 28–29 September 2000* (eds M.C. Roco and W. Bainbridge).

6 UNESCO (2006) *The Ethics and Politics of Nanotechnology*, United Nations Educational, Scientific and Cultural Organization (UNESCO), Paris.

7 Kjølberg, K., and Wickson, F. (2007) Social and ethical interactions with nano: mapping the early literature. *NanoEthics*, **1**, 89–104.

8 Strand, R. (2001) ELSA studies of nanoscience and nanotechnology, Memo to the COST Nanoscience and -Technology Advisory Group (NanoSTAG). Available at: http://www.stage-research.net/STAGE/nanotechnology/nanostag-elsa.pdf (accessed 16 November 2010).

9 Van de Poel, I. (2008) How should we do nanoethics? A network approach for discerning ethical issues in nanotechnology. *NanoEthics*, **2**, 35–38.

10 ETC Group (Action Group on Erosion, Technology and Concentration) (2003) No small matter II: The case for a global moratorium. Size matters! Available at: http://www.etcgroup.org/upload/publication/165/01/occ.paper_nanosafety.pdf (accessed 16 November 2010).

11 Friends of the Earth (2006) Nanomaterials, sunscreens and cosmetics: small ingredients, big risks. Available at: http://nano.foe.org.au/node/100 (accessed 16 November 2010).

12 Friends of the Earth (2008) Out of the laboratory and onto our plates: nanotechnology in food and agriculture. Available at: http://nano.foe.org.au/node/227 (accessed 16 November 2010).

13 Roco, M.C., and Bainbridge, W.S. (2002) Converging technologies for improving human performance: integrating from the nanoscale. *J. Nanopart. Res.*, **4**, 281–295.

14 Roco, M.C., and Bainbridge, W.S. (eds) (2002) Converging technologies for improving human performance, NSF-DOC Report, Arlington, VA.

15 Merton, R.K. (1942) The normative structure of science, in *The Sociology of Science: Theoretical and Empirical Investigations* (ed. N.W. Storer), University of Chicago Press, Chicago, IL, 1973, pp. 267–278.

16 Nordmann, A. (2004) Converging technologies – Shaping the future of European societies, Report for the European Commission via an Expert Group on Foresighting the New Technology Wave, European Commission, Brussels. Nydal and Strand 2008 probably to go in here

17 Nydal, R., and Strand R. (2008) God nanoetikk – god nanoteknologiutvikling.

Etikk i praksis (Nordic Journal of Applied Ethics), **1**, 33–51.
18 Gavelin, K., Wilson, R., and Doubleday, R. (2007) *Democratic Technologies? The Final Report of the Nanotechnology Engagement Group (NEG)*, Involve, London.
19 Pickering, A. (1995) *The Mangle of Practice*, University of Chicago Press, Chicago, IL.
20 Kjølberg, K., Delgado-Ramos, G.C., Wickson, F., and Strand, R. (2008) Models of governance for converging technologies. *Technol. Anal. Strateg. Manage.*, **20**, 83–97.
21 European Commission (2007) Taking European knowledge society seriously, Report of the Expert Group on Science and Governance to the Science, Economy and Society Directorate, Directorate-General for Research, European Commission.
22 European Commission (2008) Commission recommendation of 07/02/2008 on a Code of Conduct for Responsible Nanosciences and Nanotechnologies Research, C (2008) 424, Commission of the European Communities, Brussels. Available at: http://ec.europa.eu/nanotechnology/pdf/nanocode-rec_pe0894c_en.pdf (accessed 16 November 2010).
23 European Commission (2000) Communication from the Commission on the Precautionary Principle, COM (2000) 1, Brussels.
24 European Commission (2001) White Paper on European Governance, COM (2001) 428, Brussels.

17
Evolving Best Practice in Governance Policy – Developing Consumer Confidence in Risk Analysis Applied to Emerging Technologies

Hans J.P. Marvin, Hans Bouwmeester, Gijs A. Kleter, Lynn J. Frewer, and Meike T.A. Wentholt

17.1
Introduction

Risk governance is an important element that needs to be addressed following the development and identification of any new technology, including agri-food nanotechnology. Following a general introduction to the principles and components of the internationally acknowledged approach to food risk analysis, this chapter further discusses the "risk cycle" approach proposed by the European Union's Scientific Steering Committee. Risk analysis is supposed to bring together three interrelated risk-focused activities, comprising risk assessment, risk management, and risk communication. The European SAFE FOODS project has further devised amendments to these models for risk analysis, which have culminated in a newly developed integrated risk analysis framework. This framework considers risk analysis as an iterative, cyclic process that passes through four stages: framing, risk–benefit assessment, evaluation, and risk management. New elements contained by the integrated SAFE FOODS risk analysis framework include the increased emphasis on – and individuation of – the stages of framing and evaluation, at which the risk assessors and risk managers interact with each other and where also stakeholders can provide useful inputs and feedback. In addition, it is proposed that the risk assessment stage focuses not only on risks but also on other impacts, including human health benefits, as well as social, economic, environmental, and ethical impacts.

Risk management includes decision-making, implementation, and monitoring, after which review of the effectiveness of risk management can take place, possibly providing outcomes that feed back into the framing phase so that the cycle can start again. In general terms, increased stakeholder involvement, communication, and transparency are advocated throughout the risk analysis process.

An area in which the new integrated SAFE FOODS risk analysis framework could be relevant is that of nanotechnologies applied to food production and handling. This new technology holds great potential but can, at the same time, also pose new risks previously not encountered. The term "nanotechnology" actually covers a wide range of technological applications, which share the character-

Nanotechnology in the Agri-Food Sector: Implications for the Future, First Edition. Edited by Lynn J. Frewer, Willem Norde, Arnout Fischer, Frans Kampers.

istics of nanometer-scale size structures as functional units. The properties of nanomaterials can differ greatly from the generic properties of the same materials (i.e., those which are larger than nano-size). As a consequence, potential toxic effects cannot be predicted because of lack of extrapolation. In line with the new integrated risk analysis framework, proposals are made regarding the identification and prioritization of areas for future research in the area of nanotechnology applied to agriculture and food production.

17.2
Introduction to Food Safety Governance

17.2.1
General Principles of Risk Analysis

Governance of food safety is conducted by most governments through application of the *risk analysis framework*, which is the dominant model applied in the area of regulation associated with food safety. International harmonization of food safety regulations is being conducted through the Codex Alimentarius, an organization co-founded by two United Nations organizations: the Food and Agriculture Organization (FAO) and the World Health Organization (WHO).

According to the general principles developed within Codex Alimentarius, the risk analysis framework can be separated into three distinct interrelated activities – risk assessment, risk management, and risk communication.

17.2.1.1 Risk Assessment

Risk assessment is performed by technical risk assessors. In some institutional contexts – for example, the European Food Safety Authority (EFSA) – there is a structural separation between risk management and risk assessment, although such institutional compartmentalization does not always apply – for example, the British Food Standards Agency (FSA) has responsibility for assessment, management, and communication within its terms of reference. Indeed, some stakeholders have criticized the structural and functional separation of risk assessment and risk management as non-pragmatic [2].

The process of risk assessment in the context of food safety was largely defined in a joint expert meeting convened by FAO/WHO [3], and four components were distinguished, namely (i) hazard identification, (ii) hazard characterization, (iii) exposure assessment, and (iv) risk characterization (see Figure 17.1). Hazard identification is the identification of a substance or attribute of food that may potentially cause adverse effects on human health. Whether such adverse health effects will occur depends on the exposure to the hazard. Within risk assessment, this is determined by studies on dose–response relationships between the hazard and the harmful effects in the target organs (the hazard characterization step), the

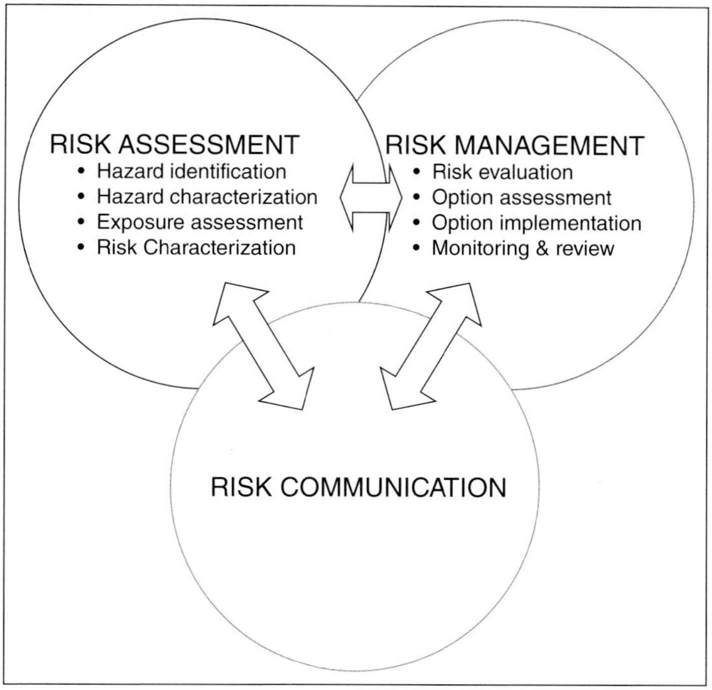

Figure 17.1 Risk analysis framework [1].

actual exposure of humans to this hazard (exposure assessment), and the risk of becoming exposed to levels of concern (risk characterization). The risk is determined as the product of the likelihood of the occurrence of the harm and the magnitude or severity of the effect.

17.2.1.2 **Risk Management**

Risk management is the decision-making process performed by risk managers in which the outcome of the risk assessment is weighed against other relevant data, and, if judged appropriate, prevention or mitigation measures are selected and implemented. Risk management generally initiates and ends the risk analysis process, and is composed of a number of elements: (i) risk evaluation (i.e., identification of food safety problem, establishing of a risk profile, ranking the hazard for risk assessment and risk management prioritization, establishing risk assessment policy, commissioning of risk assessment, and consideration of risk assessment results); (ii) risk management option assessment (i.e., identification of available management options, selection of preferred management options, and final management decision); (iii) implementation of management decisions; and (iv) monitoring and review (i.e., assessment of effectiveness of measures taken, and review of management and/or assessment as necessary) [4].

To initiate a risk assessment, the risk manager will prepare a "risk profile" by consulting all parties interested that are likely to be affected by the risk manager's decision. Once it has been decided to perform a risk assessment, this activity may be outsourced to independent risk assessors. In addition to safety, the risk manager may consider other factors as part of the risk assessment, such as the potential social, ethical, and economic impacts of the hazard or, indeed, hazard prevention or mitigation.

17.2.1.3 Risk Communication

Risk communication has been defined as the interactive exchange of information and opinions concerning risk and risk management activities among risk assessors, risk managers, consumers, and other interested parties [1]. It is assumed that interactive communication among all interested stakeholders, such as risk assessors, risk managers, industry, non-governmental organizations, primary producers, and consumers, *inter alia*, will assure transparency, facilitate the development of consistent decision-making, and improve the quality of decisions made.

17.2.2 Risk Analysis in Europe–the European Commission's Scientific Steering Committee Model

The advice to the European Commission regarding consumer health and food safety has previously been provided via its Directorate-General for Health and Consumers by a Scientific Steering Committee (SSC) and additional Scientific Expert Committees. This role is currently fulfilled by the European Food Safety Authority (EFSA). The mandate of these advisory bodies is to provide the European Commission with scientific advice or opinions about scientific and technical issues in their respective field of expertise and is based on the principles of "excellence", "independence", and "transparency". In 2000, a dedicated Working Group of the SSC published recommendations to harmonize risk assessment procedures between the various Scientific Advisory Committees [5]. The primary goal of this activity was to harmonize definitions and procedures among the various Scientific Advisory Committees, as well as to describe the underlying principles and to stimulate consistency, transparency, and quantitative approaches of the assessments.

The report also considers the "risk cycle" (see Figure 17.2), showing the various stages in the risk analysis framework, and builds further on the framework proposed by the experts consulted by FAO/WHO [1]. In a second report [6], further improvements to risk assessment were proposed, addressing, *inter alia*, issues such as the use of probabilistic modeling, the impact of emerging technologies on the risk assessment process, and the assessment of the impact on quality of life, including individual experience of quality-of-life parameters. The ultimate goal is maximizing health and well-being (considering potential impacts on human, animal, and environmental targets) as well as the quality of life experienced. This

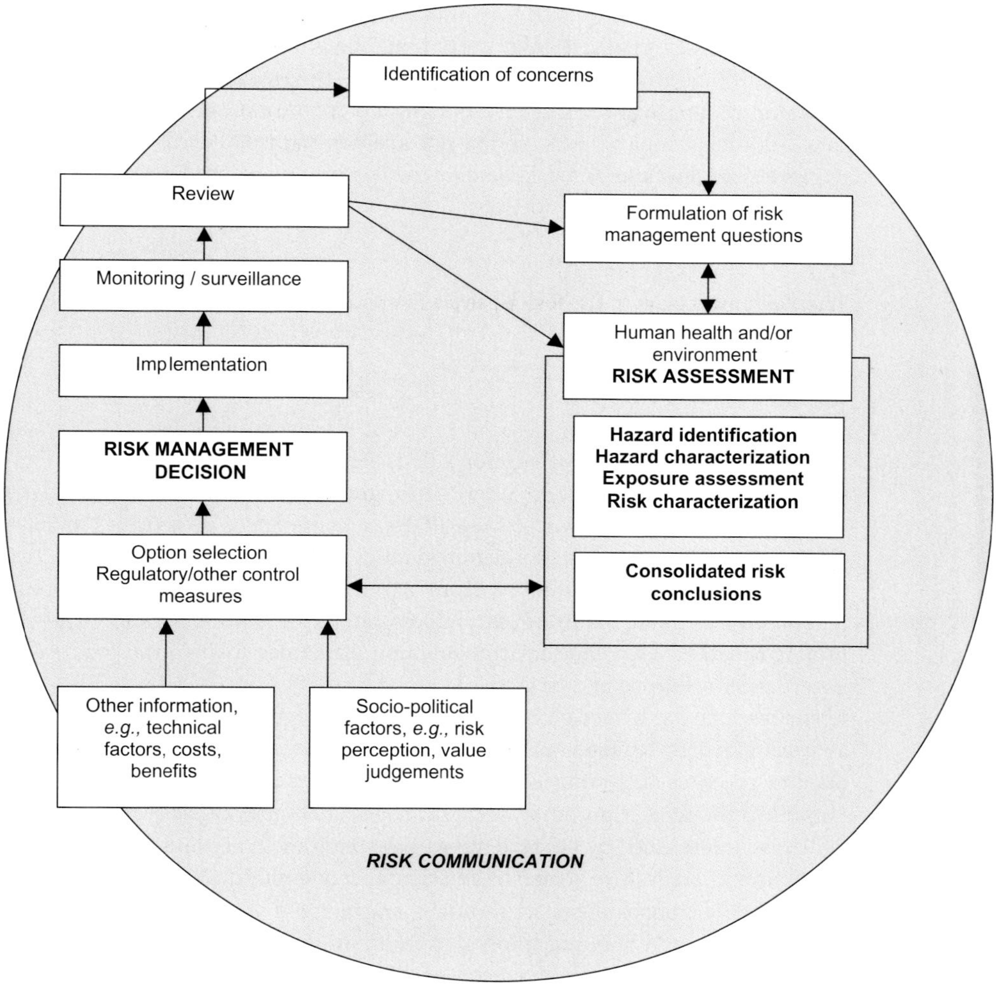

Figure 17.2 The "risk cycle" with components of risk analysis according to the EU Scientific Steering Committee [5].

is, of course, a remit reaching beyond the minimization of human health risks, requiring additional methodologies for collection of data on which decisions can be based.

The framing of appropriate questions by the Commission services is often the trigger for a new risk assessment. It is emphasized that a dialog between risk managers and risk assessors is important to achieve clear and achievable questions and should include the development of risk profiles of the putative hazard. It is also recognized that involvement of stakeholders in profiling the risk (e.g., identification of the criteria to be used, the specific issues to be addressed, and major concerns) will contribute to a more transparent and consistent risk assessment.

To facilitate improved transparency and acceptability of management decisions, all sources of data that have been used should be provided, including any limitations regarding the accessibility of potential data, the weighting of different datasets, and whether or not stakeholders have the opportunity to submit additional data. Additional improvement of the risk analysis framework will be achieved if risk assessors and relevant stakeholders are involved in the analysis.

17.3 Potential Innovations to the Risk Analysis Framework as Proposed by SAFE FOODS

17.3.1 The SAFE FOODS Project

Funded through the European Union's (EU) Sixth Framework Program, the four-year project "Promoting Food Safety Through a New Integrated Risk Analysis Approach for Foods" (known as SAFE FOODS) commenced its activities in 2004. The research has attempted to integrate natural and social science research activities, and involves 37 institutions from 21 countries (including non-European institutions in China, South Africa, and Russia). As an overarching objective, this project aimed to contribute to strengthening consumer trust in the food safety governance in Europe and beyond.

The research performed within SAFE FOODS attempted to improve current risk analysis practices for foods produced by different breeding approaches and production practices deploying high- and low-input systems. As one of the main outputs of this project, an "improved risk analysis framework" has been developed, which is underpinned by new scientific assessment methods, and embedded in a broad impact analysis of social, financial, and economic consequences, and with high levels of transparency, active public engagement, and improved risk communication. In addition, practitioners working in the field of food safety governance and other relevant stakeholders were consulted to maximize the applicability and acceptability of the framework.

The research was conducted in a number of interdependent projects, which delivered the elements for the construction of the improved risk analysis framework. The project's strategic objectives were the following.

- To design a European working procedure for early identification of emerging chemical or microbial risks in food production chains in an expanding European market.
- To develop comparative safety assessment methods for foods produced by different breeding approaches and production practices, using modern profiling techniques, and new qualitative and quantitative risk assessment models.
- To investigate consumers' confidence and/or preferences in risk analysis practices for foods.

- To understand differences in food risk perceptions of consumers, experts, and decision-makers, and to design informative risk communication strategies that directly address societal concerns.
- To investigate the role of institutions across Europe involved in risk assessment and management given the greater interest of the consumer in taking a broader impact of food production on environment, animal welfare, sustainability, and socio-economic consequences into account.
- To design a new risk analysis approach for foods, integrating scientific principles, societal aspects, and effective public participation.

17.3.2 The SAFE FOODS Risk Analysis Framework

The integrated framework describes an iterative decision process consisting of four stages: framing, risk–benefit assessment, evaluation, and risk management (for a schematic overview, see Figure 17.3 [7–8]).

At the *framing* stage, interested parties, stakeholders, experts, and officials work together to gain an initial shared understanding of the issue, objectives, and broad

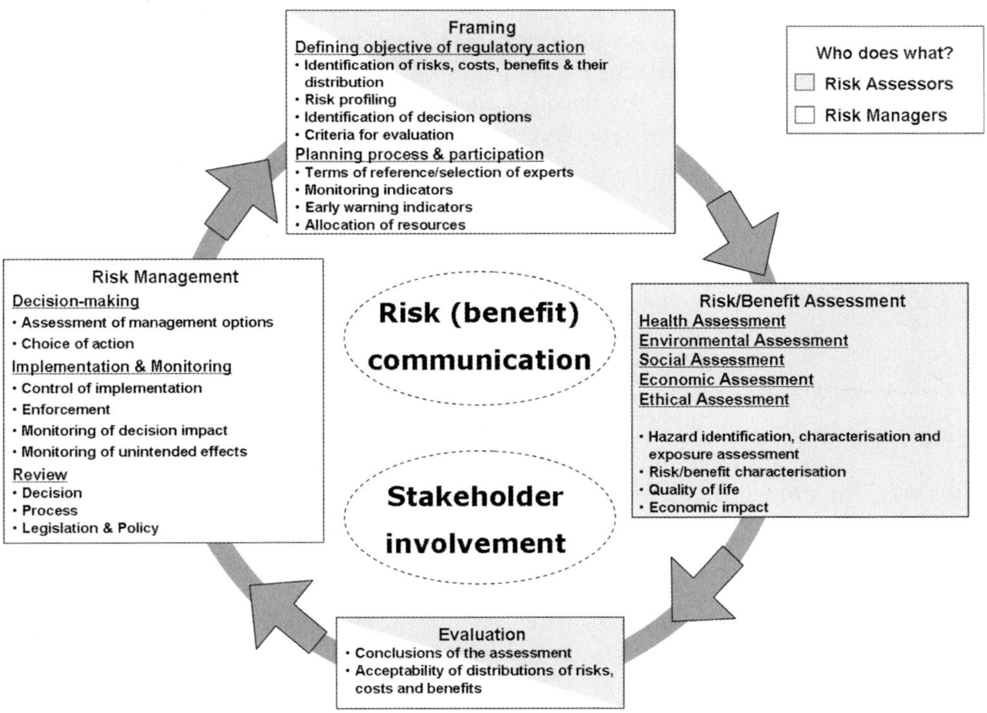

Figure 17.3 Risk analysis framework suggested by SAFE FOODS (adapted from [8]).

courses of regulatory action. Areas of consensus and dissent are documented in order to provide the basis for planning future decisions. Framing also includes defining the scope of the assessment, together with the terms of reference for those involved in the assessment process, proposing criteria for ranking regulatory options, together with monitoring indicators.

The *risk–benefit assessment* includes not only single pre-identified risks, but also human health impacts in general (including health benefits), as well as environmental, economic, social and ethical impacts, and their distribution. This reflects recent debate that has focused on extending the risk assessment paradigms applied in the process of risk analysis to include a broader assessment of the social, economic, and ethical impacts of hazards, whether this results from active risk prevention and mitigation or its omission. Social impact assessment, and its subcategory, health impact assessment, is a rapidly evolving area from the perspective of potential policy impact [9] and evolving assessment methodologies (see e.g. [10, 11]. However, inclusion of socio-economic and ethical impact aligns with the more general trend in policy to address the broader societal context in which risks are embedded.

The *evaluation* stage is proposed as an intermediate stage between risk assessment and management. Evaluation is a participatory process in which interested parties, stakeholders, experts, and officials use the assessment outcome to compare the risks, costs, and benefits and their distribution in the absence of any risk management measures. The outcomes of the evaluation are recommendations regarding which consequences are deemed acceptable, and whether risk management measures may be required.

In this context, *risk management* includes decision-making, implementation, monitoring, and review. Risk management involves the definition, ranking of alternative measures, and final selection of appropriate regulatory options in the context of the assessment outcomes and regulatory options available. Monitoring indicators are the result of proposals made at the framing stage. At the review stage, the impacts of the decision, together with the process by which the decision has been made, and the legislation under which the issue is regulated, are revisited and the effectiveness of what has been done assessed.

The three main differences from the other models described above (e.g., [1, 5], see sections 17.2.1. and 17.2.2, respectively) can be described as: (i) expansion of the scope of the formal risk assessment to include assessment of benefits and costs; (ii) more formal (and institutionalized) stakeholder participation; and (iii) improved risk communication and publicly accessible reports at each stage of the process.

17.3.3
Stakeholders' Views on the New Risk Analysis Framework

To assess the expert opinions on the new risk analysis framework, a Delphi survey [12] was used to solicit stakeholder and end-user views regarding the potential utility of the new framework. Details of this survey are reported elsewhere [7]. The Delphi approach involves a degree of interactivity and dialog, similar to the kind of interactive dialog found in group meetings, but which enables access to wider expertise than might otherwise be attainable. In addition, the approach uses questionnaires to elicit the opinions, which provides a structured dialog. The methodology essentially involves the repeated surveying of experts, the opinions from whom are used as feedback on subsequent “rounds”.

Within this Delphi study, the participants were first sent a questionnaire about the new framework, and then presented with a second survey containing similar questions, which they were asked to complete. In the second round, the experts were provided as well with anonymized feedback regarding the opinions of the whole group on the first round, either in the form of averaged results, or quotations derived from individual expert views, which may have resulted in them reconsidering their views. The views of two groups of experts in risk assessment, risk management, and risk communication were addressed, the first group with experts from within EU Member States, and the second with experts from outside of the EU.

The results suggested that most of the novel concepts in the model were acceptable to many of the experts, though those experts from within the EU seemed to be more positive than their counterparts from the international community. There was substantial support for the idea of broadening assessment to include socio-economic and ethical impacts. Furthermore, there was general support for increasing the role of other stakeholders in the overall risk analysis process. While there was broad stakeholder support for the use of these innovations, there was consensus that they should be applied on a case-by-case basis, rather than applied routinely, perhaps a decision to be made at the framing stage. Varying views existed, however, as to how stakeholders should be involved, what were the appropriate methodological approaches required to measure risks and benefits associated with the different impact factors, including health, and how these different factors should be weighted in the risk analysis process. Finally, the applicability of the new model to emerging risks, including those associated with new technologies, such as nanotechnology, required further discussion.

17.4 Risk Analysis and Nanotechnology

The case of nanotechnologies applied to food and agricultural production is presented here, as it provides an example of a new technology associated with potential new risks and benefits. The authors suggest that the integrated risk analysis framework discussed above can provide a balanced approach toward safe and prudent policies for development and integration of nanotechnology into the domain of food production and handling.

17.4.1 Background of Nanotechnology

Working at the atomic level only became within reach when key analytical tools such as the scanning tunneling microscope were developed in the 1980s. Advances like these and other analytical tools quickly spread to be utilized in many other fields of science. This has led to the development of materials showing unique properties that are dependent on their nanostructure, for example, nanoscale size. Current research is leading to the development of sophisticated and heterogeneous materials and devices, based on an increasing ability to engineer their functionality at the nanoscale [13]. In this context, it has been emphasized that the benefits that have the potential to change and improve our lives will inevitably bring with them new risks that need to be identified and managed [14], which of course emphasizes the need for the application of effective technology governance.

Nanotechnology itself and its applications are now rapidly growing, as hundreds of (claimed) nanotechnology products, including enhanced materials, electronic products and devices, and pharmaceutical products, are already on the market [15]. Nanotechnology applications are beginning to impact on the food-associated industries and are predicted to grow rapidly in the coming years. Applications in this area are already many and wide-ranging: the development of improved taste, color, flavor, texture, and consistency of foodstuffs; increased absorption and bioavailability of nutrients and health supplements; new food packaging materials with improved mechanical, barrier, and antimicrobial properties; and nanosensors for traceability and monitoring the condition of food during transport and storage [16–19].

It is this broadness of application of nanotechnologies that makes it particularly difficult to discuss potential risks in general terms. Moreover, this broadness also makes the technology very sensitive to any emerging consumer concerns about its application, because (negative) discussions about applications of nanotechnology in one sector are likely to have an effect on applications in another sector. This is also one of the reasons why the newly developed integrated SAFE FOODS risk analysis framework involving stakeholders in the framing and evaluation stages of the risk analysis process appears particularly suited for the topic of the safety of nanomaterials in food.

Nanotechnology is a collection of enabling and converging technologies, which mean that it is not a single type of technology used in a single field of science, but a great variety of techniques that have only one thing in common: the nanometer-size scale. Given that premise, it is useful to provide the definitions that have been applied to the field of nanotechnology and nanoparticles and which are used throughout this chapter:

Nanotechnologies The design, characterization, production, and application of structures, devices, and systems by controlling shape and size at the nanometer scale [20].

Nanoparticle A discrete entity that has three dimensions of the order of 100 nm or less.

Nanoparticulate matter A substance comprising of particles, the substantial majority of which have three dimensions of the order of 100 nm or less [21].

Nanoparticles as such are not new to biology. Nano-sized particles can have a natural origin, such as sand dust, and ash resulting from volcanic eruption, or can be the unintended result of human activities, such as ultra-fine particles in diesel exhaust (combustion). In the remaining part of this chapter, the discussion will be solely dedicated to engineered nanoparticles:

Engineered nanoparticle Any material that is deliberately created such that it is composed of discrete functional parts, either internally or at the surface, many of which will have one or more dimensions of the order of 100 nm or less [21].

17.4.2
Historic Picture of Nanoparticle Safety in Relation to Risk Analysis and Good Governance

Recently, Oberdörster *et al.* [22] described the roots of nanotechnology from strands of knowledge gained during the development of modern particle toxicology (fine dust, pollution particles), virology, and other sciences.

It was not before 1990 that the size of (fine dust) particles was recognized as an important factor in its translocation over the lung epithelium. Before this period, possible contributions of fine or ultra-fine particles were not considered or imagined. But in the early 1990s, it was observed that a significantly greater pulmonary inflammation and interstitial translocation occurred from a given mass of ultra-fine particles than from the same mass of fine particles [23, 24]. The same scientists, two years later, concluded that "toxic responses to new technology metal compounds may not be extrapolated from known metal toxicology" [25]. Mechanistic research on effects caused by asbestos had also been initiated, which, in the mid-1990s, led to the "oxidative stress hypothesis" explaining the toxicity of ultra-fine particles including metal nanoparticles following inhalatory exposure [26–28].

It was in this same period that results from dose (metric) and pulmonary effect studies led to the conclusion that parameters such as particle surface area, size, and surface chemistry as key dose metric parameters explained the observed effects [29]. This list of parameters was subsequently extended by Oberdörster *et al.* [30] and Warheit *et al.* [31], who concluded that "knowledge about only one or two characteristics of nanoparticles is not sufficient to interpret their biological and toxicological effects".

In the last decade, an ever-increasing number of studies with engineered nanoparticles have been published. Both positive and negative aspects were highlighted, and these results, in turn, have been compiled in numerous reviews. It was during this period that, for the first time, general concerns were raised about the unknown potential of some nanomaterials to pose a hazard to both human health and the environment. This also highlights the need for risk managers to be able to consider the benefits besides the risks of the new technology, of which the assessment is one of the new features recommended by the newly developed integrated SAFE FOODS risk analysis framework for food safety. The framework also provides for a mechanism of involving stakeholders during the framing stage of the risk analysis process, when issues to be addressed during the scientific stage of risk and benefit assessment are identified.

The challenges of nanotoxicology, that is, the branch of science focusing on the potential toxicity of nanomaterials, for science, industry, and regulators have been discussed in many conferences, workshops, and scientific committees. The need for toxicological testing of nanomaterials is clearly identified in the reports of these meetings [20, 32–35], which have also contributed to the growing awareness that an improved understanding of the hazards of nanomaterials is essential to enable a sustainable maturation of nanotechnologies. This is also reflected in the European Union's approach to the introduction of nanotechnology as being required to be "safe, integrated and responsible" [36]. In practical terms, the risk is likely to differ from one nanomaterial to another, ranging from safe and innocuous for most nanomaterials to highly toxic for some others [22].

17.4.2.1 **Risk Assessment**

Current safety and risk assessment requirements are based on knowledge gathered for conventional chemicals. In these assessments, knowledge gaps for less well-characterized chemicals may occasionally be encountered. Such uncertainties are approached on the basis of general knowledge, for example through extrapolation from a well-characterized compound to a similar, less well-characterized one. For nanoparticles, however, such a knowledge base is lacking, and, at the same time, the uncertainties in the safety assessments are also expected to be greater [37].

At this stage, the (lack of) knowledge about nanotoxicology may result in risk assessors basing their risk assessments on available, yet incomplete, information about nanoparticles and their appearance in products. Over time, it will be possible to obtain more comprehensive data and to extract the most relevant information for the risk assessment.

From a regulatory point of view, the question has been raised as to what information is additionally required for effective regulation of nanotechnology. In addition, the question has arisen as to whether the current regulatory system within the EU is suited to cope with the regulatory demands placed upon it in this context. To evaluate this, the EU has commissioned its Scientific Committees and Commission services, as well as EFSA in 2008, to perform a scientific and legislative review on the suitability of the existing regulation for nanotechnologies. The Scientific Committee on Emerging and Newly-Identified Health Risks (SCENIHR) concluded that the EU regulatory framework covers, in principle, also nanotechnologies [38]. In line with this, the Health Council of The Netherlands considered that "the best course of action would be to modify existing laws and rules as and when developments within the fields of nanoscience and nanotechnologies render such measures necessary" [39]. SCENIHR and others deemed adjustments of legislation, guidelines, and guidance documents concerning the testing of nanoparticles of the substance to be necessary [38]. However, such adaptations can currently not be made due to the lack of knowledge on this topic.

It is clear that, in the existing regulatory framework, the responsibility for the safety of the product is assigned to the producers. There is currently a need for guidance on how to approach the safety assessment of nanoparticles, and to define what information should be presented by producers to the regulatory agencies. To further elaborate such guidance, a close collaboration among all stakeholders is required, for which the newly developed integrated SAFE FOODS risk analysis framework approach can serve as a useful model. In fact, such (early) stakeholder involvement is now being arranged at EU and national level, addressing not only the risks and benefits of the technology but also ethical issues. A typical example is the broad national debate initiated in 2008 in the Netherlands (http://www.nanopodium.nl/). In addition, large research programs are funded at EU and national level aimed to provide answers for the risk managers to improve the decision-making process.

17.5 Recommendations

Discussions on the improvement of the reliability of risk assessment of nanomaterials, data requirements, and expected performance of current assays have demonstrated that it is important to focus the question on what information is additionally required to dossier requirements for conventional chemicals. Some research agendas or roadmaps try to circumvent the uncertainties that are accepted in the risk assessment of conventional chemicals. In other words, in an area where such additional research questions can be or are being raised, it is essential to define those questions that represent the "need to know" issues. This approach should be leading all roadmaps or research agendas that are developed to be applicable to the field of potential risks of nanotechnology.

To improve the existing risk assessment methodology, good governance, and regulatory framework associated with the application of nanotechnology to food and agriculture, the following issues should be addressed in line with the proposed integrated SAFE FOODS risk analysis framework described above.

- Developing analytical tools for the detection and characterization of nanoparticles in food matrices to estimate exposure, kinetics, and toxicological dose–response relationships.
- Establishing dose metrics to facilitate the interpretation of scientific studies as well as regulatory frameworks.

- Investigating kinetics (especially oral bioavailability) and (oral) toxicity of the different types of nanoparticles, with special attention to those parts of the body that are normally protected by barriers like the blood–brain barrier and placenta.
- Assess the validity of currently used toxicological assays for detecting the effects caused by nanoparticles.
- Identifying products containing nanoparticles that are currently on the market (or being developed), including the type of nanoparticles that are (or will be) used and the estimated consumption of these products.
- Investigating the potential health benefits linked to the introduction of nanomaterials (e.g., packaging with antibacterial properties or nanosensors, higher bioavailability), as well as the economic, social, ethical, and environmental impacts of the application of the various forms of nanotechnology to food production and handling.
- Involving representatives of the relevant stakeholders' parties involved with the application of nanotechnology in food (e.g., consumers, producers) as well as the risk assessors and risk managers at the framing stage so as to ascertain that all relevant aspects and viewpoints are covered in the assessments of risks and benefits. Furthermore, these stakeholders should also be involved in the evaluation of the outcomes of the assessments in order to incorporate stakeholder views and priorities into decision options.

The request for extra information is not to be considered solely as a request for additional studies using new methodologies. It can also imply that conventional approaches and methodologies need to be redesigned. The use of novel technologies (e.g., profiling approaches) and the more frequent use of *in vitro* approaches for risk assessment need to be studied and used in parallel with conventional techniques.

In conclusion, the issue of governance of nanotechnology applied to food production represents an example of technological innovation requiring broader and more inclusive governance structures to be developed and applied, in order to meet the requirements and preferences of all key stakeholders, including the general public.

Acknowledgments

The authors gratefully acknowledge funding received from the European Union's Sixth Framework Program and from the Dutch Ministry of Economic Affairs, Agriculture and Innovation.

References

1 FAO/WHO (1997) Risk management and food safety. Report of a Joint FAO/WHO Consultation, Rome, Italy, 27 to 31 January 1997. FAO Food and Nutrition Paper 65, Food and Agriculture Organization, Rome. Available at: http://www.fao.org/docrep/w4982e/w4982e00.htm (accessed 16 November 2010).

2 Wentholt, M.T.A., Fischer, A.R.H., Rowe, G., Marvin, H.J.P., and Frewer, L.J. (2010) Effective identification and management of emerging food risks: results of an international Delphi survey. *Food Control*, **21** (12) S1, 1731-1738.

3 FAO/WHO (1995) Application of risk analysis to food standards issues, Report of the Joint FAO/WHO Expert Consultation, Geneva, Switzerland, 13–17 March, 1995. Food and Agriculture Organization, Rome. Available at: http://www.who.int/foodsafety/publications/micro/en/march1995.pdf (accessed 16 November 2010).

4 Codex Alimentarius Commission (2005) Procedural Manual. Codex Alimentarius Commission, Joint FAO/WHO Food Standards Program, Food and Agriculture Organization, Rome. Available at: http://www.codexalimentarius.net/web/procedural_manual.jsp (accessed 16 November 2010).

5 SSC (2000) Opinion of the Scientific Steering Committee on Harmonisation of Risk Assessment Procedures. Scientific Steering Committee, Directorate-General Health and Consumers, European Commission, Brussels. Available at: http://ec.europa.eu/food/fs/sc/ssc/out82_en.html (accessed 16 November 2010).

6 SSC (2003) Opinion of the Scientific Steering Committee on Setting the Scientific Frame for the Inclusion of New Quality of Life Concerns in the Risk Assessment Process. Scientific Steering Committee, Directorate-General Health and Consumers, European Commission, Brussels. Available at: http://ec.europa.eu/food/fs/sc/ssc/out357_en.pdf (accessed 16 November 2010).

7 Wentholt, M.T.A., Rowe, G., Koenig, A., Marvin, H.J.P., and Frewer, L.J. (2009) The views of key stakeholders on an evolving food risk governance framework: results from a Delphi study. *Food Policy*, **34** (6), 539–548.

8 Koenig, A., Kuiper, H.A., Marvin, H.J.P., Boon, P.E., Busk, L., Cnudde, F., Cope, S., Davies, H.V., Dreyer, M., Frewer, L.J., Kaiser, M., Kleter, G.A., Knudsen, I., Pascal, G., Prandini, A., Renn, O., Smith, M.R., Traill, B.W., van der Voet, H., van Trijp, H., Vos, E., and Wentholt, M.T.A. (2010) The SAFE FOODS framework for improved risk analysis of foods. *Food Control*, **21** (12), 1566–1587.

9 Dreyer, M., Renn, O., Cope, S., and Frewer, L.J. (2010) Including social impact assessment in food safety governance. *Food Control*, **21** (12), 1620–1628.

10 Owen, R., and Handy, R. (2007) Formulating the problems for environmental risk assessment of nanomaterials. *Environ. Sci. Technol.*, **41** (16), 5582–5588.

11 Cope, S., Frewer, L.J., Renn, O., and Dreyer, M. (2010) Potential methods and approaches to assess social impacts associated with food safety issues. *Food Control*, **21** (12), 1629–1637.

12 Linstone, H.A., and Turoff, M. (1975) *The Delphi Method*, Addison-Wesley, Reading, MA.

13 Roco, M.C. (2004) Nanoscale science and engineering: unifying and transforming tools. *AIChE J.*, **50** (5), 890–897.

14 Maynard, A.D., Aitken, R.J., Butz, T., Colvin, V., Donaldson, K., Oberdörster, G., Philbert, M.A., Ryan, J., Seaton, A., Stone, V., Tinkle, S.S., Tran, L., Walker, N.J., and Warheit, D.B. (2006) Safe handling of nanotechnology. *Nature*, **444** (7117), 267–269.

15 Woodrow Wilson International Center for Scholars (2009) Consumer products: an inventory of nanotechnology-based consumer products currently on the market. Project on Emerging Nanotechnologies. Available at:

http://www.nanotechproject.org/inventories/consumer/ (accessed 8 November 2010).
16 Chen, H.D., Weiss, J.C., and Shahidi, F. (2006) Nanotechnology in nutraceuticals and functional foods. *Food Technol.*, **60** (3), 30–36.
17 Weiss, J., Takhistov, P., and McClements, J. (2006) Functional materials in food nanotechnology. *J. Food Sci.*, **71** (9), R107–R116.
18 Bouwmeester, H., Dekkers, S., Noordam, M., Hagens, W., Bulder, A., De Heer, C., Ten Voorde, S., Wijnhoven, S., and Marvin H. Sips, A. (2009) Review of health safety aspects of nanotechnologies in food production. *Regul. Toxicol. Pharmacol.*, **53**, 52–62.
19 Chaudhry, Q., Scotter, M., Blackburn, J., Ross, B., Boxall, A., Castle, L., Aitken, R., and Watkins, R. (2008) Applications and implications of nanotechnologies for the food sector. *Food Addit. Contam.*, **25** (3), 241–258.
20 Royal Society and the Royal Academy of Engineering (2004) Nanoscience and Nanotechnologies: Opportunities and Uncertainties, Royal Society, London. Available at: http://www.nanotec.org.uk/finalReport.htm (accessed 16 November 2010).
21 SCENIHR (2007) Opinion on the scientific aspects of the existing and proposed definitions relating to products of nanoscience and nanotechnologies. Scientific Committee on Emerging and Newly Identified Health Risks, European Commission, Brussels. Available at: http://ec.europa.eu/health/ph_risk/committees/04_scenihr/docs/scenihr_o_012.pdf (accessed 16 November 2010).
22 Oberdörster, G., Stone, V., and Donaldson, K. (2007) Toxicology of nanoparticles: a historical perspective. *Nanotoxicology*, **1** (1), 2–25.
23 Ferin, J., Oberdörster, G., Penney, D.P., Soderholm, S.C., Gelein, R., and Piper, H.C. (1990) Increased pulmonary toxicity of ultrafine particles? I. Particle clearance, translocation, morphology. *J. Aerosol Sci.*, **21** (3) 381–384.
24 Oberdörster, G., and Yu, C.P. (1990) The carcinogenic potential of inhaled diesel exhaust: a particle effect? *J. Aerosol Sci.*, **21** (Suppl. 1), S397–S401.
25 Ferin, J., and Oberdörster, G. (1992) Translocation of particles from pulmonary alveoli into the interstitium. *J. Aerosol Med.*, **5** (3), 179–187.
26 Gilmour, P.S., Brown, D.M., Lindsay, G.T., Beswick, P.H., MacNee, W., and Donaldson, K. (1996) Adverse health effects of PM10 particles: involvement of iron in generation of hydroxyl radicals. *Occup. Environ. Med.*, **53**, 817–822.
27 Donaldson, K., Beswick, P.H., and Gilmour, P.S. (1996) Free radical activity associated with the surface of particles: a unifying factor in determining biological activity? *Toxicol. Lett.*, **88** (1–3), 293–298.
28 Zhang, Q., Kusaka, Y., Sato, K., Nakakuki, K., Kohyama, N., and Donaldson, K. (1998) Differences in the extent of inflammation caused by intratracheal exposure to three ultrafine metals: role of free radicals. *J. Toxicol. Environ. Health A*, **53** (6), 423–438.
29 Donaldson, K., Li, X.Y., and MacNee, W. (1998) Ultrafine (nanometre) particle mediated lung injury. *J. Aerosol Sci.*, **29** (5–6), 553–560.
30 Oberdörster, G., Oberdörster, E., and Oberdörster, J. (2005) Nanotoxicology: an emerging discipline evolving from studies of ultrafine particles. *Environ. Health Perspect*, **113** (7), 823–839.
31 Warheit, D.B., Webb, T.R., Sayes, C.M., Colvin, V.L., and Reed, K.L. (2006) Pulmonary instillation studies with nanoscale TiO_2 rods and dots in rats: toxicity is not dependent upon particle size and surface area. *Toxicol. Sci*, **91** (1), 227–236.
32 ILSI (2006) Meeting of the HESI Nanomaterial Environmental, Health, and Safety Project Committee, September 15, 2006, Health and Environmental Sciences Institute, International Life Sciences Institute, Washington, DC.
33 SCENIHR (2005) Opinion on the appropriateness of existing methodologies to assess the potential

risks associated with engineered and adventitious products of nanotechnologies. Scientific Committee on Emerging and Newly Identified Health Risks, European Commission, Brussels. Available at: http://ec.europa.eu/health/ph_risk/committees/04_scenihr/docs/scenihr_o_003.pdf (accessed 16 November 2010).

34 EFSA (2010) Nanotechnology. European Food Safety Authority, Parma. Available at: http://www.efsa.europa.eu/en/sctopics/topic/nanotechnology.htm (accessed 16 November 2010).

35 OECD/Allianz (2005) Small Sizes That Matter: Opportunities and Risks of Nanotechnologies, Organization for Economic Cooperation and Development, Paris. Available at: http://www.oecd.org/dataoecd/37/19/37770473.pdf (accessed 16 November 2010).

36 European Commission (2004) Towards a European strategy for nanotechnology. European Commission, Brussels. Available at: http://ec.europa.eu/nanotechnology/pdf/nano_com_en_new.pdf (accessed 16 November 2010).

37 Morgan, K. (2005) Development of a preliminary framework for informing the risk analysis and risk management of nanoparticles. *Risk Anal.*, **25** (6), 1621–1635.

38 SCENIHR (2007) Opinion on the appropriateness of the risk assessment methodology in accordance with the technical guidance documents for new and existing substances for assessing the risks of nanomaterials. Scientific Committee on Emerging and Newly Identified Health Risks, European Commission, Brussels. Available at: http://ec.europa.eu/health/ph_risk/committees/04_scenihr/docs/scenihr_o_010.pdf (accessed 16 November 2010).

39 Health Council of The Netherlands (2006) Health Significance of Nanotechnologies. Report 2006/06E. Health Council of The Netherlands, The Hague. Available at: http://www.gezondheidsraad.nl/sites/default/files/Nanotechnologies%20eng_0.pdf (accessed 16 November 2010).

Index

Nanotechnology in the Agri-Food Sector: Implications for the Future, First Edition. Edited by Lynn J. Frewer, Willem Norde, Arnout Fischer, Frans Kampers.